D1318587

QUANTITATIVE PROBLEMS
IN BIOCHEMISTRY

QUANTITATIVE PROBLEMS IN BIOCHEMISTRY

BY

EDWIN A. DAWES
Ph.D., D.Sc., F.R.I.C.

Reckitt Professor of Biochemistry
University of Hull

Foreword by

J. NORMAN DAVIDSON
C.B.E., M.D., D.Sc., F.R.C.P.(Glasg.),
F.R.C.P.(Edin.), F.R.I.C., F.R.S.
Gardiner Professor of Biochemistry
University of Glasgow

FIFTH EDITION

BALTIMORE
THE WILLIAMS AND WILKINS COMPANY
1972

First Edition	1956
Second Edition	1962
Reprinted	1963
Third Edition	1965
Fourth Edition	1967
Reprinted	1969
Fifth Edition	1972
Spanish Edition	1959
Japanese First Edition	1960
Japanese Second Edition	1964
Japanese Third Edition	1968
Czechoslovakian Edition	1965

ISBN 0 443 00807 8

PRINTED IN GREAT BRITAIN

FOREWORD

IN the course of the last twenty years or so Biochemistry has evolved from the descriptive and qualitative stage into a more mature phase in which it ranks as probably the most exact of all the sciences included in the field of Biology. This process of evolution has necessitated a change in the outlook and training of the student who wishes to make Biochemistry his speciality. Whereas in the past it was sufficient for him to take as his main ancillary subjects physiology and organic chemistry, he is now compelled to have more than a nodding acquaintance with physical chemistry as well and to be able to understand how the principles of physical chemistry can be applied to biological systems.

This situation in itself creates a pedagogic problem, for the number of teachers of physical chemistry who have the necessary biological background to appreciate the biochemist's problems is still all too small. Consequently any attempt made to help the student of biochemistry to a better understanding of physico-chemical problems deserves the most vigorous encouragement.

Students of chemistry have available to them a number of books dealing with chemical calculations, but no equivalent volume dealing with biochemical problems has until now been on the market, although several texts on physical biochemistry have been published during the last few years. This volume by Dr. Dawes should therefore be of considerable value in filling a gap in modern biochemical education.

Dr. Dawes has had wide experience in teaching senior students of biochemistry working for the Honours Degree in Glasgow, and at an earlier date in Leeds, and his compilation of a sufficient range of numerical problems has been a task which has spread over several years. Each group of problems is gathered into a chapter containing enough explanatory text to give the funda-mental information required for their solution without at the same time converting the volume into a full fledged textbook of physical biochemistry.

Although the book has been designed to meet the specific needs of senior students of biochemistry, it may also have an

v

appeal to workers in allied fields who are anxious to keep their knowledge of biochemistry up to date. However this may be, it is certain of a very warm welcome from members of the Honours Biochemistry Class in the University of Glasgow and, it is hoped, by students taking the equivalent courses in other universities.

J. N. DAVIDSON.

THE UNIVERSITY,
 GLASGOW, 1956.

PREFACE TO THE FIFTH EDITION

THE necessity for a new edition of *Quantitative Problems in Biochemistry* has provided the opportunity to include additional material in most chapters and to revise some of the text in the light of the inexorable progress of biochemistry. It is hoped that the widely acknowledged dual function of the book, as both a teaching manual and a reference source in the laboratory, will be fully sustained by this latest revision.

Once again it is a pleasure to acknowledge the constructive criticisms and helpful suggestions of several friends and I am, as on previous occasions, especially indebted to my colleague Dr. George W. Crosbie for his unstinting and generous help. I am also grateful for the assistance rendered by Dr. R. B. Beechey, Mr. T. J. Bowen, Dr. F. M. Dickinson, Dr. I. G. Jones and Dr. R. Y. Thomson. Mr. Robin Callander and Mr. M. Girling have been responsible for the new diagrams, and Mr. C. J. Dickenson, Mr. M. P. Stephenson and Mrs. Barbara K. Roberts have kindly assisted with the checking of answers and proof reading. The various universities acknowledged in the text have generously permitted problems set in their respective Finals examinations to be reproduced and thus broadened the scope of the problems presented; their help is greatly appreciated.

As ever, I am very grateful to my publishers, whose masthead reveals some change from the previous edition but whose patience and unfailingly high standards of production command constant admiration.

Kingston upon Hull, 1972.   E. A. D.

PREFACE TO THE FIRST EDITION

' When you can measure what you are speaking about and express it in numbers, you know something about it, and when you cannot measure it, when you cannot express it in numbers, your knowledge is of a meagre and unsatisfactory kind.'—*Lord Kelvin.*

In the past the biochemist has been at a disadvantage compared with his colleagues in the more chemical and physical sciences because much of his work was, by its very nature, of a qualitative rather than a quantitative kind. Consequently the biochemist was regarded with something akin to suspicion by his colleagues in these fields, a suspicion perpetuated by the remark of that great physical chemist G. N. Lewis, when he accused living things of being 'cheats in the game of entropy'; clearly the investigator of such phenomena was himself something of an enigma !

During the past two decades the advances made in physical biochemistry and biophysics have completely revolutionized the quantitative approach to living matter and we are now in the position of having a wide field of data upon which to draw for numerical problems. Furthermore, living things are now known to 'play the game' with entropy.

This book is the outcome of a policy pursued by the author, initially at Leeds and later at Glasgow, of including numerical problems in the examinations for the Honours Degree in Biochemistry. The author feels that the greatest benefit to be derived from numerical problems is the attitude of mind engendered. All too frequently one encounters a tendency for students of biochemistry, after receiving their initial training in chemistry and physics, to become rather illogical and imprecise in thought if their course, apart from certain sections of the practical work, provides little quantitative expression. Numerical problems, properly applied, can help to combat this danger. They can also afford a clearer understanding of experimental techniques that are not usually encountered in the undergraduate laboratory course. For instance, not every biochemical laboratory possesses an analytical ultracentrifuge, and a better appreciation of the measurements that must be made before a molecular weight can be calculated is obtained if actual sample calculations are carried out.

The original plan was to assemble a collection of suitable problems alone, but it was then felt that the book might prove to be of more value to the student if some theoretical background and worked examples were included, especially since the literature on the various topics treated is somewhat diffuse and there is no single textbook which provides the required coverage. As the intention was for emphasis to be placed on problems rather than on the development of a textbook of physical biochemistry, the greatest difficulty besetting the author was the attempt to achieve a suitable balance between inclusion and exclusion of material ; the pitfalls of such a course are patent. Choice of material has been quite deliberate in order to keep the book within reasonable compass, and some basic knowledge is assumed. However, to compensate for any deficiency resulting from this treatment, references and suggestions for further reading are included at the end of each chapter.

It is not intended that the chapters should necessarily be worked through in a given order ; each chapter is reasonably self-contained and cross-references are provided where necessary. Many of the problems have either been taken or constructed from data in published papers and, in these instances, the references are cited. This enables the student to consult the original publication should difficulty arise. Some attempt has been made to provide problems of roughly graded difficulty in order that the book might be of use not only to honours students but also to those taking biochemistry to principal or subsidiary level, corresponding to the Double and Single Science courses of the University of Glasgow.

It is a pleasure to record my gratitude to several friends who have assisted during the preparation of this book. Professor J. N. Davidson has offered constant encouragement and advice throughout, and kindly consented to write a Foreword. I am greatly indebted to my former colleagues Dr. S. Dagley and Mr. T. J. Bowen, both of whom have given generously of their time to read the entire manuscript, and who not only criticized freely but also contributed many valuable suggestions which have been incorporated in the text. Helpful suggestions in connexion with certain portions of the text were also made by Dr. H. N. Munro, Dr. W. C. Hutchison, Mr. W. H. Holms and Dr. J. C. Speakman.

Through the kindness of Professor Eric G. Ball, Dr. Fred Richards and Dr. Frank Gurd of Harvard University Medical School I have been able to include some problems devised for use in the Harvard Medical Sciences 201 ab course. Many of the diagrams are the work of Mr. R. Callander, for whose help I am extremely grateful. Thanks are due to authors and publishers for permission to reproduce certain diagrams, and to Mr. T. J. Bowen for Figures 7.1 and 7.2. Answers to some of the problems were kindly checked by Mr. A. Fleck and Dr. Shelagh M. Foster, but I alone must accept responsibility for errors which may still remain. I should also like to thank the authors and publishers of the papers quoted in this book for permission to use their data. Mr. W. H. Holms, Mr. D. Murray and my wife all rendered valuable assistance with proof reading, and their help is greatly appreciated. Mr. C. Macmillan and Mr. J. Parker, of Messrs. E. & S. Livingstone Limited, have given kindly guidance and have helped in many ways to facilitate the production of the book. Finally, but by no means least, I am indebted to my wife, who, starting as a novice, finished as a skilful typist in the preparation of the manuscript for the printer.

Glasgow, 1956.   E. A. D.

CONTENTS

xiii

LOGARITHMIC NOTATION

THE following convention with regard to logarithms is employed throughout this book.

$$\log = \log_{10}$$
$$\ln = \log_e$$

They are related by the expression :

$$\ln x = 2{\cdot}303 \log x$$

NOMENCLATURE OF THE COENZYMES

In accordance with the recommendations of the Commission on Enzymes of the International Union of Biochemistry, the nicotinamide nucleotide coenzymes formerly referred to as diphosphopyridine nucleotide (DPN) and triphosphopyridine nucleotide (TPN) are given as nicotinamide-adenine dinucleotide (NAD) and nicotinamide-adenine dinucleotide phosphate (NADP) respectively. Two different notations for the oxidation-reduction of these coenzymes are employed, e.g. for NAD :

$$NAD + 2H \rightleftharpoons NADH_2$$
$$NAD^+ + 2H \rightleftharpoons NADH + H^+$$

CHAPTER I

DETERMINATION OF MOLECULAR WEIGHTS

THE molecular weight of substances of small molecular size may be determined by measurement of the colligative properties of their solutions. These include the elevation of boiling point, depression of freezing point and of vapour pressure of the pure solvent, and osmotic pressure. Many compounds of biological origin, such as proteins and polysaccharides, are macromolecules and have very large molecular weights. They can be studied only in very dilute solution; this point is made clear by an example. Molecular weight determinations of small molecules are usually carried out in 0·01-0·1 M solution. A solution 0·01 M with respect to a protein having a molecular weight of 75,000 would have to contain 750 grams of the protein per litre and this is a physical impossibility. The most highly soluble proteins (certain albumins) will not dissolve to this extent in water, and even if this were possible to achieve, the resulting solution would be quite unsuitable for molecular weight determination. This is because the colligative properties of solutions depend on the number of molecular units present per unit volume and not on their size. Consequently a very small amount of a compound of low molecular weight may exert an effect equal to or even greater than that of the protein itself. Of the colligative properties, only osmotic pressure is used for such determinations, because in this case, by suitable choice of conditions, the effect of small molecules or ions associated with the protein may be eliminated. Further, osmotic pressure gives a measurable pressure where depression of freezing point, if applicable, would give a depression too small to measure accurately by the usual Beckmann thermometer. It is clear, therefore, that in general, to determine the molecular weights of macromolecules, other methods must be employed. These may be divided into two main groups :

1. Those based on the chemical composition, by analysis of certain elements or amino acids, or by combining weights.
2. Physico-chemical methods which include sedimentation, diffusion, flow birefringence, osmotic pressure and light scattering.

Additionally, viscosity measurements can be used if prior calibration with a similar compound of known molecular weight has been made. These methods find the widest application at the present time.

Some comment is necessary at this point on the ways in which a molecular weight may be recorded in the biochemical literature for within recent years the use of the *dalton* as a unit of mass has gained currency, particularly with molecular biologists, although this unit has not yet been recognized officially by the appropriate international committees.[1] The dalton may be defined by the statement that one atom of the carbon isotope ^{12}C has a mass of 12 daltons. Consequently one dalton equals N^{-1} g $= 1.663 \times 10^{-24}$ g, where N is the Avogadro number. This is clearly identical with the unified atomic mass unit.

Since molecular weight is defined as the relative molecular mass of a substance, i.e. the ratio of the mass of one molecule of the substance to one-twelfth the mass of an atom of ^{12}C, it is a pure number and therefore dimensionless.

Molar mass, M, is defined as the amount of substance containing as many elementary units (properly specified by some formula, e.g. $C_6H_{12}O_6$) as there are carbon atoms in exactly 0.012 kg of ^{12}C. Thus molar mass is commonly expressed in g mole^{-1} which, in fact, is the way in which molecular weights determined by sedimentation, osmotic pressure and other colligative properties have been customarily expressed. A mass in daltons is obviously numerically identical with molar mass in g mole^{-1}, but, as Edsall has pointed out, it is clearly at variance with the accepted definition of molecular weight, and clarification by the international nomenclature committees is urgently needed.

One advantage of the dalton is that it enables the biochemist to report the size of structures for which the term 'molecular weight' is manifestly inappropriate, e.g. cellular organelles such as mitochondria and ribosomes, viruses, etc., which are complex, organized structures containing many different kinds of molecule.

1. Calculation of Molecular Weight from Chemical Composition

(a) ELEMENTARY AND AMINO ACID ANALYSIS

The molecular weight of a compound such as a protein may be calculated if its elementary composition is known. Since a

---

[1] See, for example, Edsall, J. T. (1970). *Nature*, **228**, 888.

compound must contain in its molecule at least one atom of every element shown by analysis to be present, the mass of the compound which contains one gram-atom of such an element will be the minimal molecular weight. In other words, the molecular weight cannot be smaller than this amount. The compound may contain more than one atom of the element, and the molecular weight is then an integral multiple of the minimal value based on the assumption that only one atom is present. Thus, if the percentage of the element present in the compound is known, the minimal molecular weight is given by the expression :

Minimal molecular weight

$$= \frac{\text{atomic weight of element}}{\text{percentage of element in compound}} \times 100 \quad . \quad . \quad (1.1)$$

and true molecular weight

$$= n \times \text{minimal molecular weight} \quad . \quad . \quad (1.2)$$

where n is the number of atoms of the element present in the molecule.

Example 1.1.—The nitrogen content of serine is 13·33 per cent. Calculate the minimal molecular weight.

$$\text{Min. mol. wt.} = \frac{14}{13 \cdot 33} \times 100 = 105.$$

Serine contains one nitrogen atom so that $n = 1$ and the true molecular weight is also 105.

Example 1.2.—Lysine contains 19·17 per cent. nitrogen. What is the minimal molecular weight?

$$\text{Min. mol. wt.} = \frac{14}{19 \cdot 17} \times 100 = 73.$$

Lysine contains two nitrogen atoms and therefore the true molecular weight is 146.

To obtain the true molecular weight it is necessary, therefore, to know the value of n. This is usually obtained by some other method such as the measurement of one of the colligative properties of a solution of the compound. As the value of n increases, its evaluation becomes more difficult. This is because the value for the minimal molecular weight becomes correspondingly smaller, until, eventually, it may be within the range of experimental error in the measurement of the colligative property being used to determine n. For accuracy it is necessary to select an element which is present in the molecule in a relatively small amount and which can be determined with accuracy.

Further information may be obtained if analytical figures are available for more than one element which is present in the compound. For instance, if in Example 1.1 the additional information had been given that serine contains 45·71 per cent. oxygen, we should have been able to obtain the minimal molecular weight of $\frac{16}{45·71} \times 100 = 35$ from the oxygen data and 105 from the nitrogen value. If the serine molecule contains n_1 atoms of nitrogen and n_2 atoms of oxygen, then the molecular weight of serine must be represented by the equation

$$n_1 \times 105 = n_2 \times 35.$$

This is satisfied by $n_1 = 1$ and $n_2 = 3$, and hence the molecular weight is 105 or some multiple of it. The data given do not permit a decision on the latter point, and in the absence of further evidence the molecular weight might be 210 or 315.

For the determination of the molecular weight of proteins, the elements normally used are sulphur and, in the case of metalloproteins, the metal present in the prosthetic group. The minimal molecular weight of haemoglobin, for example, has been determined by analysis of its iron content. The sulphur present in a protein may exist as disulphide, thiol or thio-ether, and the sulphur determination may be expressed in these terms or simply as the percentage total sulphur. Elements such as carbon and nitrogen are present in too large a percentage for minimal molecular weight determinations and they give very large values of n.

Some proteins do not contain any element in sufficiently small quantity to be used in these calculations, but in many cases the percentage content of certain amino acids has been used for the determination of the minimal molecular weight. The procedure is exactly analogous to that used when the basis is the elementary composition. Choice is made of an amino acid present in small percentage ; the minimal molecular weight must contain at least one of these amino acid molecules. Provided the amino acid analysis is accurate, this method offers advantages, because, unlike an element, it is not likely that an amino acid will be present as a contaminant of the protein. Amino acids finding greatest application in this way are tyrosine, tryptophan and cystine, for they are usually present in small amount and can be accurately determined.

There are many data now available on the composition of various proteins, but unfortunately not all of them are reliable for the calculation of minimal molecular weights because of uncertainty in the state of purity of the protein analysed. There is also the possibility that in hydrolysing the protein prior to analysis some of the amino acids may be destroyed.

(b) COMBINING WEIGHTS

Minimal molecular weights may be determined as the weight of the compound which combines or reacts with 1 gram-molecule of a suitable chemical reagent such as a monovalent acid or base. Proteins, for example, contain a number of free carboxyl and amino groups in their molecules and these may be titrated with base and acid respectively. In this way the maximal base and acid-binding capacities are determined. If there are, say, x carboxyl groups per molecule, then one gram-molecule of protein will combine with x equivalents of base and the minimal molecular weight will be $1/x$ of the true molecular weight. x may be evaluated in a similar manner to n as described in the preceding section. The main disadvantage of this method is the large number of free acidic and basic groups which are present in most protein molecules and which thus yield very small minimal molecular weights with attendant difficulties in the evaluation of x.

Titration with acidic and basic dyes has also been used to determine the combining weights of proteins, and in the special case of respiratory proteins, combination with oxygen has been accurately measured (see Problem 1.6).

The minimal molecular weight of insulin has been determined by reacting the hormone with 1-fluoro-2, 4-dinitrobenzene under conditions which permitted the introduction of 1·5 moles of reagent per insulin molecule, based on an assumed molecular weight of 12,000. The reaction products were subjected to countercurrent distribution and this enabled the mono-substituted derivative to be separated from unsubstituted and di-substituted entities. Colorimetric analysis for the dinitrophenyl group of a known concentration of the mono-substituted insulin then

indicated a molecular weight of about 6000, corresponding to one
A and one B chain per insulin unit of this molecular weight.

Example 1.3.—The maximal acid-combining capacity of egg albumin is
8.7×10^{-4} equivalents per gram protein. Calculate the minimal molecular
weight. The molecular weight of this protein, as determined by diffusion and
sedimentation velocity measurements, is 43,800. Determine the approximate
number of basic groups per molecule.

8.7×10^{-4} equivalents of acid combine with 1 g protein

$\therefore$ 1 equivalent of acid combines with $\dfrac{1}{8.7 \times 10^{-4}}$

$= 1149$ g protein.

Hence the minimal molecular weight is 1149.
The number of basic groups per egg albumin molecule will therefore be

$$\frac{43800}{1149} \simeq 38.$$

(c) END-GROUP ANALYSIS

This method presupposes a known or postulated structure of
the molecular units on which basis the results are interpreted.
Not unnaturally, this constitutes a major limitation to its usefulness. Furthermore, where a linear chain type of structure is
assumed, any branching, unless quantitatively assessed, will
introduce error. The classical example of end-group analysis is
the estimation of the chain length of cellulose by Haworth and
his collaborators. Assuming that cellulose consists of linear
chains of glucose molecules joined by 1, 4 linkages, the cellulose
is completely methylated by the gentlest possible means and then
hydrolysed under conditions which permit breakage of glucose-
glucose links but not hydrolysis of the methyl groups. At one
end of each cellulose chain will be a 2,3,4,6-tetramethyl glucose
molecule, whereas all the other members of the chain will be
2,3,6-trimethyl glucose units. Accordingly analysis of the
products of hydrolysis permits an estimate of the chain length.
In this way the value of 100 to 200 β-glucose units, corresponding
to a molecular weight of 20,000 to 40,000, was obtained. A
further difficulty encountered in this work is the possibility of
degradation of the chain during the methylation procedure so
that the values must be regarded as minimal limits.

2. Physico-chemical Methods

These are based on molecular kinetic theory and fall into two
main categories depending on (1) the colligative properties of

solutions, i.e. dependent on the *number* of molecular units present per given volume and (2) the *weight* of the units present. The number-average methods include measurement of osmotic pressure, spread monolayers, as well as end-group assay, whilst weight-average methods embrace sedimentation equilibrium, viscosity and light-scattering.

The number-average molecular weight, M_n, is given by the expression

$$M_n = \frac{\sum\limits_i n_i M_i}{\sum\limits_i n_i} = \frac{C_i \sum\limits_i}{\sum\limits_i n_i} \quad . \quad . \quad . \quad (1.3)$$

and the weight-average value, M_w, by

$$M_w = \frac{\sum\limits_i n_i M_i^2}{\sum\limits_i n_i M_i} = \frac{\sum\limits_i C_i M_i}{\sum\limits_i C_i} \quad . \quad . \quad . \quad (1.4)$$

where n_i is the number of gram molecules of molecular weight M_i, and C_i is their concentration, equal to $n_i M_i$, and the summation is taken over all values of i.

The number-average and weight-average molecular weights afford information as to the dispersity of high molecular weight substances in solution. If a protein consists of molecular units all of the same size, it is said to be *monodisperse* and the molecular weight determined by number-average and weight-average methods will be the same. But if it consists of molecular units of different sizes, i.e. if it is *polydisperse*, the determined molecular weights are not the same, the weight-average figure always being greater than the number-average value. The ratio of these averages is a rough indication of the *polydispersity* of the protein.

Example 1.4.—The composition of a protein corresponds to 5 moles of molecular weight 15,000 and 10 moles of molecular weight 30,000.
The number-average molecular weight

$$= \frac{5 \times 15000 + 10 \times 30000}{5 + 10} = 25,000,$$

and the weight-average molecular weight

$$\frac{5 \times (15000)^2 + 10 \times (30000)^2}{5 \times 15000 + 10 \times 30000} = 27,000.$$

In certain cases, such as with synthetic polymer samples, which usually consist of varying molecular chain lengths, it is often desirable to place greater emphasis on the higher molecular weight species in the distribution. A Z-average molecular weight, M_z, is then calculated from the expression

$$M_z = \frac{\sum\limits_i n_i M_i^3}{\sum\limits_i n_i M_i^2} = \frac{\sum\limits_i C_i M_i^2}{\sum\limits_i C_i M_i} \qquad . \qquad . \qquad . \qquad (1.5)$$

For some purposes a $(Z+1)$ average is used, given by

$$M_{z+1} = \frac{\sum\limits_i n_i M_i^4}{\sum\limits_i n_i M_i^3} \qquad . \qquad . \qquad . \qquad (1.6)$$

Values for the molecular weight obtained by viscometry are expressed as a *viscosity average*, M_v.

$$M_v = \left[\frac{\sum\limits_i n_i M_i^{1+a}}{\sum\limits_i n_i M_i} \right]^{\frac{1}{a}} \qquad . \qquad . \qquad . \qquad (1.7)$$

where a is the exponent in the Mark-Houwink equation (1.48) which relates intrinsic viscosity to molecular weight ($[\eta] = K'M^a$, p. 43). K' and a are constants for the particular solute-solvent-temperature combination; a depends on the shape of the solute molecule and usually has values ranging from 0·5 to 0·8 for flexible random coils. M_v varies accordingly but generally is much closer to the weight-average than to the number-average molecular weight. When $a = 1·0$ then $M_v = M_w$.

The number-average and weight-average molecular weights are the most frequently used. The relationship between the different averages for a condensation polymer such as a polyester is shown in Fig. 1.1 where the weight fractions of the polymers are plotted as a function of the molecular weight. The weight fraction, w_i, is given by $W_i / \sum\limits_i W_i$, where W_i is the weight of the ith species. It will be noted that M_v is represented by a broad band rather than a single line, in consequence of equation 1.7.

The different molecular weight averages are related in the following way

$$M_n < M_v \leqslant M_w < M_z < M_{z+1}$$

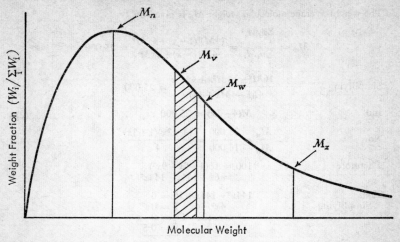

FIG. 1.1

Molecular weight distributions and averages for a condensation polymer. The values are superimposed on the 'most probable' molecular weight distribution for such a polymer. The polydispersity ratio (M_w/M_n) is 2·0 in this distribution.

(ROWLAND, 1967.)

Example 1.5.—A macromolecule P_4 dissociated into the monomer P at low concentrations

$$P_4 \rightleftharpoons 4P.$$

The system, in equilibrium, at a concentration of 1 g per litre at 25° C, was examined by light scattering and osmotic pressure.

The molecular weight by light scattering = 25,000
The molecular weight by osmotic pressure = 16,000

Calculate the free energy change of the dissociation.

(Hull Honours Course Finals, Part II, 1967.)

Let M be the molecular weight of the monomer P and x the degree of dissociation. Then

$$\begin{array}{cc} P_4 & \rightleftharpoons \quad 4P \\ (1-x) & 4x \end{array}$$

and the molecular weight of P_4 is $4M$. Osmotic pressure and light scattering measurements yield, respectively, number- and weight-average molecular weights. The number-average molecular weight M_n is given by

$$M_n = \frac{\sum\limits_i n_i M_i}{\sum\limits_i n_i} = \frac{4M\,(1-x)+4Mx}{(1-x)+4x} = 16,000$$

Therefore

$$\frac{4M}{1+3x} = 16,000 \ . \quad . \quad . \quad . \quad . \quad (1)$$

The weight-average molecular weight M_w is given by

$$M_w = \frac{\sum\limits_{i} n_i M_i^2}{\sum\limits_{i} n_i M_i} = \frac{(4M)^2(1-x)+4xM^2}{4M(1-x)+4xM} = 25,000$$

Simplifying

$$\frac{16M^2-16M^2x+4M^2x}{4M-4Mx+4Mx} = 25,000$$

and

$$M(4-3x) = 25,000 \quad . \qquad . \qquad . \qquad . \qquad (2)$$

Now

$$\frac{M_w}{M_n} = \frac{25,000}{16,000} = \frac{(4-3x)(1+3x)}{4}$$

Therefore

$$100 = 16(4+9x-9x^2)$$
$$= 64+144x-144x^2$$

Thus

$$144x^2-144x+36 = 0$$

Simplifying

$$4x^2-4x+1 = 0$$
$$(2x-1)^2 = 0$$
$$x = 0.5$$

Since P_4 is at 1 g per litre, 0.5 g per litre of it is dissociated.
Put $x = 0.5$ in equation (2)

$$M.\ 2.5 = 25,000$$
$$M = 10,000$$

Thus the molecular weights are $P = 10,000$ and $P_4 = 40,000$.

The equilibrium constant $K = \dfrac{[P]^4}{[P_4]}$

The molar concentrations at equilibrium are

$$[P_4] = \frac{0.5}{40,000} \quad \text{and} \quad [P] = \frac{0.5}{10,000}$$

Whence

$$K = \frac{\left(\dfrac{0.5}{10,000}\right)^4}{\left(\dfrac{0.5}{40,000}\right)} = \frac{(0.5)^3 \times 4}{(10,000)^3} = \frac{0.5}{10^{12}} = 5 \times 10^{-13}$$

$$\Delta G = -RT \ln K = -2.303\ RT \log K$$
$$= -2.303 \times 1.987 \times 298 \times \log (5 \times 10^{-13})$$
$$= -2.303 \times 1.987 \times 298 \times (-12.301)$$
$$\Delta G = 16,770 \text{ cal mole}^{-1}$$

Molecular weight determinations on proteins are complicated by the shape and the electrical charge of the molecules. The former factor has a profound effect on molecular movement such as that measured in sedimentation and diffusion studies. Spherical molecules behave in a normal manner under these experimental conditions, but elongated, thread-like molecules of fibrous proteins deviate from the normal due to increased frictional and also hydration effects ; as a consequence their rate of diffusion is reduced.

The possibility of aggregation of molecules is increased in concentrated solution. Colligative phenomena, with the exception of osmotic pressure, are of little value for the determination of the molecular weights of macromolecules. Not only are the elevation of boiling point and depression of freezing point or vapour pressure too small to be measured, but, as already mentioned, the presence of traces of low molecular weight compounds, such as salts, would produce an effect equal to or even greater than that of the macromolecule itself. By correct choice of conditions the effect of such compounds can be eliminated in osmotic pressure measurements, the basis of which is now described.

(a) OSMOTIC PRESSURE

If a solution is separated from the pure solvent by a membrane permeable to molecules of solvent but impermeable to those of solute, then solvent molecules pass through the membrane into the solution until equilibrium is reached. The pressure which must be exerted on the solution to prevent the net passage of solvent molecules across the membrane is a measure of the osmotic pressure of the solution. Osmotic pressure was first studied by the botanist Pfeffer in 1877, which perhaps emphasizes the importance of the phenomenon in biological systems. Van't Hoff showed that for very dilute solutions

$$\Pi V = RT \quad . \quad . \quad . \quad . \quad (1.8)$$

where Π is the osomtic pressure, V the volume containing 1 gram-molecule of solute, R the gas-constant and T the absolute temperature. Alternatively,

$$\Pi = CRT \quad . \quad . \quad . \quad . \quad (1.9)$$

where C is the concentration of solute in moles per litre. The value of R is 0·082 when Π is expressed in atmospheres and V in litres. The molar concentration C is equal to c/M, where c is the concentration in grams per litre and M the molecular weight of the solute. Thus :

$$M = \frac{cRT}{\Pi} \quad . \quad . \quad . \quad (1.10)$$

and by measuring the osmotic pressure of a solution of known concentration at a given temperature the molecular weight of the solute may be obtained.   The concentration of a solute producing an osmotic pressure equal to that of a molar solution of a perfect solute is referred to as *osmolar*.   Solutions which exert the same osmotic pressure are said to be *iso-osmotic*.   It should be noted that the term *isotonic* is not synonymous with iso-osmotic.   Tonicity is defined in terms of the response of a cell immersed in a solution ; if a cell neither shrinks nor swells the solution is said to be isotonic with it.   While an isotonic solution is generally also iso-osmotic, this is not necessarily so, e.g. an iso-osmotic solution of sodium chloride is isotonic with sea urchin eggs whereas an iso-osmotic solution of calcium chloride is hypotonic to them.

The van't Hoff equation holds only for infinitely dilute solutions and should therefore be written as

$$M = \frac{RT}{(\Pi/c)_{c \to 0}} \qquad . \qquad . \qquad . \qquad (1.10a)$$

In practice, therefore, the osmotic pressures determined experimentally are often considerably above those computed from equation 1.10.   Now, since $\Pi/c = RT/M$, it follows that Π/c

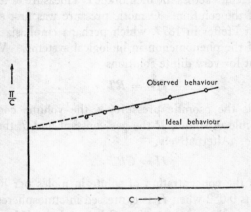

Fig. 1.2

Determination of the osmotic pressure Π at infinite dilution, by extrapolation to zero concentration, for molecular weight calculation.

(sometimes called the reduced osmotic pressure) should be constant for all concentrations if the solution behaves ideally. Accordingly, Π/c is measured at several different concentrations and then Π/c is plotted as a function of c.   In many cases this gives an almost linear curve which may be extrapolated to zero concentration, i.e. infinite dilution, and the Π/c intercept thus obtained used in the van't Hoff equation to evaluate M. This is shown in Fig. 1.2. The slope of the line is a measure of the interaction between solute and solvent and becomes greater when the solvent has a large solvating effect. It is also dependent on the shape of the molecule ; the greater the asymmetry, the greater is the deviation from ideal behaviour because the elongated molecules, by solvation, immobilize solvent molecules. The general equation which represents such behaviour (of which equation 1.10 is the limiting expression) may be written as

$$\frac{\Pi}{c} = RT\left[\frac{1}{M} + Ac + Bc^2 + \quad\right] \qquad . \qquad . \quad (1.10b)$$

and for dilute solutions this can be approximated to

$$\frac{\Pi}{c} = RT\left[\frac{1}{M} + Ac\right] . \qquad . \qquad . \qquad . \quad (1.10c)$$

The slope of the Π/c versus c plot gives a quantitative measure of the energy of interaction, expressed as the second virial coefficient, A in equation 1.10b, which is called the *interaction constant* and is thus a measure of the deviation from ideal behaviour.

With protein solutions there is the additional complication of the Donnan equilibrium (p. 94) whereby the free diffusion of inorganic ions through the membrane is restricted. Here the observed osmotic pressure is produced both by protein molecules and inorganic ions. This arises because proteins are ampholytes and their ionization is affected by pH ; on the acid side of their isoelectric point they exist as cations and on the alkaline side as anions. The Donnan effect may be minimized or even eliminated by making osmotic pressure measurements at the isoelectric point of the protein, but difficulty is often encountered because the protein displays its minimum solubility at its isoelectric point. Thus the observed osmotic pressure of protein solutions is dependent on the pH and is higher in acid and alkaline solution than at the isoelectric point. Also, to reduce the Donnan effect and to eliminate errors caused by traces of salts present in the protein solution, a concentrated salt solution is often used as solvent for the protein and a salt solution of the same concentration is placed on the other side of the semi-permeable membrane. Suitable corrections can be made for the ion effect in these cases.

A difficulty encountered in the determination of molecular weights of macromolecules by osmotic pressure measurements is the long periods of time required for the attainment of equilibrium.

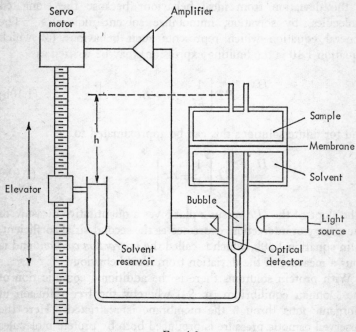

FIG. 1.3

Diagram of a dynamic membrane osmometer with optical detector system and servo-mechanism for reducing the hydrostatic pressure head on the solvent side of the membrane.

(By courtesy of HEWLETT-PACKARD LIMITED)

There is a danger that proteins may be either denatured or contaminated by bacterial growth. To overcome this time factor various dynamic osmometers have been devised. In the original type the rate of flow of the solvent through the membrane, in response to a varied external pressure on the solution side of the membrane, is measured. By plotting the rate of flow against the external pressure and extrapolating to zero rate of flow, the osmotic pressure is obtained. It is equal to the intercept on the pressure axis. An example of this type of experiment is given in Problem 1.25. A recent commercial instrument illustrated diagrammatically in Fig. 1.3, employs the principle of opposing the movement of the solvent through the membrane by reducing the hydrostatic head on the solvent side of the membrane until it exactly equals the osmotic pressure of the solution, and thus prevents net solvent flow across the membrane. Movement of the solvent is detected by an optical system consisting of a light source, focused on a bubble in the capillary leading to the solvent reservoir, in conjunction with a photocell detector. The amplified photocell output activates a servo motor which adjusts the height of the solvent reservoir by means of a screw elevator to return the bubble to its original position. The hydrostatic head of the solvent is thus exactly opposed to the osmotic pressure of the solution and this is indicated on a digital counter. The counter reading at equilibrium, which may be reached in as little as ten minutes, is equivalent to the osmotic pressure of the solution. Such instruments permit the rapid determination of number-average molecular weights within the range of 10^4 to 10^6.

To summarize, the determination of molecular weights from osmotic pressure measurements involves the following assumptions :

1. The system is at equilibrium.
2. The membrane is permeable to solvent molecules but not to solute molecules.
3. The osmotic pressure is proportional to the concentration of the solute (van't Hoff's Law).
4. The molecules of the solute are all of the same size.

With aqueous solutions of proteins and using collodion membranes, assumptions 1 to 3 are normally justified, subject to the provisions already discussed, but 4 may or may not be true and cannot be ascertained by osmotic measurements. It can be decided only by sedimentation and diffusion measurements. Finally, highly purified proteins must be used if the results are to be of any value. The method is only practicable for molecular weights of up to one million, above which the observed osmotic pressure is too small to be measured accurately.

Example 1.6.—An aqueous solution containing 0·388 g of a sugar per 100 ml exerted an osmotic pressure of 380 mm Hg at 10° C. Determine the molecular weight of the sugar.

$$M = \frac{cRT}{\Pi} \qquad \qquad \Pi = 380 \text{ mm} = \frac{380}{760} \text{ atm}$$

$$T = 283°$$

$$c = 3·88 \text{ g per litre}$$

$$M = \frac{3·88 \times 0·082 \times 283 \times 760}{380} = \underline{180}.$$

Example 1.7.—The osmotic pressure of horse haemoglobin in 0·2 M-phosphate was measured by Gutfreund at various concentrations of protein. The following results were obtained at 3° C:

Haemoglobin concentration (g/100 ml)	0·65	0·81	1·11	1·24	1·65	1·78	2·17	2·54
Osmotic pressure (cm H_2O)	2·5	3·1	3·9	4·7	5·7	6·8	8·3	8·9

Haemoglobin concentration (g/100 ml)	2·98	3·52	3·90	4·89	6·06	8·01	8·89
Osmotic pressure (cm H_2O)	11·2	13·4	14·6	19·6	23·9	34·2	38·7

Use these data to determine the molecular weight of haemoglobin.

(After Gutfreund (1949), p. 197, in *Haemoglobin*, ed. Roughton & Kendrew. London Butterworth.)

This problem is solved graphically by plotting the reduced osmotic pressure Π/c against the concentration c. The values of Π/c are worked out as follows and then plotted in Fig. 1.4 :

c	0·65	0·81	1·11	1·24	1·65	1·78	2·17	2·54
Π/c	3·84	3·82	3·51	3·79	3·46	3·82	3·82	3·50
c	2·98	3·52	3·90	4·89	6·06	8·01	8·89	
Π/c	3·76	3·80	3·74	4·00	3·94	4·27	4·36	

(b) Sedimentation

Svedberg, in 1925, invented the ultracentrifuge, an instrument which enables very high centrifugal fields of up to 250,000*g* to be attained routinely. If solutions of macromolecules are subjected to such fields, the molecules sediment owing to their large mass. Their concentration therefore increases from the centre of the centrifuge to the periphery. The tendency to sediment is opposed by the diffusion of the protein back from the highly concentrated region in the periphery to the more dilute region in

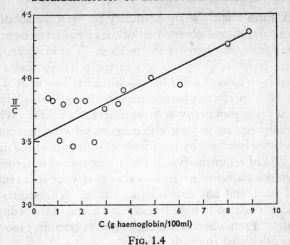

FIG. 1.4

The best straight line drawn through the experimental points is extrapolated to zero concentration and the intercept on the Π/c axis, which is equal to 3·51, is used in equation 1.7 to evaluate M. Note particularly that since Π is expressed in cm H_2O the appropriate value of the gas constant R must be used. This is $8·471 \times 10$ litre cm H_2O per mole per degree (see Appendix 2), and since c in equation 1.7 refers to grams per litre, whereas the concentrations given are in grams per 100 ml, the value obtained for Π/c must be divided by 10 to obtain the correct units, i.e. Π/c becomes 0·351.

$$\text{Hence } M = \frac{8·471 \times 10 \times 276}{0·351} = \underline{66{,}600}.$$

the centre. At relatively low speeds an equilibrium is reached, when sedimentation is exactly opposed by the diffusion, and the distribution of protein in the centrifuge tube is then in a steady state. This occurs after a fairly long period of time and is the basis of the *sedimentation equilibrium* method of determining molecular weights. Alternatively, if high centrifugal fields are employed, the rate of sedimentation of macromolecules is greatly in excess of their rate of diffusion and the latter factor becomes of minor importance. This is the *sedimentation velocity* method. To these two earlier methods must now be added the technique employing transient states or the *approach to sedimentation equilibrium*. These techniques will now be treated separately.

SEDIMENTATION VELOCITY.—This is the most widely used method. It consists of the measurement of the rate at which macromolecules move in centrifugal fields of such force that the process of sedimentation greatly exceeds that of free diffusion. The solute molecules, which move outward from the axis of rotation, form a fairly sharp boundary between the solution and the pure solvent, and observation of the rate of movement of this boundary forms the basis of the method. The movement of the boundary is followed by changes in the refractive index and is observed by means of a Schlieren cylindrical lens system similar to that used in electrophoretic measurements. Movements are recorded photographically at given intervals of time. When the solute molecules are moving with constant velocity, the centrifugal force is being balanced by the frictional resistance of the medium.

The rate of sedimentation is usually expressed in terms of the *sedimentation coefficient s*, which is the velocity for unit centrifugal field of force and has dimensions of time. The sedimentation coefficient is related to both the size and shape of the sedimenting molecules. From observation of the rate of boundary movement, *s* can be calculated from the formula

$$s = \frac{dx}{dt} \cdot \frac{1}{\omega^2 x},$$

where x is the distance from the centre of rotation and ω the angular velocity in radians per second.[1] For proteins studied to date, s has values lying between 1 and 200×10^{-13} sec. A sedimentation coefficient of 10^{-13} sec. is termed one Svedberg unit (S) so that a value of 6×10^{-13} would be denoted by 6S.

A solute in a centrifugal field sediments with a centrifugal force per gram-mole of $M(1 - \bar{v}\rho)\omega^2 x$ exactly opposed by the frictional force per gram-mole $F dx/dt$. M is the molecular weight of the solute, $\bar{v}$ the partial specific volume (this is the increment in volume when 1 gram of dry solute is added to a large volume of solution and for most proteins has a value near 0.74), and ρ the density of the solvent (see below). The molar frictional coefficient F is usually obtained from the diffusion coefficient D by the relationship $F = RT/D$, where R is the gas constant in ergs per mole per degree and T the absolute temperature. Hence the molecular weight from sedimentation velocity and diffusion measurements is obtained by equating the centrifugal and frictional forces, and is given by the expression

$$M = \frac{RT}{D(1 - \bar{v}\rho)} \cdot \frac{dx}{dt} \cdot \frac{1}{\omega^2 x}$$

[1] See Appendix 4.

$$M = \frac{RTs}{D(1 - \bar{v}\rho)} \quad . \quad . \quad . \quad . \quad (1.11)$$

Equation 1.11 is referred to as the Svedberg equation. It will be noticed that an independent value for the diffusion coefficient is required which is not necessary in the sedimentation equilibrium method. D is defined as the quantity of material diffusing per second across a surface area of 1 cm² when the concentration gradient is unity. Its measurement is discussed in the next section. ρ in equation 1.11 is usually taken as the density of the solvent and not of the solution although there is no universal agreement about this. In actual practice the distinction is not important.

Example 1.8.—Diphtheria toxin has been subjected to ultracentrifugal analysis and the following data obtained at 20°. The average sedimentation coefficient was $4{\cdot}6 \times 10^{-13}$ sec. and the average diffusion coefficient $5{\cdot}96 \times 10^{-7}$ cm²/sec. These analyses were carried out on solutions of the protein at various concentrations over the range 0·38 to 0·61 per cent. The density of water at 20° is 0·998. Obtain a value for the molecular weight of diphtheria toxin. The toxin has a partial specific volume of 0·736.

(After PETERMANN & PAPPENHEIMER, Jr. (1941), *J. phys. Chem.*, 45, 1.)

The data given are substituted in equation 1.11.

$$M = \frac{RTs}{D(1 - \bar{v}\rho)}$$

R, in this case, has the value $8{\cdot}314 \times 10^7$ ergs per mole per degree.

Thus,
$$M = \frac{8{\cdot}314 \times 10^7 \times 293 \times 4{\cdot}6 \times 10^{-13}}{5{\cdot}96 \times 10^{-7} \times (1 - 0{\cdot}736 \times 0{\cdot}998)}$$
$$= \underline{70{,}850.}$$

SEDIMENTATION EQUILIBRIUM.—The centrifuge is operated at a relatively low speed, e.g. about 8000 r.p.m. for a protein with a molecular weight of 60,000. During the early stages of the experiment the concentration of solute decreases at the meniscus and increases at the bottom of the cell, due to sedimentation. However, on account of back diffusion, there is no region of pure solvent formed and the concentration of solute will remain finite at the meniscus provided the centrifuge is not run at too high a speed. For example, under conditions which are ideal for this type of work, the concentration of solute at the bottom of the cell is about double and that at the meniscus about one-half of its original value. When the centrifuge has been operated for a sufficient length of time, equilibrium is attained and the concentration of the

B

solute as a function of the distance from the axis of rotation undergoes no further change, i.e. there is no net movement of solute in the cell.

The amount of solute transported by sedimentation across a given surface in the ultracentrifuge cell in unit time, $(dm/dt)_S$, is equal to the product of the concentration, c, of solute at the surface, the area of the surface, A, and the velocity of sedimentation of the solute. The velocity of sedimentation is given by the product of the sedimentation coefficient s and the centrifugal field, namely $s\omega^2x$, therefore

$$\left(\frac{dm}{dt}\right)_S = cAs\omega^2x.$$

The amount of solute transported across the same surface in the opposite direction as a result of diffusion, $(dm/dt)_D$, is, by Fick's First Law of Diffusion (see page 31), equal to the product of the diffusion coefficient D, the area of the surface and the concentration gradient, dc/dx. Thus

$$\left(\frac{dm}{dt}\right)_D = -DA\frac{dc}{dx}$$

where dc is the change in solute concentration over distance dx. At equilibrium the amount of solute driven in time dt by the centrifugal force in the direction of the periphery through unit surface is exactly the same as the increment moved by diffusion in the opposite direction. Hence

$$\left(\frac{dm}{dt}\right)_S = \left(\frac{dm}{dt}\right)_D$$

and

$$c s\omega^2x = D\frac{dc}{dx}.$$

Therefore

$$D = \frac{cs\omega^2x}{\dfrac{dc}{dx}}.$$

If this value for D is now substituted in the Svedberg equation (1.11) we obtain

$$M = \frac{RT(dc/dx)}{(1 - \bar{v}\rho)\omega^2cx}$$

or
$$M = \frac{2RT}{(1 - \bar{v}\rho)\omega^2} \cdot \frac{d \ln c}{d(x^2)} \qquad . \qquad . \quad (1.12)$$

Thus if c_2 and c_1 are the concentrations of solute at distances x_2 and x_1 from the centre of rotation

$$M = \frac{2RT \ln(c_2/c_1)}{\omega^2(1 - \bar{v}\rho)(x_2{}^2 - x_1{}^2)} \qquad . \qquad . \quad (1.12a)$$

Hence, at equilibrium and if the solute is monodispersed, measurement of the relative concentrations of solute at two distances from the centre of rotation enables the molecular weight to be evaluated. Inhomogeneity with respect to molecular weight will be revealed by differences in the value of M if a series of values of c are determined as a function of x.

It should be noted that, unlike the sedimentation velocity method, knowledge of the diffusion coefficient is not required. The great disadvantage of this method is the long period of time required to attain equilibrium which, in the case of proteins, is several days. However, since the time to attain equilibrium varies as the square of the depth of liquid in the ultracentrifuge cell, modern technique employs a depth of 1 to 3 mm instead of about 15 mm and thus a great saving of time is achieved. A layer of silicone oil is placed in the cell to define the cell bottom clearly and a Rayleigh interference optical system is used to show the concentration change.

Yphantis (1964) has described a short column equilibrium method variously referred to as the high speed or *meniscus depletion method*. In principle it is the same as the method just described but the centrifuge is operated at speeds about three times greater than usual in order that the macromolecular solute is sedimented out of the region of the cell near the meniscus and the meniscus concentration can be taken as zero. Concentrations in the cell are then determined as being proportional to the refractive index differences between the meniscus region and the radii of revolution of interest. This method is extremely useful for the determination of the molecular weights of monodisperse solutes on account of rapidity, simple computation and the fact that much lower initial concentrations can be studied, e.g. 15-100 μg. With polydisperse systems number-, weight- and Z-average molecular weights can be determined and, in many cases, the molecular weight of the smallest macromolecular species present can be obtained. The lower limit for the meniscus depletion technique is a molecular weight of about 10,000 because the ultracentrifuge cannot be run at a speed which would be necessary to secure the condition of zero solute concentration at the meniscus for molecular weights lower than this value.

Example 1.9.—The outer surface of the protozoon *Paramecium aurelia* is covered with a layer of a soluble protein, the immobilization antigen, which constitutes about 1 per cent. of the total cell mass. In studies of the genetic control of the primary structure of the immobilization antigen several different types have been isolated in pure form by ion exchange chromatography. One of these, antigen 90 D, was examined in a meniscus depletion type of experiment at 10,500 rev/min and the following values of a concentration function were measured:

x(cm)	$f(c)$
6·65	0·0000
6·70	0·0135
6·72	0·0245
6·74	0·0430
6·76	0·0760
6·78	0·1380

where x is the distance from the centre of rotation. The experiment was carried out at a temperature of 27°. Calculate the molecular weight.

$$R = 8·31 \times 10^7 \; ; \; (1 - \bar{v}\rho) = 0·30 \; ; \; \pi = 3·142$$

(Data of I. G. JONES, used for part of a question in the Hull Honours Course Finals, Part II, 1969.)

The appropriate equation for sedimentation equilibrium (equation 1.12) is

$$M = \frac{2RT}{(1 - \bar{v}\rho)\omega^2} \cdot \frac{d \ln c}{d(x^2)}$$

Consequently a plot of $\ln f(c)$ versus x^2 will give a slope equal to $d \ln c/d(x^2)$ from which value and the other data provided M may be evaluated. For convenience $\log f(c)$ is plotted.

x(cm)	x^2	$f(c)$	$\log f(c)$
6·65	44·22	0·0000	
6·70	44·89	0·0135	$\bar{2}$·1303
6·72	45·16	0·0245	$\bar{2}$·3892
6·74	45·43	0·0430	$\bar{2}$·6335
6·76	45·70	0·0760	$\bar{2}$·8808
6·78	45·97	0·1380	$\bar{1}$·1399

Log $f(c)$ is plotted versus x^2 in Fig. 1.5 from which the slope is determined as 0·93.

Thus $\dfrac{d \log c}{dx^2} = 0·93$ and $\dfrac{d \ln c}{dx^2} = 2·303 \times 0·93$

The angular velocity of the ultracentrifuge ω is given by

$$\omega = 2\pi \times \text{rev/sec}$$
$$= 2 \times 3·142 \times \frac{10500}{60}$$

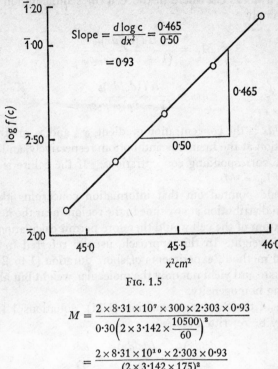

$$\text{Slope} = \frac{d \log c}{dx^2} = \frac{0.465}{0.50}$$

$$= 0.93$$

FIG. 1.5

Hence

$$M = \frac{2 \times 8.31 \times 10^7 \times 300 \times 2.303 \times 0.93}{0.30 \left(2 \times 3.142 \times \frac{10500}{60}\right)^2}$$

$$= \frac{2 \times 8.31 \times 10^{10} \times 2.303 \times 0.93}{(2 \times 3.142 \times 175)^2}$$

$$M = 294,500.$$

APPROACH TO SEDIMENTATION EQUILIBRIUM.—On account of the long periods necessary for sedimentation equilibrium to be achieved, considerable attention has been directed to the theoretical analysis of the transient states before equilibrium is reached and while the solute is being redistributed. At sedimentation equilibrium the condition holding throughout the cell is that there is no net movement of solute. Archibald (1947) drew attention to the fact that, as the meniscus and the bottom of the cell are both impervious to the flow of solute, there can be no net transport of solute *across* either the meniscus or the bottom. Consequently at these bounding surfaces the situation is exactly analogous to conditions of sedimentation equilibrium (i.e. no net transport of solute) and therefore at these radii of revolution the formal equation which holds throughout the experiment is identical with that which holds elsewhere in the cell for sedimentation equilibrium, namely

$$M_m = \frac{RT(dc/dx)_m}{(1 - \bar{v}\rho)\omega^2 x_m c_m} \qquad . \qquad . \quad (1.12b)$$

$$M_b = \frac{RT(dc/dx)_b}{(1 - \bar{v}\rho)\omega^2 x_b c_b} \qquad . \qquad . \quad (1.12c)$$

where dc/dx is the concentration gradient, x_m and x_b are the radii of revolution at the meniscus and bottom respectively, and c_m and c_b are the corresponding concentrations. If the solute is homogeneous $M_m = M_b$.

Archibald pointed out that information concerning the concentration distribution at any time in the regions near the meniscus or the bottom of the cell would therefore permit calculation of the molecular weight. In this approach, usually referred to as 'the Archibald method', experiments of short duration (1 to 2 hours) are adequate and yield not only the molecular weight but also data concerning homogeneity.

By use of the Svedberg equation (1.11), equations 1.12b and 1.12c may be rewritten as :

$$\frac{(dc/dx)_m}{x_m c_m} = \frac{(dc/dx)_b}{x_b c_b} = \frac{\omega^2 s}{D} \qquad . \qquad . \quad (1.13)$$

if it is assumed that the solute is homogeneous. This equation reveals that the concentration of solute at the bounding surfaces changes continuously as dc/dx changes, in such a manner that, at constant rotor speed, $(1/cx)(dc/dx)$ remains constant.

The Archibald method can also be applied to polydisperse systems, giving the average molecular weight and information concerning the homogeneity or heterogeneity of the system. Equations 1.12b and 1.12c can be written for each of the molecular species present and the equations rearranged to give $C_i M_i$. If these equations are now summed and divided by $\sum_i C_i$, which is equal to the observed concentration C^{obs}, the weight average molecular weight M_w is obtained for the meniscus

$$M_w = \frac{RT}{(1 - \bar{v}\rho)\omega^2}\left(\frac{dc}{dx}\right)_m^{obs} \frac{1}{x_m c_m^{obs}} \qquad . \qquad . \quad (1.14)$$

A similar equation may be obtained for the bottom of the cell. Now M_w at both the meniscus and the bottom of the cell is a function of time and represents the molecular weight at a given instant. As the experiment proceeds, the larger molecules concentrate at the cell bottom and deplete the meniscus so that M_w increases and decreases respectively in these regions. The weight average molecular weight of the original solution is obtained by extrapolation to zero time of the data obtained for the meniscus and bottom before redistribution has occurred.

The operating velocity follows the approximate formula of 15,000 r.p.m. for a protein of molecular weight of 10,000 and 3000 r.p.m. for a protein of molecular weight of 200,000. A heavy rotor, weighing 22 lb., is available for low-speed Archibald runs down to 2000 r.p.m. This rotor obviates the disadvantage of the low speed wobble usually associated with the customary 7 lb. rotor when running slowly. The centrifuge is operated and photographs of the boundaries are taken at intervals. Fig. 1.6 illustrates a typical example. Measurement of the heights z, on enlarged photographs of the Schlieren curves for a series of, say, ten equally spaced values of r, defines the curve at the meniscus. Information obtained from these data is used in conjunction with the area under the curve (hatched in Fig. 1.6), suitably measured, e.g. by planimetry, to evaluate $(dc/dx)_m/x_m c_m$, after taking into account the relevant magnification factors. A value for $(dc/dx)_b/x_b c_b$ can be obtained in a similar manner.

DENSITY GRADIENT CENTRIFUGATION.—A relatively new technique in sedimentation, and one which has led to considerable advances in recent years, is that of density gradient centrifugation. A preparative ultracentrifuge is used with a swing-out bucket rotor and the separation is carried out in plastic centrifuge tubes. Two types of gradient are employed, namely *continuous gradients*, produced automatically by gradient building devices, and *discontinuous gradients*, achieved by carefully layering solutions of different density by hand, one above another, usually with the heaviest at the bottom of the tube. Density gradients can be used in two distinct ways:

1. *Zone or velocity centrifugation.* In this method a sucrose gradient is used and the solution of macromolecular particles is introduced as a thin layer at the top of the tube and centrifuged into the gradient column. As the centrifugal field is

applied, the particles migrate as distinct bands through the gradient, each zone consisting of one type of molecule. The bands are prevented from spreading as a result of convection by the density gradient. The sedimentation process is in accordance with the equations already derived for sedimentation velocity in the ultracentrifuge (p. 18). After a given

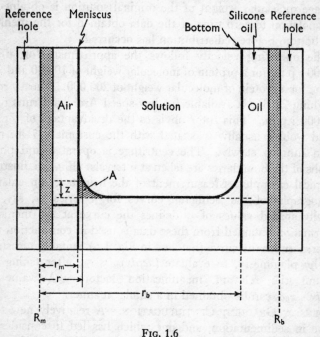

FIG. 1.6

Schlieren pattern obtained during approach to sedimentation equilibrium. The reference edges at the meniscus end (R_m) and the bottom end (R_b) of the diagram represent points of standard distance from the axis of rotation. The distances between R_m and the meniscus (r_m), and between R_m and the bottom (r_b) can be measured on the diagram and converted to x_m and x_b, the distances of the meniscus and the bottom respectively, from the axis of rotation. Sedimentation is from left to right.

length of time, the centrifuge is stopped, the tubes removed and the various fractions collected from either the bottom or top of the tube. In the former instance the bottom of the tube is pierced with a hypodermic needle and the solution collected in pre-determined fractions and monitored. In the latter case a denser fluid is forced up through the bottom of the tube and the fractions are again collected and monitored.

2. *Equilibrium density or isopycnic centrifugation.* The macromolecules are allowed to sediment through a density gradient which is constructed so as to include their own density. When equilibrium is established the particles will have come to rest in a zone centred around a position corresponding to their own buoyant density, the isopycnic point. These bands can be made quite narrow by careful attention to conditions. In some cases the gradient may be formed by the centrifugal field during the course of the experiment. Initially the low molecular weight solute (often in these experiments inorganic salts such as caesium chloride and rubidium chloride are used) and the macromolecules are distributed uniformly throughout the solution. The duration of these experiments is of necessity longer than with pre-formed gradients.

Density gradient centrifugation has been widely used for preparative and analytical purposes and has been applied to proteins, nucleic acids, ribosomes, mitochondria, microsomes, nuclei and viruses. It is particularly suitable for subcellular fractionation and has been extensively developed by de Duve and his colleagues.

Frequently in density gradient centrifugation it is convenient to make a comparison of an unknown macromolecule with a marker or standard of known s and molecular weight which is centrifuged in the same tube. For proteins, various enzymes now find general application as standards, e.g. catalase, $M = 250,000$, $s_{20, w} = 11·36S$; lysozyme, $M = 17,200$, $s_{20, w} = 2·15S$.

The sedimentation process is in accordance with the equations already discussed in connexion with sedimentation velocity in the ultracentrifuge (p. 18). If we consider the general case of a particle of volume V and density P_ρ (rather than a macromolecule of molecular weight M and partial specific volume $\bar{v}$) the equation may be written as

$$s = \frac{dx}{dt} \cdot \frac{1}{\omega^2 x} = \frac{V(P_\rho - P_m)}{f}$$

where P_m is the density of the medium. If the particle is assumed to be spherical and Stokes' Law (p. 32) is applied then, since $V = 4\pi r^3/3$,

$$s = \frac{2(P_\rho - P_m)r^2}{9\eta} \qquad \cdot \qquad \cdot \qquad \cdot \qquad \cdot \qquad (1.15)$$

In the case of an ellipsoid of revolution a corrected form of equation 1.15 is applied by introducing a numerical shape factor, θ, tabulated by Svedberg and Pederson, for ellipsoids of different axial ratios

$$s = \frac{2(P_\rho - P_m)r^2}{9\eta\theta} \qquad . \qquad . \qquad . \qquad (1.15a)$$

A theoretical treatment of particle distribution in density gradient sedimentation has been presented by de Duve, Berthet and Beaufay (1959). Since these experiments are normally carried out in cylindrical centrifuge tubes (as opposed to the sector shaped cells of the analytical ultracentrifuge) and the centrifugal field is radial from the centre of rotation, its direction is not parallel to the walls of the tube, but diverges progressively down the tube. The lateral component of the field causes many more particles to strike the side of the tube than would be the case with a parallel field and the behaviour of these particles after striking the walls will profoundly influence their eventual distribution. Two extreme situations exist (1) the particles may adhere to the wall and be removed from the medium, resulting in a progressive depletion of the suspension during sedimentation and (2) all the particles may rebound into the medium, i.e. a situation analogous to the application of a centrifugal field parallel to the walls of the tube with little depletion until the particles reach the bottom of the tube.

de Duve *et al.* compensated for this effect by introducing into the sedimentation equation a factor γ, defined as the fraction of particles adhering to the walls and sedimenting immediately. $1 - \gamma$ is thus the fraction rebounding from the walls into the medium and the two limiting cases correspond to $\gamma = 1$ and $\gamma = 0$ respectively.

Variation of particle concentration c with t and radial distance x can be expressed by the partial differential equation

$$\frac{\delta c}{\delta t} + \frac{\delta c}{\delta x} = -c\left(\gamma \frac{dx}{dt} \cdot \frac{1}{x} + \frac{\delta(dx/dt)}{\delta x}\right) \qquad . \qquad (1.16)$$

If particle interactions can be neglected, dx/dt is a function of x only and

$$\frac{dx}{dt} = s\omega^2 \phi x \qquad . \qquad . \qquad (1.17)$$

where ϕx is a function determined principally by the density gradient. If equation 1.17 is introduced into equation 1.16 the latter can be integrated to give

$$\int_{x_i}^{x} \frac{dx}{\phi x} = s \int_{0}^{t} \omega^2 dt = sW \simeq s\omega^2 t . \qquad . \qquad (1.18)$$

$$\frac{c}{c_i} = \left(\frac{x_i}{x}\right)^{\gamma} \frac{\phi(x_i)}{\phi(x)} \qquad . \qquad . \qquad . \qquad (1.19)$$

where c_i and x_i are the initial concentration and position of the particles at $t = 0$, and c and x are the corresponding values after time t. W is the time integral of the squared angular velocity as defined by equation 1.18.

Equation 1.18 defines the position of the particles after time t and equation 1.19 defines the profile of particle concentration in the centrifuge tube. To apply these equations a knowledge of the value of ϕx is demanded. The simplest case is that of a monodisperse suspension in a homogeneous medium, where $s\phi x = sx$ and consequently

$$sW = \ln \frac{x}{x_i} . \qquad . \qquad . \qquad . \qquad (1.20)$$

and

$$\frac{c}{c_i} = \left(\frac{x_i}{x}\right)^{1+\gamma} \qquad . \qquad . \qquad . \qquad . \qquad (1.21)$$

Eliminating x_i between the equations gives

$$\ln \frac{c}{c_i} = - (1+\gamma)sW \qquad . \qquad . \qquad (1.22)$$

If the radial distances of the meniscus and the bottom of the tube are respectively x_m and x_b, sedimentation is complete when

$$\frac{x_b}{x_m} = e^{sW} \qquad . \qquad . \qquad . \qquad . \qquad (1.23)$$

and this equation may be used to calculate the amount of material of sedimentation coefficient s deposited after any given time t of sedimentation at an angular velocity of ω radians/sec.

In the case of a monodisperse suspension in a density gradient, de Duve *et al.* showed that the function $\phi(x)$, which is non-linear in this instance, could be expressed approximately by a polynomial of the type

$$\phi(x) = \frac{B}{n + px + qx^2 + rx^3} \qquad . \qquad . \quad (1.24)$$

Insertion of this function into equations 1.18 and 1.19, with numerical evaluation, enables values of x and c to be obtained from given values of sW and x_i. For more complex situations involving polydisperse suspensions in a density gradient the reader is referred to de Duve *et al.* (1959).

(c) DIFFUSION

When a system is at equilibrium, the distribution of any molecular species is uniform throughout any single phase. Thus, if a protein solution is placed in contact with pure solvent, the protein molecules diffuse from the region of high concentration into the solvent until equilibrium is achieved. This molecular movement or diffusion stems from the thermal energy of the molecules. The speed with which a molecule moves is characterized by its *diffusion coefficient* which is a function of the size and shape of the molecule. Knowledge of the diffusion coefficient enables the molecular weight to be determined ; it is also necessary for ultracentrifuge studies and affords information as to the shape of molecules in solution.

Two types of diffusion may be recognized—translational and rotational. For the former, free diffusion is permitted to occur, or it may be modified by the application of an external force (such as centrifugal force in the sedimentation equilibrium technique). Rotational diffusion is characteristic of the type of molecule. A spherical molecule is characterized by a single rotary diffusion coefficient, an ellipsoid of revolution by two coefficients, one for each axis, and a general ellipsoid molecule by three coefficients, one corresponding to each of the axes. Rotary diffusion coefficients are determined by flow birefringence and, in the case of molecules that are dipolar, electrical methods.

A little explanation is, perhaps, advisable at this point on the subject of ellipsoids. Ellipsoids of revolution are termed prolate if they are elongated and oblate if flattened. This is made clear in Fig. 1.7, where it will be noticed that the semi-axis of revolution a is greater than the equatorial semi-axis b for a prolate ellipsoid but less than b in the case of an oblate one. The axial ratio a/b is usually employed to describe these ellipsoids. In most work asymmetric molecules are assumed to be ellipsoids of revolution on account of the complexity of treatment involved if they are considered as general ellipsoids.

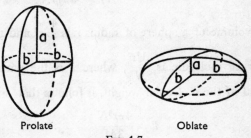

Prolate   Oblate

FIG. 1.7
Ellipsoids of revolution.

TRANSLATIONAL DIFFUSION.—Fick, assuming analogy with the conduction of heat, stated that the mass of solute ds diffusing in time dt across an area A is proportional to the concentration gradient dc/dx at that point and to the area A. This is Fick's First Law and is expressed thus :

$$ds = - DA \frac{dc}{dx} dt,$$

where dc is the change in solute concentration over distance dx and D is the diffusion coefficient. D is defined as the quantity of material diffusing per second across a surface area of one square centimetre when the concentration gradient is unity. The minus sign indicates that diffusion is in the direction of the lower concentration. For amino acids in water at 20°, D is of the order of 10^{-5} cm²/sec and for proteins 10^{-6} to 10^{-7} cm²/sec. With dilute solutions of non-electrolytes, D is related to the *molar frictional coefficient* F, which is the force which must act on 1 gram-molecule of the diffusing substance to give it a velocity of 1 cm/sec, in the following way

$$D = \frac{RT}{F} \qquad . \qquad . \qquad . \qquad . \qquad (1.25)$$

The molar frictional coefficient $F = Nf$, where f is the force that acts *per molecule* and N is the Avogadro number ($6 \cdot 06 \times 10^{23}$).

For spherical molecules of radius r Stokes' Law, which shows the relationship to the viscosity η of the medium, may be applied.

$$f = 6\pi\eta r \qquad \qquad (1.26)$$

Thus,
$$D = \frac{RT}{F} = \frac{RT}{Nf} = \frac{RT}{6\pi\eta rN} \qquad (1.27)$$

Since the volume of a sphere of radius r is $\frac{4}{3}\pi r^3$ and the volume of a spherical molecule is $\frac{M\bar{v}}{N}$, where $\bar{v}$ is the partial specific volume and M the molecular weight, it follows that

$$\bar{v} = \frac{4\pi r^3 N}{3M} \qquad (1.28)$$

and
$$r = \left(\frac{3M\bar{v}}{4\pi N}\right)^{\frac{1}{3}} \qquad (1.29)$$

Hence the theoretical diffusion coefficient for a spherical molecule D_0 is given by the relationship

$$D_0 = \frac{RT}{Nf_0} = \frac{RT}{6\pi\eta N\left(\frac{3M\bar{v}}{4\pi N}\right)^{\frac{1}{3}}} \qquad (1.30)$$

and this enables M to be evaluated. Furthermore, the frictional force per spherical molecule f_0 may be expressed by:

$$f_0 = 6\pi\eta\left(\frac{3M\bar{v}}{4\pi N}\right)^{\frac{1}{3}} \qquad (1.31)$$

Experimentally determined values of D are usually smaller than those calculated from molecular volumes on the assumption that the proteins are spherically shaped. The ratio D_0/D, which is equal to f/f_0, is a measure of the extent to which the protein molecule deviates in shape from a perfect sphere. This ratio is known as the *dissymmetry constant* or *frictional ratio* and is usually assumed to be related to the ratio of the major and minor axes of an ellipsoid of revolution. Of the proteins investigated at the present time, ribonuclease most closely approaches a spherical molecule with a frictional ratio of 1·04, whereas tobacco mosaic virus, with a ratio of 3·12, is a 'cigar-shaped' molecule.

The study of the shape of protein molecules is complicated by the contribution of hydration which exerts its greatest effect on D_0 values for spherical molecules, the evaluation of which is usually based on the assumption that the protein is anhydrous. Hydration causes a swelling of the molecule by binding water, and this increases the frictional effect. The hydration contribution becomes negligible, however, when the frictional ratio is greater than 2·5.

There are two main methods available for the measurement of translational diffusion, namely analytical and refractometric. In the former, chemical analysis of two solutions of different solute concentration, separated by a porous disc, is carried out over an interval of time. Usually one starts with a fairly concentrated protein solution on one side of the disc and pure solvent on the other. The diffusion coefficient is then given by the expression

$$D = \frac{mh}{ctA} \qquad . \qquad . \qquad . \qquad . \qquad (1.32)$$

where m is the amount of material diffusing in time t, c the concentration of the solution, h the effective distance through which the solute diffuses and A the effective area of the pores of the disc. The ratio h/A is a constant for any given disc and is known as the membrane constant. It may be evaluated by calibration with low molecular weight substances of known diffusion coefficient. Furthermore, it is only necessary to know the percentage of quantity of material which has traversed the membrane ; knowledge of absolute quantities is not needed. Equation 1.32 can be further simplified if the amount of solute contained in 1 ml of the concentrated solution is taken as unity and the amount which has diffused is expressed in this same unit (i.e. as the number of ml. of the concentrated solution containing the quantity diffused) and becomes

$$D = \frac{Q_{ml} \cdot h}{tA} \qquad . \qquad . \qquad . \qquad . \qquad (1.33)$$

where Q_{ml} is the number of millilitres of concentrated solution that contains the amount of substance diffused. For instance, if 0·005 mole of glycerol diffused into water from a 0·1 M solution Q_{ml} would be 50, since 0·005 mole glycerol is contained in 50 ml. of 0·1 M solution.

In the refractometric method the diffusing boundary is observed by means of a Schlieren cylindrical lens system and is recorded photographically. Other methods are, however, available.

Example 1.10.—Northrop and Anson carried out experiments to determine the diffusion coefficient of haemoglobin at 5° by means of the porous disc method. A 1 per cent. solution of haemoglobin was permitted to diffuse for 2·06 days and the amounts of haemoglobin passing through the disc were estimated colorimetrically at intervals of time. The membrane constant was evaluated by the use of hydrochloric acid, lactose and several salts ; the same value of 0·150 was obtained in each case.

Results from one such experiment were as follows :

| Time (days) | 0·75 | 0·92 | 2·06 |
| Haemoglobin diffused (ml concentrated solution) | 0·202 | 0·264 | 0·558 |

Calculate the average diffusion coefficient.

(Data from NORTHROP & ANSON (1929), *J. gen. Physiol.*, 12, 543.)

$$D = \frac{Q\text{ml}}{t} \cdot \frac{h}{A} \qquad \text{and} \; \frac{h}{A} = 0\cdot150$$

$$\text{(i)} \; D = \frac{0\cdot202 \times 0\cdot150}{0\cdot75} = 0\cdot0405$$

$$\text{(ii)} \qquad \frac{0\cdot264 \times 0\cdot150}{0\cdot92} = 0\cdot0430$$

$$\text{(iii)} \qquad \frac{0\cdot558 \times 0\cdot150}{2\cdot06} = 0\cdot0407$$

Hence the average diffusion coefficient is 0·0414 cm² per day.

Example 1.11.—The diffusion coefficient of a 2·5 per cent. solution of haemoglobin in 0·05 M-phosphate buffer pH 6·8 and at 5° had an average value of 0·0420 cm² per day, according to Northrop and Anson. The viscosity coefficient of water at 5° is 0·01519 erg seconds per cm³ and the partial specific volume of dry haemoglobin is 0·75. Taking the molecular weight of haemoglobin to be 67,000 calculate the degree of hydration of the protein.

(Data from KUNITZ, ANSON & NORTHROP (1934), *J. gen. Physiol.*, 17, 365.)

The volume of hydrated molecules in solution is, per mole, $\frac{4}{3}\pi r^3 N$ and the volume of one mole of dry protein $M\bar{v} = 67,000 \times 0\cdot75 = 50,260$. The volume of water of hydration per mole will be given by

$$\frac{4}{3}\pi r^3 N - M\bar{v}.$$

The radius of a hydrated molecule r may be obtained from the diffusion coefficient by means of equation 1.14.

$$r = \frac{RT}{6\pi\eta ND}.$$

Care must be exercised in the units employed for the diffusion coefficient ; it is given as cm^2 per day which must be converted to cm^2 per second before substitution in the above equation, i.e. $0.0420/60 \times 60 \times 24$

$$r = \frac{8.314 \times 10^7 \times 278 \times 60 \times 60 \times 24}{6 \times 3.142 \times 0.01519 \times 6.06 \times 10^{23} \times 0.0420}$$

$$= 2.74 \times 10^{-7} \text{ cm.}$$

This value for r can now be substituted in the first equation and the volume of water of hydration per mole will be

$$\frac{4 \times 3.142 \times (2.74 \times 10^{-7})^3 \times 6.06 \times 10^{23}}{3} - 50,260 \text{ cm}^3$$

$$= 52210 - 50260 = \underline{1950 \text{ cm}^3}.$$

And the water of hydration per gram dry protein $= \dfrac{1950}{67000} = \underline{0.029 \text{ cm}^3}.$

Gel Filtration

Gel filtration is a technique which offers a method for the separation of molecules based on differences in their hydrodynamic radius. The principle employed is the preparation of gel granules which have a variable range of accessible volumes for molecules of different sizes ; this accessibility depends upon the gel porosity. When a solution passes through a given gel small solute molecules can diffuse relatively freely through the network structure of the grains, whereas diffusion of large molecules will be restricted according to the porosity of the network, and those of very large size will be excluded completely. The grains of the gels commercially available, such as Sephadex, are obtained by synthetically cross-linking the bacterial polysaccharide dextran to give a three dimensional network of non-ionic character ; the polar properties of the gel are almost entirely due to the high content of hydroxyl groups. Different degrees of cross-linking are available and these determine the porosity of the polysaccharide network. A low degree of cross-linkage gives a highly porous structure whereas high cross-linkage gives a structure of low porosity.

The grains have a great affinity for water on account of their hydrophilic nature and swell to a size determined by the degree of cross-linking. Thus a gel of given cross-linkage always holds a definite quantity of water expressed as *water regain*, i.e. grams of water per gram of dry substance. The gel is insoluble in water and salt solutions and stable in alkaline solutions and weak acids, although strong acids effect hydrolysis of the dextran.

The water contained in a gel column consists of internal and external water. The volume of internal or inner water, V_i, is equal to the sum of the water content of all gel grains in the column and can be obtained from knowledge of the weight of dry material taken and its water regain. Part of this internal water is water of hydration, which is firmly bound and inactive as solvent. The total volume of external water is referred to as the *void volume* (V_0). If V_g is the volume of the gel matrix then the total volume of the packed bed, V_t, is given by

$$V_t = V_0 + V_i + V_g \quad . \qquad . \qquad (1.34)$$

A solute will be distributed between water in the grains and the surrounding water according to its *distribution coefficient* K_D. When $K_D = 0$ the solute is completely excluded, and when its value lies between 0 and 1 it is partially excluded from the gel. A value of K_D greater than unity indicates that the solute has interacted with the gel matrix, probably by adsorption.

Suppose a sample containing two solutes A and B, with K_D values of 0 and 1 respectively, is introduced into the column. After a volume V_0 has been added solute A will appear in the effluent (Fig. 1.8a). After an additional volume V_i (assuming that the water of hydration is very small in comparison with V_i) the non-excluded solute B appears, and so the effluent then contains A and B in the same concentration ratio as the original sample. The addition of sample is then discontinued and water introduced. Solute A ceases to emerge after a volume V_0, and B after a volume ($V_0 + V_i$) has passed. If the sample volume added is less than V_i, complete separation of the solutes is possible (Fig. 1.8b).

The general expression relating the appearance of a solute in an effluent to its distribution coefficient and the appropriate volumes is

$$V_e = V_0 + K_D V_i \quad . \qquad . \qquad (1.35)$$

where V_e is the elution volume. To achieve complete separation of two solutes with distribution coefficients K^1_D and K^2_D they must therefore be introduced to the column in a volume less than $(K^2_D - K^1_D) V_i$.

The distribution coefficient K_D is primarily determined by the size of the solute molecule and very large molecules have $K_D = 0$

whereas small molecules, such as glucose and ions of sodium chloride, have K_D values between 0·7 and 1.

Gel filtration, in addition to enabling the separation of solutes, permits the determination of approximate molecular weights. For

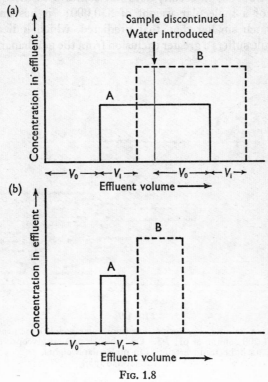

Fig. 1.8

Principles of gel filtration. The sample introduced to the column contains solutes A and B having distribution coefficients $K_D^A = 0$ and $K_D^B = 1$ respectively. In (a) the sample volume is greater than the internal volume V_i of the gel and complete separation cannot be effected. In (b) the sample volume is less than V_i and complete separation of A and B is achieved.

this technique it is necessary to ascertain the elution volumes for molecules of known molecular weight and to construct a graph relating elution volume to the logarithm of the molecular weight (Fig. 1.9). The elution volume of the unknown substance is then determined and the molecular weight obtained by interpolation from the graph. This procedure has the advantages of being rapid and of permitting the molecular weights of different molecules in

the same solution to be estimated in one experiment. It suffers
from the disadvantage that it depends on the physical properties
of the molecules in solution and anomalous results are obtained
with asymmetric molecules. For example fibrinogen, which is a
highly asymmetric molecule, does not conform to the behaviour
expected of a molecular weight of 330,000; it has an elution
volume much smaller than that predicted, which indicates that
the molecule suffers a greater exclusion from the gel than molecules

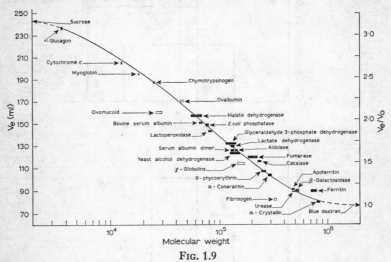

FIG. 1.9

Plot of elution volume, V_e, against log (molecular weight) for proteins on a
Sephadex G-200 column at pH 7·5. Open bars represent glycoproteins. The
lengths of bars indicate uncertainties in molecular weights.
(After ANDREWS, 1965.)

of comparable molecular weight which approximate more closely
to spherical shape in solution. Clearly this consideration limits
the general application of the method to unknown proteins where
problems of asymmetry, hydration or other factors might contribute to behaviour divergent from that of spherical molecules.

In use, the void volume of the column is first obtained by placing
on the top of the column a narrow zone of a solution of a high
molecular weight substance, which will be excluded by the gel,
e.g. dextran blue, and eluting with the solvent. The elution
volume is equal to the void volume. The column is then calibrated
with reference proteins of known molecular weight, the elution

of which is monitored by extinction measurements at 280 nm. In this way the elution volumes of the reference proteins are measured and can be plotted against the logarithm of the molecular weight. The use of small amounts of enzymes having relatively little absorption at 280 nm, and monitored by enzyme assay, is a useful adjunct. Andrews (1965) has carried out considerable work in this manner (Fig. 1.9) and has used the technique to estimate the molecular weights of various enzymes and also of the sub-units of glutamate dehydrogenase.

Inspection of Fig. 1.9 reveals certain anomalies of behaviour. Apoferritin, the protein moiety of ferritin, displays the same elution volume as ferritin, despite their respective molecular weights of 480,000 and 750,000. This may be attributed to the fact that apoferritin has the form of a hollow, approximately spherical shell with an external diameter of about 12 nm and this dimension is unchanged when the central cavity is occupied by a micelle of hydrated ferric oxide to yield ferritin. The anomalous behaviour of fibrinogen has already been mentioned. Glycoproteins do not conform to the general relationship and the presence of carbohydrate appears to produce an expanded form of structure in comparison with globular proteins.

A relationship between the gel-filtration behaviour of proteins and their equivalent hydrodynamic radii, r, has been deduced by various workers. Since, if Stokes' Law is applicable, r is related to the diffusion coefficient D by equation 1.27, then

$$r = \frac{\text{Constant}}{D}$$

and the first relationship may be tested by plotting the elution values of proteins against $1/D$ (Fig. 1.10). Andrews has thus shown that glycoproteins which failed to fit the V_e versus log M relationship now conform very well.

(d) VISCOSITY

Viscosity affords a secondary method for determining molecular weights and depends for its success upon calibration against primary methods of the types previously discussed. The approach used is essentially of an empirical nature. The molecular weight obtained is usually expressed as the *viscosity-average*, M_v, which

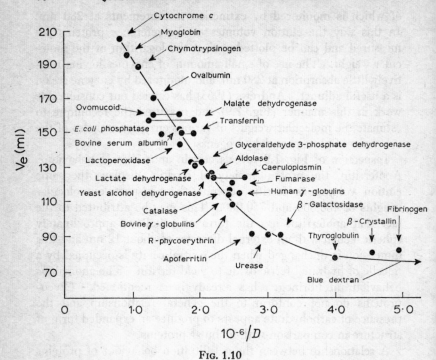

FIG. 1.10

Plot of elution volume, V_e, against reciprocal of diffusion coefficient, D, for gel filtration of proteins on a Sephadex G-200 column at pH 7·5. Different values of $1/D$ for the same protein are joined by a horizontal line.
(After ANDREWS, 1965.)

approximates more closely to the weight-average than the number-average molecular weight (p. 8).

Viscosity is usually determined by one of two methods.

1. THE CAPILLARY FLOW METHOD.—The rate of flow of liquid through a capillary tube of known radius and length and under a known pressure is measured. The method is based on Poiseuille's Law.

2. THE COUETTE OR CONCENTRIC FLOW METHOD.—This consists of two coaxial cylinders with liquid between them; the outer cylinder is rotated at constant speed, and the inner, suspended freely by a torsion wire, has a turning moment exerted on it, the angle through which it is deflected being measured by means of a light beam, mirror and scale.

The Ostwald viscometer is an example of the capillary flow

method where the liquid flows under its own head of pressure. For these conditions

$$\eta = C\rho t \quad . \qquad . \qquad . \qquad . \quad (1.36)$$

where η is the viscosity coefficient, C a constant depending upon the dimensions of the capillary, ρ the density of the liquid and t the time of flow of the liquid between two fixed points on the viscometer. The value of η in c.g.s. units is expressed in dynes per square centimetre per unit velocity gradient, i.e. g cm^{-1} sec^{-1}; these units are referred to as *poise*. The use of poise will be discontinued under the SI and viscosity will be recorded in units of kg m^{-1}s^{-1}, i.e. units tenfold greater than poise. If the times are compared for the flow of a protein solution and for water (viscosity coefficients η and η_0 respectively), then, since the density of water ρ_0 is unity, the *relative viscosity* η_r is given by the expression

$$\eta_r = \frac{\eta}{\eta_0} = \frac{C\rho t}{C\rho_0 t_0} = \frac{\rho t}{t_0} \quad . \qquad . \qquad . \quad (1.37)$$

For very dilute solutions ρ is often also equated with unity.

The *specific viscosity* η_{sp} is defined as

$$\eta_{sp} = \frac{\eta - \eta_0}{\eta_0} = \frac{\eta}{\eta_0} - 1 = \eta_r - 1 \quad . \qquad . \quad (1.38)$$

For spherical particles and very dilute solutions Einstein formulated the equation

$$\frac{\eta}{\eta_0} = \eta_r = \frac{1 + 0 \cdot 5\phi}{(1 - \phi)^2} \quad . \qquad . \qquad . \quad (1.39)$$

which can be expanded into a power series, only the first term of which is significant, and equation 1.39 thus becomes

$$\eta_r = 1 + 2 \cdot 5\phi \quad . \qquad . \qquad . \quad (1.40)$$

where ϕ is the volume fraction of the solute, i.e. the volume occupied by solute molecules relative to the volume of the system. Thus, if n molecules of volume v are present in a total volume of solution V, the volume fraction $\phi = nv/V$, and if n is known it is possible to calculate the volume occupied by each particle or molecule and hence its degree of hydration. If ϕ is divided by the concentration of the solute, the *specific volume fraction* is obtained.

Equation 1.40 may be written in the form

$$\frac{\eta_r - 1}{\phi} = \frac{\eta_{sp}}{\phi} = 2 \cdot 5 \qquad . \qquad . \qquad . \qquad (1.41)$$

so that a plot of η_{sp}/ϕ against ϕ should give a straight line parallel to the ϕ axis and cutting the η_{sp}/ϕ axis with an intercept of 2·5. This affords a method of calculating the hydration of spherical molecules. The intercept should be 2·5, and therefore the additional volume (equal to the hydration) which must be added to the volume of the suspended particles in the dry state to obtain this intercept can be calculated. Equation 1.41 indicates that the relative viscosity at low concentration is independent of both particle size and the nature of the solute, due to the fact that the increased viscosity is caused by indirect action of the particles on solvent flow. Particle size cannot therefore be obtained by use of this equation.

For asymmetric molecules the Einstein equation (1.39, 40) does not hold, and many workers have introduced empirical equations to fit better the observed data. Kunitz proposed the expression

$$\eta_r = \frac{1 + 0 \cdot 5\phi}{(1 - \phi)^4} \qquad . \qquad . \qquad . \qquad (1.42)$$

which, when expanded, approximates to

$$\eta_r = 1 + 4 \cdot 5\phi \qquad . \qquad . \qquad . \qquad (1.43)$$

This approximation proves satisfactory for small values of ϕ. A strict hydrodynamic derivation for hydrated spherical molecules where f/f_0 departs from unity is the following, due to Polson

$$D = \frac{9 \cdot 35 \times 10^{-6}}{M^{\frac{1}{3}} \eta_{sp}^{\frac{1}{3}}} \qquad . \qquad . \qquad . \qquad (1.44)$$

Staudinger studied the viscosity of long chain polymers and related viscosity to molecular weight by the equation

$$\frac{\eta_{sp}}{c} = K_m M \qquad . \qquad . \qquad . \qquad (1.45)$$

where K_m is a constant characteristic of each polymer system and c the concentration, usually in grams of solute per 100 ml solution.

On account of the concentration dependence of η_{sp} the limiting value of η_{sp}/c for infinitely dilute solution is obtained by plotting

η_{sp}/c (termed the reduced viscosity, c.f. reduced osmotic pressure, p. 12) against c and extrapolating to zero concentration. This value, called the *intrinsic viscosity* and denoted by $[\eta]$, is thus defined as

$$\lim_{c \to 0} \left(\frac{\eta_{sp}}{c} \right) = [\eta] \quad . \qquad . \qquad . \quad (1.46)$$

Because of the variation of η_{sp}/c with c, equation 1.45 should be replaced by

$$\lim_{c \to 0} \left(\frac{\eta_{sp}}{c} \right) = [\eta] = K_m M \quad . \qquad . \quad (1.47)$$

An alternative form of equation 1.47 is the Mark-Houwink equation

$$[\eta] = K' M^a \quad . \qquad . \qquad . \qquad . \quad (1.48)$$

where K' and a are empirical constants for the particular solute-solvent-temperature combination. The constant a is a function of the shape of the solute molecule and for flexible random coils has values ranging from 0.5 to 0.8, and for rods is 2.0.

The volume intrinsic viscosity is given by $\lim_{\phi \to 0} \left(\frac{\eta_{sp}}{\phi} \right)$ and for unhydrated spheres is 2.5.

Scheraga and Mandelkern (1953) introduced a new treatment of the hydrodynamic properties of proteins by assuming that the molecule possesses some degree of flexibility and solvation. The configurations of globular type protein molecules in solution are represented in terms of an *effective* hydrodynamic ellipsoid whose axial ratio and size may be determined from accurate measurements of sedimentation coefficient, intrinsic viscosity and molecular weight, all made in the same solvent. The size and shape of this effective or equivalent hydrodynamic ellipsoid are assumed to be such that it can be treated as a rigid ellipsoid which exhibits the same hydrodynamic behaviour as the solvated protein molecule in solution. These workers showed that the axial ratio of this effective ellipsoid of revolution in solution may be calculated from a single function, β, defined by

$$\beta = N\eta \left(\frac{s[\eta]}{1 - \bar{v}\rho} \right)^{\frac{1}{3}} \left(\frac{D}{RT} \right)^{\frac{2}{3}} \quad . \qquad . \quad (1.49)$$

where η is the viscosity of the solvent, ρ the density of the solution, $[\eta]$ the intrinsic viscosity of the solute, and N the Avogadro number. The β coefficient is related to the axial ratio of the ellipsoid. An alternative expression which eliminates the diffusion coefficient is

$$\beta = \frac{Ns\eta[\eta]^{\frac{1}{3}}}{M^{\frac{2}{3}}(1 - \bar{v}\rho)} \qquad \cdot \qquad \cdot \qquad \cdot \quad (1.50)$$

It will be seen that this treatment unites viscosity and sedimentation measurements and the results apply to hydrated molecules. Johnson and Rowe (1961) have used this method in a study of myosin.

Example 1.12.—The relative viscosity of lactose solutions of varying concentration was measured by Kunitz at 25° with the following results :

Lactose concentration g/100 ml.	2·8	5·9	12·1	18·3	25·7
Relative viscosity η_r	1·086	1·187	1·450	1·779	2·371

What is the effect of lactose concentration on the specific volume fraction? Use the empirical equation derived by Kunitz.

(Data from KUNITZ (1926), *J. gen. Physiol.*, 9, 715.)

The expanded form of the Kunitz equation 1.43 is

$$\eta_r = 1 + 4 \cdot 5\phi,$$

whence $\phi = \dfrac{\eta_r - 1}{4 \cdot 5}$ and can be calculated for each value of η_r. Thus for 2·8 g lactose/100 ml $\phi = \dfrac{1 \cdot 086 - 1 \cdot 0}{4 \cdot 5} = 0 \cdot 0191$ and the specific volume fraction $= \dfrac{0 \cdot 0191}{2 \cdot 8} = 0 \cdot 00682$.

Other values are recorded in the following table :

Concentration (c) g/100 ml	Volume fraction (ϕ)	Specific volume fraction (ϕ/c)
2·8	0·0191	0·00682
5·9	0·0415	0·00703
12·1	0·1000	0·00826
18·3	0·1664	0·00909
25·7	0·3046	0·01185

From these figures it would appear that the specific volume fraction increases with increase in lactose concentration, indicating that there is a hydration effect. If, however, the full Kunitz equation 1.42 is used instead of the expanded approximation, reasonably constant values of the specific volume fraction are obtained. This is because equation 1.43 holds only for small values of ϕ. Solving a fourth-power equation is not an easy process however and Kunitz prepared a graph which enabled him to read off values of ϕ from the specific viscosity of the solution.

(e) LIGHT SCATTERING

When light passes through a medium containing suspended particles, a certain proportion of it is scattered in various directions and the incident beam passes through with weakened intensity. If the original light beam is monochromatic, and the scattered radiation is examined spectroscopically, it is found that most of the scattered light has the same wavelength as the incident beam (Rayleigh scattering). A certain amount of the scattered light is, however, of different wavelength as a result of the Raman effect, but the Raman spectrum is very much weaker than the Rayleigh scattering and can be safely ignored in light scattering studies. The Raman effect arises from vibrational and rotational energy changes in the molecule which are associated with a changing polarizability. In the present discussion of light scattering it is assumed that the wavelength of the incident light does not coincide with an absorption maximum of the particles under investigation. Consequently photochemical effects involving chemical changes in the molecules, fluorescence, phosphorescence etc., are assumed to be absent.

If the incident beam of light is passed through a colloidal solution the beam is clearly visible to an observer in a darkened room because of the scattered light (Tyndall effect). This effect has been known for many years and in the ultramicroscope the effect shows the presence of particles too small to be seen by ordinary microscope procedures. Individual particles can be counted by means of the haloes of light scattered from them. Zsigmondy counted the number of particles in a given volume by counting the haloes seen in the ultramicroscope. He calculated the molecular weight of the particles from knowledge of their concentration by weight and Avogadro's number.

The theory of light scattering has been developed over many years and considerable work has been carried out on light scattering by gases. The biochemist is, however, much more concerned with light scattering by macromolecules after correcting for scattering by the solvent itself. The application of the method to macromolecules in solution is largely due to Debye (1944), who pointed out that turbidity measurements of high polymer solutions could be used as an absolute method of determining molecular weights. Light scattering determinations not only yield molecular weights but also information about size, shape and solute-solute and solute-solvent interactions.

Certain experimental considerations favour light scattering measurements. Thus they can be made very rapidly and reaction kinetics (e.g. of polymerization) studied where other methods fail. The information derived from light scattering studies supports and extends that obtained by other methods. The technique has been applied, for example, to the equilibrium between the different molecular weight forms of insulin and also for ovalbumin. Molecular weights obtained for mixtures are weight average figures and the only other technique which gives true weight average figures is that of sedimentation equilibrium in the ultracentrifuge.

The proportion of incident light which is scattered increases as the number and size of the particles in the solution increases. The scattering of light may be defined in terms of τ, the turbidity, which is the extinction coefficient due to scattering, after subtracting the scattering due to solvent alone, and is analogous to the extinction coefficient of a light absorbing molecule. It is defined by

$$I = I_0 e^{-\tau l}$$

where I_0 and I are the intensities of the incident and transmitted light respectively, and l is the length of the light path through the scattering solution. It has been shown that the turbidity τ is related to the molecular weight, M, of light scattering particles by the expression

$$\tau = \frac{32\pi^3 n_0^2 [(n - n_0)/c]^2}{3\lambda^4 N} cM \qquad . \qquad . \quad (1.51)$$

where λ is the wavelength of the radiation, n and n_0 are the refractive indices of solution and solvent respectively, c is the concentration of the scattering particles in grams per ml and N is the Avogadro number.

Equation 1.51 is frequently written in the form

$$\tau = HcM \qquad . \qquad . \qquad . \quad (1.52)$$

where
$$H = \frac{32\pi^3 n_0^2 [(n - n_0)/c]^2}{3\lambda^4 N} \qquad . \qquad . \quad (1.53)$$

Debye found that equation 1.51 can be applied to solutions if the molecules are randomly orientated (the ideal case) and if they are small and isotropic. Ideal behaviour is, of course, manifest only at infinite dilution. The scattering due to the pure solvent is subtracted from that due to the solution in order to apply the equations to the scattering due to solute molecules.

For a given system a plot of Hc/τ against c yields a curve which when extrapolated to zero concentration (i.e. infinite dilution) gives an intercept equal to $1/M$ and thus enables M to be evaluated. This approach may be compared with the plot of the reduced osmotic pressure as a function of c (p. 12). Light scattering molecular weights refer to the unsolvated molecule since the solvation layer has a refractive index very close to n_0.

It will be apparent from equation 1.51 that in addition to careful measurements of the scattered light, the specific refraction increment $(n - n_0)/c$ must be precisely determined since it is squared in the numerator of the equation. It is determined by means of the differential refractometer technique which enables refractive index differences of the order of $0 \cdot 001$ to be measured within $0 \cdot 5$ per cent. error. If n and n_0 are close in value, less light is scattered and advantage is taken of this in infrared spectroscopy by the technique of suspending samples in Nujol to reduce the scattered radiation which would occur if the dry material were investigated. A further practical consideration concerns large particles such as dust, which produce considerable scattering of light and must be removed by suitable procedures. In the case of protein solutions this is usually achieved by a combination of centrifugation and filtration.

Steric effects must be considered when the molecules are highly asymmetric, as in the case of tobacco mosaic virus, and additionally there are the repulsive forces due to the electrical charges of the molecules and the accompanying ionic atmospheres. Thus the magnitude of the charge on proteins will affect the results. When measuring the turbidity of proteins, reducing the net charge by bringing the protein nearer to the isoelectric point reduces the slope of the Hc/τ against c plot but, even so, extrapolation to zero concentration must be carried out.

The equations so far cited are applicable to solutions where the dimensions of the particles are less than about one-twentieth of the wavelength of the incident light, e.g. for visible light scattering this corresponds to molecular weights of up to about five million. For large particles with a dimension of the order of the wavelength of light, each portion of the particle scatters light in accordance with the Rayleigh relation. As these portions are in fixed positions relative to one another, the scattered waves will interfere, giving rise to constructive and destructive interference. These correspond respectively to cases where the scattered waves tend to reinforce or annul each other. Where the angle (θ) of the scattered light relative to the path of the transmitted light is large, destruc-

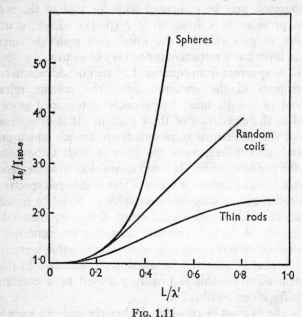

FIG. 1.11

Dissymmetry as a function of the shape of the molecule. L is the length of a thin rod, the root mean square distance between the ends of a randomly coiled molecule, and the diameter of a sphere.

tive interference occurs; constructive interference occurs at small angles. Consequently there is a higher scattered light intensity at low angles, i.e. in the forward direction, where the destructive interference is smaller. The ratio of the intensity of light scattered at some forward angle θ to that scattered at its supplementary angle $180° - \theta$ ($I_\theta / I_{180-\theta}$) is defined as the *dissymmetry of scattering*. Usually the ratio of intensities at 45° and 135° is termed the dissymmetry.

The dissymmetry will be greater for a given pair of angles the more pairs of scattering points there are on the particle and hence the larger the dimension, L, of the particle relative to the wavelength of the light, λ', the more pronounced the dissymmetry. λ' is the wavelength of light in the medium and is equal to the wavelength of the incident light divided by the refractive index of the medium. Since a spherical particle of diameter L has more pairs of scattering elements than a thin rod of length L, it displays a greater dissymmetry. This is illustrated in Fig. 1.11 from which it will be observed that the dissymmetry for very small particles $(L/\lambda' \rightarrow 0)$ approaches unity or, in other words, the angular scattering of a Rayleigh scatterer is symmetrical. In more precise terms, a Rayleigh scatterer exhibits a scattering proportional to $(1 + \cos^2 \theta)$ but this function is symmetrical about 90° and $I_\theta/I_{180-\theta}$ is equal to unity for all values of θ. Thus information concerning particle shape can be ascertained from dissymmetry measurements and, unlike estimates from flow birefringence studies, distortion of the molecule as a result of the velocity gradient in the solution will not occur.

In 1948 Zimm introduced an alternative procedure for determining particle shape based on the fact that destructive interference disappears when θ is zero. Light scattering measurements are made over a considerable range of angles and concentrations and then a plot of

$$\frac{2\pi^2 n_0^2[(n - n_0)/c]^2}{\lambda^4 N R_\theta} \cdot c \quad \text{against} \quad \sin^2 \frac{\theta}{2} + Kc$$

is carried out. R_θ is the 'Rayleigh ratio' given by $I_\theta d^2/I_0$ where I_θ is the intensity of radiation measured at angle θ and at a distance d from the scattering particles, and K is an arbitrary constant to spread the data. The intercept is

$$\frac{2\pi^2 n_0^2[(n - n_0)/c]^2}{\lambda^4 N R_0} \cdot c$$

and equals $1/M_w$ if the incident light is vertically polarized or $1/2M_w$ if it is unpolarized (because of the $(1 + \cos^2 \theta)$ term). The value of the interaction constant is found from the slope of the zero angle line ($\theta = 0$ in Fig. 1.12). A typical Zimm plot showing the double extrapolation of $\theta \rightarrow 0$ at constant c and $c \rightarrow 0$ at constant θ is given in Fig. 1.12.   The common intercept of both extrapolated curves on the ordinate yields $1/M_w$.

It should be noted that whereas dissymmetry measurements include the hydration shell of the molecules, turbidity measure-

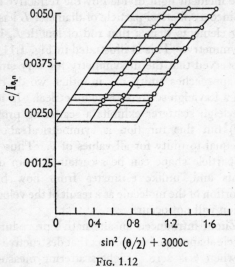

FIG. 1.12

Zimm extrapolation of light scattering data for thymus genoprotein in 0·015 M phosphate at pH 7·2. The ordinate is in grams/ml divided by the reduced intensity of scattering at angle θ.

(After STEINER (1952), *Trans. Far. Soc.*, **48**, 1185.)

ments ignore it and this can give rise to an apparent discrepancy. For example, hyaluronic acid, by dissymmetry measurements consists of spheres of 210 nm radius, and the molecular weight by turbidity measurements is 8×10^6. This corresponds to swollen spheres of radius 210 nm containing 400 parts of water to one part of hyaluronic acid.

(f) OTHER METHODS

Alternative physico-chemical methods of molecular weight determination include flow birefringence and dielectric dispersion measurements. For details of these procedures the student is referred to Mommaerts (1957), Leach (1969) and Alexander & Johnson (1949).

REFERENCES AND SUGGESTED READING

ALEXANDER, A. E. & JOHNSON, P. (1949). *Colloid Science*, vol. I. London: Oxford University Press.

ALEXANDER, P. & BLOCK, R. J., editors (1960, 1961). *A Laboratory Manual of Analytical Methods of Protein Chemistry*, vols. 2 and 3. Oxford: Pergamon Press.

ANDREWS, P. (1965). The gel-filtration behaviour of proteins related to their molecular weights over a wide range. *Biochem. J.*, **96**, 595.

BOWEN, T. J. (1970). *An Introduction to Ultracentrifugation.* London: Wiley-Interscience.

BULL, H. B. (1971). *An Introduction to Physical Biochemistry*, 2nd ed. Philadelphia: F. A. Davis.

CHERVENKA, C. H. (1969). *A Manual of Methods for the Analytical Ultracentrifuge.* Palo Alto: Spinco Division of Beckman Instruments.

DE DUVE, C., BERTHET, J. & BEAUFAY, H. (1959). *Prog. Biophys. biophys. Chem.*, **9**, 325.

EDSALL, J. T. (1953). The size, shape and hydration of protein molecules. In *The Proteins*, vol. 1, Part B, 549, ed. H. Neurath & K. Bailey. New York: Academic Press.

GOSTING, L. J. (1956). Measurement and interpretation of diffusion coefficients of proteins. *Adv. Protein Chem.*, **11**, 429. New York: Academic Press.

JOHNSON, P. & ROWE, A. J. (1961). The intrinsic viscosity of myosin and the interpretation of its hydrodynamic properties. *Biochem. J.*, **79**, 524.

KUPKE, D. W. (1960). Osmotic pressure. *Adv. Protein Chem.*, **15**, 57.

LEACH, S. J., editor (1969). *Physical Principles and Techniques of Protein Chemistry*, Part A. New York: Academic Press.

MOMMAERTS, W. F. H. M. (1957). Flow birefringence. In *Methods in Enzymology*, IV, 166, ed. S. P. Colowick & N. O. Kaplan. New York: Academic Press.

OSTER, G. (1955). Light scattering. In *Physical Techniques in Biological Research*, **1**, 51, ed. G. Oster & A. W. Pollister. New York: Academic Press.

PETERLIN, A. (1959). Determination of molecular dimensions from light scattering data. In *Progress in Biophysics*, **9**, 175, ed. J. A. V. Butler & B. Katz.

ROWLAND, F. W. (1967). *Facts and Methods Measuring for Chemistry*, **8**, 3. Avondale, Penn.: Hewlett-Packard.

SCHACHMAN, H. K. (1957). Ultracentrifugation, diffusion and viscometry. In *Methods in Enzymology*, IV, 32, ed. S. P. Colowick & N. O. Kaplan. New York: Academic Press.

SCHACHMAN, H. K. (1969). *The Ultracentrifuge in Biochemistry.* New York: Academic Press.

SCHERAGA, H. A. & MANDELKERN, L. (1953). Consideration of the hydrodynamic properties of proteins. *J. Am. Chem. Soc.*, **75**, 179.

STACEY, K. A. (1956). *Light Scattering in Physical Chemistry.* London: Butterworths Scientific Publications.

STEIN, W. D. (1962). Diffusion and osmosis. In *Comprehensive Biochemistry*, vol. 2, 283, ed. M. Florkin & E. H. Stotz. Amsterdam: Elsevier.

YANG, J. T. (1961). The viscosity of macromolecules in relation to molecular conformation. *Adv. Protein Chem.*, **16**, 323.

YPHANTIS, D. A. (1963). Equilibrium ultracentrifugation of dilute solutions. *Biochemistry*, **3**, 297.

PROBLEMS

1.1. A compound contained 10·66 per cent. sulphur and 15·5 per cent. phosphorus. Determine the minimal molecular weight and the number of sulphur and phosphorus atoms present in this molecular weight.

C

1.2. The following analytical data, in percentages, are available for horse haemoglobin : iron, 0·335 ; disulphide sulphur, 0·190 ; sulphur, 0·390. For pig haemoglobin the iron and sulphur content is respectively 0·40 and 0·48 per cent. Compare the minimal molecular weights of the haemoglobins from these two species and deduce the number of atoms of these elements present in the molecule.

(After Cohn, Hendry & Prentiss (1925), *J. biol. Chem.*, 63, 721.)

1.3. Amino acid analysis of the wheat protein glutenin has given the following percentage composition for three amino acids present in small amounts : tryptophan, 1·68 ; tyrosine, 4·5 ; β-hydroxyglutamic acid, 1·8. Use this information to calculate the minimal molecular weight and deduce the number of molecules of each amino acid present in the protein molecule.

(After Cohn, Hendry & Prentiss (1925), *J. biol. Chem.*, 63, 721.)

1.4. The copper content of haemocyanins isolated from various species has been determined. The figures given are the percentage of copper present : *Cancer*, 0·32 ; *Homarus*, 0·34 ; *Octopus vulgaris*, 0·38 ; *Helix pomatia*, 0·29 ; *Limulus polyphemus*, 0·173. Compare the minimal molecular weights of these haemocyanins of different origin.

(After Redfield, Coolidge & Shotts (1928), *J. biol. Chem.*, 76, 185.)

1.5. Analysis of ox haemoglobin gave the following percentages : iron, 0·336 ; sulphur, 0·48 ; arginine content, 4·24. Calculate the minimal molecular weight and the number of these atoms or molecules assumed to be present.

(After Cohn, Hendry & Prentiss (1925), *J. biol. Chem.*, 63, 721.)

1.6. Hüfner determined experimentally that one gram of haemoglobin combines with a maximum of 1·34 ml of oxygen at N.T.P. Subsequently it was shown by Cohn and co-workers that horse haemoglobin has an iron content of 0·335 per cent. Use these data to deduce the stoicheiometric relationship between oxygen combination and iron content for haemoglobin.

(After Hüfner (1894), *Arch. Anat. Physiol.*, 130 ; and Cohn, Hendry & Prentiss (1925), *J. biol. Chem.*, 63, 721.)

1.7. The maximal acid- and base-combining capacities per gram for myogen are $1·35 \times 10^{-3}$ and $1·28 \times 10^{-3}$ equivalents respectively. Compare the minimal molecular weights obtained by these two titrations.

1.8. Gelatin has a maximal acid-combining capacity of 96×10^{-5} equivalents per gram, according to Hitchcock. Determine the minimal molecular weight and compare this figure with the one obtained from Simms' value of 7×10^{-4} equivalents base combining per gram gelatin.

(Data from Hitchcock (1931), *J. gen. Physiol.*, 15, 125 ; and Simms (1928), *J. gen. Physiol.*, 11, 629.)

1.9. Serum albumin has been reported to possess maximal acid- and base-combining capacities of 72 and 70×10^{-5} equivalents per gram respectively. If the figure of 67,100 for the molecular weight, obtained by diffusion and sedimentation velocity measurements, is accepted, calculate the approximate number of acidic and basic groups per serum albumin molecule.

(Data from Prideaux & Woods (1932), *Proc. roy. Soc.*, B, 111, 201 ; and Lamm & Polson (1936), *Biochem. J.*, 30, 528.)

1.10. Edestin has a maximal acid-combining capacity of 134×10^{-5} equivalents per gram and the molecular weight obtained by sedimentation measurements is 309,000. Ascertain the approximate number of basic groups in the edestin molecule.

(Data from Hitchcock (1930), *J. gen. Physiol.*, 14, 99 ; and Polson (1939), *Kolloid Z.*, 87, 149.)

1.11. Fibrin was titrated with an acidic dye and gave a combining capacity of $1·475 \times 10^{-3}$ equivalents per gram protein. Calculate the equivalent combining weight of the protein cation.

(Data from Chapman, Greenberg & Schmidt (1927), *J. biol. Chem.*, 72, 707.)

1.12. Gelatin has been titrated with acid, base, acidic and basic dyes. The combining capacities in equivalents per gram protein were as follows:

Acid:	$9{\cdot}6 \times 10^{-4}$	Acidic dye:	$1{\cdot}04 \times 10^{-3}$
Base:	$7{\cdot}0 \times 10^{-4}$	Basic dye:	$7{\cdot}0 \times 10^{-4}$

Compare the equivalent combining weights of the protein cation and anion and comment on the agreement between acid-base and dye titrations.

(Data from PRIDEAUX & WOODS (1932), *Proc. roy. Soc.*, B, **110**, 353 ; RAWLINS & SCHMIDT (1929), *J. biol. Chem.*, **82**, 709 ; and CHAPMAN, GREENBERG & SCHMIDT (1927), *J. biol. Chem.*, **72**, 707.)

1.13. The composition of a macromolecule corresponds to the following:

Number of Moles	*Molecular Weight*
10	17,000
10	34,000
20	68,000

Calculate the number-average and the weight-average molecular weights.

1.14. Calculate the osmotic pressure of a 1 per cent. (w/v) solution of glycerol at 22° C.

1.15. A solution containing 0·5 gram of urea per 100 ml exerted an osmotic pressure of 2·037 atmospheres at 25° C. Determine the molecular weight of urea.

1.16. The following figures were obtained in an investigation of the osmotic pressure of a high molecular weight substance at 27° C.

Concentration (g/100 ml)	2·0	2·75	3·4	5·0	6·0	7·25	8·3
Osmotic pressure (mm Hg)	6·6	9·1	11·6	17·9	21·9	27·4	32·8

By use of a graphical method, determine the molecular weight of the compound.

1.17. Pfeffer obtained the following results with dilute solutions of sucrose at 0° C.

Concentration of sucrose (g per litre solution)	10·03	20·14	40·60	61·38
Osmotic pressure (atmospheres)	0·686	1·34	2·75	4·04

Use these figures to demonstrate that the osmotic pressure of a solution at constant temperature is proportional to the concentration.

The relationship between the osmotic pressure and absolute temperature of a 1 per cent. solution of sucrose was investigated with the following results.

Absolute temperature	273·0	279·8	286·7	287·2	288·5	295·0	305·0
Osmotic pressure (atmospheres)	0·648	0·664	0·691	0·671	0·684	0·721	0·716

Deduce the relationship between osmotic pressure and absolute temperature from these data.

(After PFEFFER (1877), *Osmotische Untersuchungen*, Leipzig.)

1.18. What special principles are involved when osmometry is applied to high molecular weight charged particles such as proteins? Two different proteins A and B give the following osmotic data at 5° C:

Concentration (g/litre)		10	20	30	40
Osmotic pressure in atmospheres	A	0·0038	0·0078	0·0120	0·0160
	B	0·0075	0·0220	0·0423	0·0720

Comment on the data and calculate molecular weights.
$(R = 0.08207$ litre atmospheres/mole/°K)
(Leeds Honours Course Finals, 1953.)

1.19. Berkeley and Hartley determined the osmotic pressure of sucrose solutions at 0° C with the following results :

Concentration (g/litre)	2·02	10·0	20·0	45·0	93·75
Osmotic pressure (atmospheres)	0·134	0·66	1·32	2·97	6·18

Show that these findings are in accordance with Boyle's Law as applied to dilute solutions.

(Data from BERKELEY & HARTLEY (1906), *Phil. Trans. roy. Soc.*, A, 206, 481.)

1.20. The following osmotic pressure data were obtained for serum albumin in 0·05 M-acetate buffer at the isoelectric point pH 4·8 and 0°.

Concentration (g/100 ml solvent)	0·78	1·25	2·79	2·82	3·38	4·18	8·98	12·45
Observed osmotic pressure (cm water)	2·39	4·01	8·53	8·68	10·80	13·01	27·84	37·69

Use these data to determine the molecular weight of serum albumin.

(After BURK (1932), *J. biol. Chem.*, 98, 353.)

1.21. The effect of hydrogen ion concentration on the osmotic pressure of dialysed sheep haemoglobin was investigated by Adair. Solutions of varying haemoglobin concentration were used and the experimental temperature was 23°. Below, the osmotic pressure is expressed in terms of the pressure in mm of mercury per 1 per cent. protein at 0°.

pH	5·0	5·4	6·5	6·7	6·8	6·8	6·8	7·2	9·6	10·2
Osmotic pressure (mm Hg)	21·5	13·4	3·2	2·4	2·4	3·5	4·5	5·0	15·6	21·4

Express these data graphically, comment on the results obtained and deduce the isoelectric point of the haemoglobin.

(After ADAIR (1925), *Proc. roy. Soc.*, A, 109, 292.)

1.22. The osmotic pressure of solutions of serum albumin in 0·0667 M phosphate buffer pH 7·4 and 0° has been determined and the results are tabulated below.

Concentration (g/100 ml solvent)	0·578	0·684	1·05	1·59	2·27	2·47	2·95
Osmotic pressure (cm H_2O)	2·04	2·35	3·56	5·62	8·55	9·88	10·84

Determine the molecular weight of serum albumin.

(After ROCHE & MARQUET (1935), *C.R. soc. biol.*, 118, 898.)

1.23. The osmotic pressure of carbon monoxide haemoglobin prepared from horse blood was measured at various concentrations at pH 7·38 and 0° C. The results obtained are tabulated below.

Concentration of HbCO (g/100 ml)	3·028	3·015	2·140	2·025	2·010	0·976	0·978
Osmotic pressure (mm Hg)	8·50	8·43	5·81	5·54	5·46	2·55	2·59

Determine the molecular weight of haemoglobin from these measurements.

(Data from ADAIR (1949), in *Haemoglobin*, p. 191, ed. Roughton & Kendrew. London : Butterworth.)

1.24. The osmotic pressure of human haemoglobin in 0·2 M-phosphate has been measured by Gutfreund. The following table presents his data obtained at 3°.

Concentration g/100 ml	Osmotic pressure cm H_2O
0·47	1·6
0·56	2·0
0·60	2·2
1·29	4·7
1·66	6·5
2·39	9·5
3·09	11·3
3·25	12·8
3·47	13·2
3·75	14·9
4·34	17·6
4·83	19·9
5·40	22·6
7·01	30·6

Use these figures to calculate the molecular weight of human haemoglobin.

(Data from GUTFREUND (1949), in *Haemoglobin*, p. 197, ed. Roughton & Kendrew. London : Butterworth.)

1.25. The action of ultraviolet light on cellulose nitrate has been studied by measuring the osmotic pressure of the solution of cellulose nitrate in acetone. The solution, containing 10·28 g of cellulose nitrate per litre, was exposed to ultraviolet irradiation for 10·2 days and osmotic measurements were made by a dynamic method after 2·08 days and at the end of the experiment. The pressure on the solution was altered and the rate of movement of a meniscus, indicating the movement of solvent into or out of the solution, measured. Data obtained are recorded below. Direction of meniscus movement is indicated by + and −.

Time of exposure (days)	Temperature °C	Pressure cm acetone	Rate of movement of meniscus cm per sec.
2·08	25	46	+ 0·0048
		56	+ 0·0067
		67	+ 0·0096
		77	+ 0·0116
10·2	22	36	− 0·0168
		46	− 0·0068
		67	+ 0·0108
		77	+ 0·0200

From these figures ascertain the effect of ultraviolet irradiation on the molecular weight of cellulose nitrate. The density of acetone is 0·78 and that of mercury 13·59.

(After MONTONNA & JILK (1941), *J. phys. Chem.*, 45, 1374.)

1.26. The molecular weight of the dye congo red has been determined by the sedimentation equilibrium method. The initial concentration of the dye was 0·1 g per litre dissolved in 0·1 M-NaCl, the density of the resulting solution being 1·0023. The speed of rotation of the centrifuge was 299·6 revolutions per second. The following data were obtained at 20°.

Distance from axis of rotation (cm)		Relative concentration of congo red	
x_2	x_1	c_2	c_1
5·87	5·84	53·60	50·46
5·81	5·78	47·57	44·79
5·75	5·72	42·18	39·76
5·66	5·63	35·36	33·36

The partial specific volume of congo red is 0·60. Calculate the mean molecular weight from these figures.

(Data from SVEDBERG & PEDERSON (1940), *The Ultracentrifuge.* Oxford University Press.)

1.27. The sedimentation equilibrium technique has been used to determine the molecular weight of the erythrocruorin of *Planorbis*. The experiment was carried out at 20° and the density of the solvent was 1·0034. The centrifuge speed was 2280 r.p.m. and the distance of the outer end of the solution from the axis of rotation was 5·95 cm. Experiments were conducted for 48, 53 and 77 hours and below are tabulated some of the experimental findings.

Distance from axis of rotation (cm)		Concentration ratio
x_2	x_1	c_2/c_1
5·85	5·80	1·289
5·70	5·65	1·195
5·60	5·55	1·237
5·55	5·50	1·223

The partial specific volume of erythrocruorin was found to be 0·745. Determine the mean molecular weight.

(After SVEDBERG & ERIKSSON-QUENSEL (1934), *J. Amer. chem. Soc.*, 56, 1700.)

1.28. The effect of pepsin on a diphtheria antitoxic pseudoglobulin from the horse has been examined by means of the ultracentrifuge. The sedimentation coefficients before and after pepsin treatment were 7·2 and 5·7 × 10⁻¹³ sec. respectively and the corresponding diffusion coefficients were 3·9 and 5·8 × 10⁻⁷ cm²/sec (all values corrected to values for water at 20°). The partial specific volume of the protein was assumed to be the same as normal horse globulin and to have a value of 0·745. The density of water at 20° is 0·9982. Determine the effect of the pepsin treatment on the molecular weight of the antitoxin.

(After PETERMANN & PAPPENHEIMER, Jr. (1941), *J. phys. Chem.*, 45, 1.)

1.29. The maximum molecular weight of insulin has been determined by sedimentation and diffusion studies. The sedimentation coefficient, corrected to the value in water at 20°, is 3·12 × 10⁻¹³ sec. and the diffusion coefficient, similarly corrected, is 8·2 × 10⁻⁷ cm²/sec. The partial specific volume of insulin was found to be 0·735, and the density of water at 20° is 0·9982. Evaluate the molecular weight.

(After CREETH (1953), *Biochem. J.*, 53, 41.)

1.30. Studies on the sedimentation and diffusion of human albumins yielded the following coefficients (corrected to their values in water at 20°) : S_{20} 4·24 × 10^{-13} sec and D_{20} 6·32 × 10^{-7} cm² sec⁻¹. The density of water at 20° is 0·9982 and the partial specific volume of the albumins was taken as 0·733. Calculate the molecular weight.

(After CHARLWOOD (1952), *Biochem. J.*, **51**, 113.)

1.31. The Lewis blood-group substance was found to have the following characteristics. At a concentration of 1g/100 ml the sedimentation coefficient was 5·44 × 10^{-13} sec and the diffusion coefficient 1·37 × 10^{-7} cm²/sec, both values being corrected to values in water at 20°. The partial specific volume was found to be 0·643. The density of water at 20° is 0·9982. Calculate the molecular weight of this substance.

(After KEKWICK (1952), *Biochem. J.*, **50**, 471.)

1.32. 20 g crystalline trypsin were dissolved in 0·25 saturated $(NH_4)_2SO_4$ acetate buffer pH 4·0 and placed in a rocking osmometer in the same solvent at 5°. The osmotic pressure was determined after 24 hours and the solution analysed for protein nitrogen.

Protein concentration (mg/ml)	72	71	50	49	22·5	18
Pressure (mm Hg)	39	38	22	21·5	11	9

The diffusion coefficient of trypsin was found to be 0·020 cm²/day and the viscosity coefficient of the solvent was 0·0303 erg sec/cm³ at 5°. Determine the molecular weight of trypsin and the average molar volume of the hydrated protein.

(After NORTHROP & KUNITZ (1933), *J. gen. Physiol.*, **16**, 295.)

1.33. In an experiment to determine the diffusion coefficient of crystalline trypsin, the following results were obtained by the porous-disc method. Diffusion was followed by both trypsin activity measurements and by determination of nitrogen.

Original solution mg N/ml	Time days	Quantity of original solution diffused by activity ml	by nitrogen content ml
3·45	0·156	0·062	——
	0·708	0·281	0·274
	2·00	0·565*	0·806

* In calculating the value of D from this figure a correction factor of 1·438 should be applied to compensate for the fact that the activity of the sample fell during the latter part of the experiment.

The solvent (0·5 saturated $MgSO_4$, 0·1 M-acetate) had a specific gravity of 1·115 and its viscosity coefficient was 0·0303 erg sec/cm³. All measurements were carried out at 5° and the membrane constant had been evaluated as 0·054.

Calculate the average diffusion coefficient in cm²/day and compare the values obtained by activity measurements with those obtained by nitrogen determinations.

What is the average radius of the hydrated trypsin molecule in solution? Determine the average molar volume of the protein.

(After SCHERP (1933), *J. gen. Physiol.*, **16**, 795.)

1.34. The viscosity of trypsin solutions of various concentrations was measured by means of an Ostwald viscometer. The protein was in 0·5 saturated magnesium sulphate and 0·1 M-acetate buffer pH 4·0, and at 5° the following results were obtained.

Trypsin concentration (g/100 ml)	0	0·8	1·6	2·4	3·2	4·0
Time of outflow (sec)	203·4	212·6	221·5	231·0	241·4	257·0

Neglecting any changes in the density of the trypsin solutions, determine the effect of concentration on the degree of hydration of trypsin. The partial specific volume of trypsin may be taken as 0·75. Use the Kunitz equation $\eta_r = 1 + 4\cdot5\phi$ to evaluate the volume fraction.

(After KUNITZ, ANSON & NORTHROP (1934), *J. gen. Physiol.*, **17**, 365.)

1.35. The van't Hoff equation for osmotic pressure does not hold for solutions that are not dilute, neither for solutes that are hydrated in solution. A suggested correction for the equation is the following, where the volume fraction ϕ represents the actual volume occupied by the hydrated molecules

$$\Pi = \frac{RT}{M} \frac{1000c}{(1000 - \phi)}$$

and where c is the concentration in grams per litre.

Compare the average value for the molecular weight of gelatin obtained by the use of the above equation with that obtained graphically by the plot of the reduced osmotic pressure as a function of the gelatin concentration. Use the Kunitz equation to evaluate the volume fraction. Temperature 35°.

Gelatin concentration g/100 ml	Relative viscosity	Osmotic pressure mm Hg
1	1·43	3·5
2	2·06	7·5
3	2·96	12·0
4	4·24	17·0
5	6·00	23·0
6	8·20	29·5

(After KUNITZ (1927), *J. gen. Physiol.*, **10**, 811.)

1.36. Kunitz proposed the following empirical equation to replace the Einstein viscosity equation for spheres

$$\eta_r = \frac{1 + 0\cdot5\phi}{(1 - \phi)^4}.$$

This equation can be expanded and approximates to $\eta_r = 1 + 4\cdot5\phi$. Compare the volume fractions calculated by the Kunitz and Einstein equations from the following data, obtained with saccharose and glucose.

Saccharose (g/100 ml)	1·0	2·0	4·9	10·3	15·6	21·7
Relative viscosity (η_r)	1·026	1·054	1·141	1·329	1·570	1·917
Glucose (g/100 ml)	2·1	4·7	10·6	16·6	21·7	26·4
Relative viscosity (η_r)	1·062	1·130	1·316	1·619	1·901	2·216

(Data from PULVERMACHER (1920), *Z. anorg. allg. Chem.*, **113**, 147, quoted by KUNITZ (1926) *J. gen. Physiol.*, **9**, 715.)

1.37. The viscosity of carbon monoxide haemoglobin was investigated at pH 6·8 and 5°, when the following data were obtained :

Protein concentration (g/100 ml)	2·10	4·20	6·30	8·36	10·45
Relative viscosity (η_r)	1·084	1·175	1·290	1·445	1·610

The partial specific volume of dry haemoglobin is 0·75. Determine the volume of water of hydration per gram haemoglobin in order to ascertain the relationship between hydration and protein concentration. Use the Kunitz equation $\eta_r = 1 + 4·5\phi$ to evaluate the volume fraction.

(After KUNITZ, ANSON and NORTHROP (1934), *J. gen. Physiol.*, **17**, 365.)

1.38. The following values of the diffusion coefficient for glucose at a series of concentrations were obtained at 25° :

Concentration (moles/litre)	0·100	0·200	0·250	0·300	0·355	0·400	0·500	0·600
Diffusion coefficient (cm²/day)	0·566	0·556	0·552	0·550	0·546	0·543	0·539	0·535

Use these data to obtain a value for the diffusion coefficient at infinite dilution and hence calculate the molecular weight of glucose. A linear graph is obtained if the diffusion coefficient is plotted against the square root of the concentration. The viscosity of water at 25° is 0·008937 poise and the density of glucose may be taken as 1·548.

(After FRIEDMAN & CARPENTER (1939), *J. Amer. chem. Soc.*, **61**, 1745.)

1.39. The following data were obtained for crystalline horse-liver alcohol dehydrogenase :

The sedimentation coefficient was found in two runs to be $S_{20} = 4·86$ and 4·90 Svedberg units, and the diffusion coefficient, D_{20}, was found to be $6·5 \times 10^{-7}$ cm² sec⁻¹. A determination of the partial specific volume gave a value of 0·751. The density of the solution was 0·998.

Use these data to calculate the molecular weight of the enzyme.

(After THEORELL & BONNICHSEN (1951), *Acta chem. scand.*, **5**, 1105.)

1.40. Extracts of disrupted baker's yeast cells contain a ribonucleoprotein which comprises some 1·5 per cent. of the weight of the cells and has a sedimentation coefficient of 80s.

In the presence of 0·01 M-phosphate the ribonucleoprotein boundary at 80s disappeared and two other ribonucleoprotein boundaries, corresponding to 60s and 40s, appeared in the ultracentrifuge. Verify that the 60s and 40s components have arisen by dissociation of the 80s particles.

It may be assumed that all the particles are spherical and have the same partial specific volume.

(After CHAO (1957), *Arch. Biochem. Biophys.*, **70**, 426.)
(Glasgow Honours Course Finals, 1959.)

1.41. The viscosities of solutions of polyvinyl chloride have been measured by Fuoss and Mead who used the equation $\underset{c \to 0}{\text{limit}} \left(\frac{\eta_{sp}}{c} \right) = K_m M$ in the equivalent form $\text{limit} \frac{\ln \eta_r}{c} = \lambda_0$, whence $K_m = \frac{\lambda_0}{M}$. They evaluated the constant

$K_m = \lambda_0/M$ on two ultracentrifuge runs (sedimentation velocity) and one diffusion coefficient determination using methyl amyl ketone as solvent to obtain

$$M = \frac{RTs}{D(1 - \bar{v}\rho)} = 102{,}000.$$

The value of λ_0 for the same sample was obtained by viscosity measurements in cyclohexanone at 25° C, and found to be equal to 7·19. Thus $K_m = \lambda_0/M = 7 \times 10^{-5}$.

(1) Using this value of K_m determine the molecular weight (weight average) of the six polyvinyl chloride samples for which viscosity data in cyclohexanone at 25° C are given below. These samples were obtained by carrying out the polymerization at various temperatures and with various methods of initiating the chain reactions.

	I			II		
η_r	1·149	1·290	1·473	1·303	1·612	2·023
$100c$	3·06	5·70	8·78	3·20	5·98	9·18
	III			IV		
η_r	1·125	1·267	1·435	1·278	1·555	1·917
$100c$	2·79	5·70	9·00	3·12	5·81	8·94
	V			VI		
η_r	1·094	1·179	1·289	1·092	1·181	1·295
$100c$	3·12	5·83	9·00	3·05	5·82	9·04

The concentration c was conveniently expressed in monomoles/litre (Staudinger's " Grundmole ").

(2) Verify Fuoss and Mead's observations that

(a) For any particular fraction $\lambda = \lambda_0(1 - \gamma c)$, where γ is a constant for that fraction.

(b) For the different fractions γ increases as λ_0 increases linearly so that we may write $\lambda = \lambda_0(1 - \beta\lambda_0 c)$, where β is a constant independent of molecular weight.

(c) Evaluate β.

The authors suggest that the equation in (b) can be used to save much work. λ_0 may be calculated from a knowledge of β and measurement of λ at a particular concentration c. When c is less than 5 g/l this equation may be solved by successive approximations. Furthermore, since λ_0 is proportional to molecular weight, then $c\lambda_0$ is proportional to ϕ, the volume concentration of solute.

Hence, the equation $\lambda = \lambda_0(1 - \beta\lambda_0 c)$ becomes $\ln\eta_r = k\phi(1 - k^1\phi)$.

This must be compared with Einstein's (1906) argument that it is the relative volume of solute and solvent that really matters in viscosity.

(After Fuoss & Mead (1942), *J. Am. chem. Soc.*, **64**, 277.)

1.42. (a) The osmotic pressure of bovine plasma albumin was measured in phosphate buffer at various concentrations of protein. The results obtained at 24° C were :

Protein concentration (g/100 ml)	Osmotic pressure (cm H_2O)
2·997	13·71
1·958	8·38
1·468	6·05
0·979	3·93
0·734	2·88
0·489	1·90

Use these data to calculate the molecular weight of bovine plasma albumin.

$$(R = 84 \cdot 71 \text{ litre cm } H_2O/\text{mole}/^\circ K)$$

(b) A molar solution of the potassium salt of a non-diffusible substance is separated from an equal volume of $0 \cdot 8$ M-KCl solution by a semi-permeable membrane. Calculate the final concentrations of potassium and chloride ions each side of the membrane after equilibrium has been attained.

(Leeds Honours Course Finals, 1958.)

1.43. Derive any expression for the particle weight of a protein from ultra-centrifuge data.

A protein has the following properties :

$$S^\circ{}_{20, w} = 4 \cdot 86 \times 10^{-13} \text{ sec}, \quad D^\circ{}_{20, w} = 6 \cdot 73 \times 10^{-7} \text{ cm}^2 \text{ sec}^{-1}, \quad \bar{v} = 0 \cdot 733.$$

Calculate the particle weight, assuming that water at 20° C has a density of $0 \cdot 9982$.

$$(R = 8 \cdot 314 \times 10^7 \text{ ergs/mole per degree } K)$$

(Leeds Honours Course Finals, 1960.)

1.44. The following data were obtained in an experiment to determine the diffusion coefficient of the citrase (citrate lyase) of *Aerobacter aerogenes*. A synthetic boundary cell was used at low speed in the ultracentrifuge, to reduce sedimentation to a minimum, and the area-maximum height method of evaluation was used.

Time (*min*)	H_{max}. (mm)
9	11·66
13	10·32
17	9·22
21	8·38
29	7·22
37	6·29
45	5·68
53	5·12
75	4·30
150	3·10

The area under the curves was $0 \cdot 1438 \text{ cm}^2$ and this remained constant throughout the run. The magnification factor, F, was $2 \cdot 04$. The following relationship holds

$$D_t = \frac{(\text{Area})^2}{H_{.max.}{}^2} \cdot \frac{1}{4F^2 \pi t}$$

where D_t is the diffusion coefficient at time t. Determine the diffusion coefficient from these data.

Note.—Consultation of Bowen (1954) *Laboratory Practice*, **3**, 369, may help in the solution of this problem.

(Unpublished data of BOWEN & ROGERS.)

1.45. Outline the special considerations that apply to the determination of physical properties of a charged macromolecule such as a protein as opposed to an uncharged macromolecule.

The osmotic pressure of a protein was measured in phosphate buffer. The following results were obtained at 20°.

Protein concentration (g/100 ml)	Osmotic pressure (cm H_2O)
3·01	13·82
1·96	8·40
1·46	6·04
1·00	4·00
0·73	2·86
0·50	1·91

Calculate a molecular weight for the protein.

$$(R = 84·71 \text{ litre cm } H_2O/\text{mole}/°K)$$

(Leeds Honours Course Finals, 1964.)

1.46. Experiments were carried out by the porous disk technique to determine the diffusion coefficient of a corpuscular protein which was judged to be of approximately spherical shape.

A solution (0·8 per cent.) of the protein was allowed to diffuse for 2 days and the quantities of protein passing through the disk were determined by the biuret method at suitable intervals of time. The membrane constant of the disk was evaluated as 0·205 by the use of surcose and various salts. The following data were obtained at 5° C.

Time (days) . . .	0·60	0·95	2·00
Protein diffused . . . (ml of concentrated solution)	0·154	0·243	0·512

The viscosity coefficient of water at 5° C is 0·01519 erg seconds per cm^3 and the partial specific volume of the protein is 0·74. The molecular weight of the protein was estimated to be 36,000.

From these data calculate the degree of hydration of the protein, assuming that Stokes' Law is applicable to the molecules of the protein.
(The Avogadro number $N = 6·06 \times 10^{23}$, $R = 8·314$ joules/mole/°K; $\pi = 3·142$.)

(Hull Honours Course Finals, Part II, 1966.)

1.47. What happens to bacteria when they undergo 'thymineless death'? The sedimentation of E. coli DNA has been investigated by u.v. photography in the ultracentrifuge after thymineless death either with or without heat denaturation.

The following data were obtained:

DNA Sample	Distance of the boundary from the axis of rotation (cm)		
	After 40 min	After 20 min	After 0 min
Before thymineless death, native . . .	5·980	5·870	5·775
Before thymineless death, heat-denatured . .	5·963	5·864	5·765
After thymineless death, native . . .	5·790	5·715	5·628
After thymineless death, heat-denatured . .	5·895	5·810	5·725

Calculate the sedimentation coefficients of the DNA samples given that in each analysis the rotor has a speed of 24,700 rev/min. Do the data support a proposal that thymineless death is accompanied by single-strand scissions in DNA?

What other experiments could you carry out to investigate this possibility?

(University of St. Andrews, Queen's College, Honours Course Finals, 1966.)

1.48. A rat liver cell contains 10^{-11} gram of DNA. Assuming this to be distributed equally amongst all 42 chromosomes, that all the DNA of the chromosomes is present in one molecule and that each base pair in the DNA double helix occupies 3·4 Å, calculate the length of DNA double helix present in each chromosome.

Avogadro's No. $= 6 \times 10^{23}$.

(Glasgow Biochemistry [Higher Ordinary] Course, 1965.)

1.49. Chang and Liener have described new chromophoric azomercurials for labelling the thiol groups of proteins. The following data are adapted from their paper.

Azomercurial reacting with amino acid μmoles	Cysteine μmoles	Mixture of 18 amino acids excluding cysteine μmoles
0·00	0·00	0·00
0·20	0·20	0·00
0·40	0·39	0·01
0·60	0·62	0·02

Ficin activity, per cent.	100	76	53	30	13	2
Molar ratio azomercurial/ficin	0	0·2	0·4	0·6	0·76	0·92

It was found that 0·65–0·70 mole of mercury is incorporated per mole of ficin. Interpret the data and appraise the value of the reagent.

(After CHANG & LIENER (1964), Nature, 203, 1065. Leeds Honours Course Finals, 1965.)

1.50. During the investigation of the molecular structure of E. coli ribosomes a ribonucleic acid sub-unit was isolated and shown to be homogeneous by DEAE-cellulose chromatography, gel electrophoresis and by end-group analysis. When heated in 0·2 M-NaCl this RNA exhibited a well defined helix-coil transition leading to an overall increase in ultraviolet absorption of 25 per cent. Upon solution in distilled water the same hyperchromicity was observed and no further change occurred on heating. The partial specific volume of the RNA was found to be 0·5 in 0·2 M-NaCl and 0·65 in distilled water.

In order to gain support for the theory that this RNA consisted of two strands combined in a single double helix, ultracentrifugal analyses were performed, both in 0·2 M-NaCl and in distilled water.

Sedimentation velocity studies at 20° and 50,740 rev/min (i.e. 5314 radians/sec) yielded the following data:

Time (sec)	Radius of boundary (cm)	
	In 0·2 M-NaCl	In distilled water
480	5·778	5·765
960	5·911	5·826
1440	6·044	5·887
1920	6·177	5·948
2400	6·310	6·009
2880	6·443	6·070

Free diffusion studies, using a synthetic boundary cell at low speed, gave a diffusion coefficient of $5\cdot3 \times 10^{-7}$ cm^2/sec in $0\cdot2$ M-NaCl and $31\cdot6 \times 10^{-7}$ cm^2/sec in distilled water.

By means of the foregoing data and a knowledge of the densities of $0\cdot2$ M-NaCl ($1\cdot0072$ g/ml) and water ($0\cdot9981$ g/ml) at $20°$, the molecular weight of the RNA under both conditions was calculated using the Svedberg equation. (R, the gas constant equals $8\cdot314 \times 10^7$ ergs/degree/mole.)

Interpret the information obtained and explain its relevance to the structure of the RNA.

(Glasgow Honours Course Finals, 1965.)

1.51. A globular protein was found to have the following sedimentation properties when examined in water at $20°$ in the ultracentrifuge.

Concentration (per cent. w/v)	S
$2\cdot0$	$2\cdot45$
$1\cdot5$	$2\cdot50$
$1\cdot0$	$2\cdot58$
$0\cdot5$	$2\cdot65$
$0\cdot25$	$2\cdot68$

The diffusion coefficient at infinite dilution in water at $20°$ was $9\cdot34 \times 10^{-7}$ cm^2 sec^{-1}. Assuming a value of $0\cdot72$ cm^3 g^{-1} for the partial specific volumes, calculate the molecular weight of the protein. R, the gas constant, is $8\cdot314 \times 10^7$ erg mole^{-1} deg^{-1}, the density of water at $20°$ is $0\cdot998$ g cm^{-3}. Describe briefly one method for the determination of the diffusion coefficient. What other molecular parameter might have been used instead of the diffusion coefficient in order to derive the molecular weight from sedimentation data and what assumptions would then have been necessary?

(University of Manchester, Biological Chemistry Ordinary Degree, 1967.)

1.52. The β-galactosidase of *Escherichia coli* has been purified and found to be a protein of molecular weight 540,000. After reduction with 2-mercaptoethanol in the presence of 8 M-urea and treatment with sodium iodoacetate, the molecular weight was determined to be 45,000–55,000. Amino acid analysis of the reduced, carboxymethylated protein gave the following data (in residues/100 residues).

Lysine	$2\cdot48$
Arginine	$6\cdot30$
S-Carboxymethyl cysteine	$1\cdot64$

[The average residue molecular weight may be taken as 115]

Digestion of the reduced, carboxymethylated protein with trypsin yielded at least 86 separable peptides. When the protein was treated with sodium iodoacetate in 10 M-urea, subsequent amino acid analysis showed no cystine but about $1\cdot6$ moles per cent. of S-carboxymethyl cysteine. After sodium iodoacetate treatment in buffer at pH $8\cdot5$, $0\cdot92$ mole per cent. of S-carboxymethyl cysteine and $0\cdot6$ mole per cent. of half-cystine were detected. On the basis of the above data discuss the structure of the β-galactosidase molecule.

(Hull Honours Course Finals, Part II, 1967.)

1.53. (a) Describe briefly three physical methods for determining the molecular weight of a soluble protein and assess their relative merits.

(b) If you were given a supply of D$_2$^{18}O (sp. gr. $1\cdot217$) and an ultracentrifuge, suggest how you might determine the partial specific volume of a protein.

(c) The buoyant density of native *Escherichia coli* DNA determined in a caesium chloride gradient at pH 7·0 was 1·704 g/ml. After heat denaturation the buoyant density in the neutral caesium chloride gradient was 1·719 g/ml. When either native or heat denatured DNA was centrifuged in a caesium chloride gradient at pH 12·0 a buoyant density of 1·766g/ml was found. Discuss these observations.

(Hull Honours Course Finals, Part II, 1968.)

1.54. An 0·025 per cent. solution of bovine pancreatic ribonuclease was examined in the ultracentrifuge by the meniscus depletion method. The speed was 45,000 rev/min at a temperature of 28° C. The following values of a concentration function were measured:

x(cm)	$f(c)$
6·45	0·000
6·54	0·006
6·56	0·010
6·58	0·016
6·60	0·028
6·62	0·042
6·64	0·066

The term $(1 - \bar{v}\rho) = 0.305$.
Determine the molecular weight.

1.55. A pure preparation of alkaline phosphatase in solution in water at a concentration of 1 mg/ml at 20° was examined in the ultracentrifuge. The speed was 60,000 r/m and the position of the boundary at various times was:

x(cm)	*Time* (sec)
6·152	100
6·229	600
6·307	1100
6·385	1600
6·465	2100

In another experiment the diffusion coefficient was estimated as 6.6×10^{-7} cm^2/sec. The factor $(1 - \bar{v}\rho)$ can be taken as 0·29. Calculate the molecular weight.

By amino acid analysis alkaline phosphatase was found to contain 0·00875 residue of tryptophan per 100 g protein. After digestion with the enzyme trypsin, three peptides containing tryptophan were found. What does this suggest about the structure of alkaline phosphatase?

$$(R = 8.31 \times 10^7 \; ; \; \pi = 3.142)$$

(Hull Joint Biological Chemistry Finals, Part II, 1969.)

1.56. What special problems are involved in determining sedimentation coefficients of DNA solutions?

Three samples of DNA of different molecular weights were analysed by the zone sedimentation method, (a) as native DNA in 1 M-NaCl, pH 8·0 ('native'), (b) as denatured DNA in 1 M-NaCl, pH 8·0 ('neutral denatured'), and (c) in 0·9 M-NaCl 0·1 M-NaOH, pH 13·0 ('alkaline'). The following sedimentation coefficients were obtained:

Double stranded molecular weight $\times 10^{-6}$	$s^{\circ}_{20, w}$(s)		
	'Native'	'Alkaline'	'Neutral denatured'
16·3	27·7	30·7	64·7
26·4	32·0	37·2	84·8
32·6	35·1	40·1	95·8

Derive relationships between sedimentation coefficient and molecular weight for the three forms. What can you infer about the conformation of DNA in the different solutions from these relations?

A sample of DNA was partially degraded with either pancreatic deoxyribonuclease or endonuclease I of *Escherichia coli* and sedimentation coefficients determined in neutral solution and in alkaline solution:

Treatment	$s^\circ_{20, w}$(s) Neutral	Alkaline
Untreated	35	40
Pancreatic DNase	35	10
Endonuclease I	20	25

What do these results suggest about the mode of action of the two enzymes?

(Hull Honours Course Finals, Part II, 1970.)

1.57. The Svedberg equation defines the molecular weight, M, of a molecule in terms of certain of its hydrodynamic properties. Given that R is the gas constant ($8\cdot314 \times 10^7$ erg mole^{-1} deg^{-1}), T is the absolute temperature and ρ the solvent density, rigorously define the remaining terms of the equation.

Describe in detail how you would calculate a sedimentation coefficient from the photographic record of an analytical ultracentrifuge experiment.

An enzyme in solution at pH 6·2 and 20° yielded the following sedimentation data:

Concentration per cent. (w/v)	S
3·50	3·00
3·00	3·05
2·50	3·15
2·00	3·20
1·50	3·15
1·00	2·90
0·50	2·68
0·25	2·60

Comment upon these results. The density of the solvent was 0·998 g cm^{-3}, the partial specific volume was 0·72 cm^3 g^{-1} and the translational diffusion coefficient in water at 20° is $9\cdot34 \times 10^{-7}$ cm^2 sec^{-1}.

(University of Manchester, Biological Chemistry Honours Course Finals, 1967.)

1.58. A globular protein on treatment with a denaturing agent was investigated by ultracentrifugation and diffusion techniques which gave the following data at 20° C.

Molarity of denaturing agent	$s^\circ_{20} \times 10^3$ (sec)	$D^\circ_{20} \times 10^7$ cm^2 sec^{-1}
1	4·32	5·96
2	4·31	5·85
3	4·25	5·86
4	3·75	5·17
4·5	2·25	3·06
5	1·30	1·74
6	1·03	1·39
7	1·01	1·39
8	1·00	1·36

Deduce a relationship, without going to first principles, between s° and M, where s° is the sedimentation coefficient, D° the diffusion coefficient, and M is the molecular weight.

Using the derived relationship and graphs, discuss the structural change which the protein has undergone in the denaturing solvent.

What is the value of the molecular weight of the protein in 2 M and 7 M concentrations of the denaturing agent?

Very briefly suggest one other physical technique which you might use to confirm your deductions, and give a rationale for your choice.

$$R = 8 \cdot 314 \times 10^7 \text{ ergs/mole/degree}$$
$$\bar{v} = 0 \cdot 734$$
$$\rho = 0 \cdot 998$$

(Glasgow Honours Course Finals, 1969.)

1.59. Comment upon each of the following phenomena, suggesting briefly ways in which your explanations might be verified.

(a) Horse spleen ferritin has a molecular weight of $700\text{-}800 \times 10^3$ while the apoprotein (iron-containing prosthetic group removed) has a molecular weight of $460\text{-}490 \times 10^3$, both values being obtained by conventional hydrodynamic means. Ferritin and apoferritin, however, are eluted from Sephadex G-200 columns with very nearly the same volume of eluent, and well away from the void volume of the column.

(b) A mixture of fibrinogen and γ-globulin in 0·5 M-tris buffer (pH 8·5) containing 0·1 M-NaCl is not resolved when examined in the analytical ultracentrifuge at 52,640 rev/min, but is resolved under the same conditions in the presence of $6 \cdot 8 \times 10^{-3}$ g/ml hyaluronic acid.

(c) Solutions of the optically inactive dye acriflavine have an absorption spectrum with a maximum at 4,570 Å. Although polyglucose 6-sulphate exhibits only a plain rotatory-dispersion curve, its complex with acriflavine shows a Cotton effect centred around 4,570 Å.

(University of Manchester, Biological Chemistry Honours Course Finals, 1967.)

1.60. Outline the advantages and disadvantages of gel-filtration as a method for the determination of molecular weights of proteins in solution.

Discuss how the following data might be interpreted :

Substance	Molecular weight	Elution volume (ml)
Sucrose	342	240
myoglobin	18,000	195
malate dehydrogenase	60,000	160
serum albumin	?	150
aldolase	145,000	125
apoferritin	480,000	95
ferritin	750,000	95
β-galactosidase	520,000	95
fibrinogen	330,000	85
Dextran blue	approx. 10,000,000	70

(University of Manchester, Biological Chemistry, Second B.Sc. Examination, 1968.)

1.61. The elution volume of a number of proteins of known molecular weight was determined on a Sephadex gel column. The values obtained were as follows :

Protein molec. wt.	10,000	20,000	40,000	60,000	80,000	200,000	300,000
Elution volume (ml)	210	188	169	150	140	120	105

A proteolytic enzyme was studied using the same column under the following conditions, with the elution volumes observed. Account for these observations. How would you confirm your interpretation of the data?

Enzyme alone at low concentration	160 ml
Enzyme alone at high concentration	135 ml
Enzyme at low concentration plus a factor isolated from soy bean	125 ml
Enzyme in 1 per cent. sodium dodecyl sulphate	205 ml

(University of Manchester, Biological Chemistry, Second B.Sc. Examination, 1968.)

1.62. A preparation of an enzyme was obtained as a single peak off an ion exchange chromatography column. What other criteria would you apply to confirm the homogeneity of this preparation?

State briefly what conclusions you can draw from the following data concerning the structure of the enzyme:

(a) In 0·1 M-phosphate, pH 7, the weight average molecular weight $M_w = 24,500$.
In 5 M-guanidine hydrochloride, $M_w = 24,500$.
In 5 M-guanidine hydrochloride and 0·1 M-mercaptoethanol $M_w = 11,400$, $M_z = 12,500$.
(b) The intrinsic viscosity, $[\eta] = 3\cdot1$ ml g^{-1} in 0·1 M-phosphate, pH 7.
(c) Fluorodinitrobenzene was reacted with the performic acid oxidized protein and the following amino acid derivatives were detected by chromatography after protein hydrolysis: DNP-Alanine, ε-DNP-lysine, imidazole-DNP-histidine, DNP-isoleucine, DNP-cysteic acid and O-DNP-tryosine. (DNP = dinitrophenyl—).
(d) The enzyme hydrolyses certain peptide bonds, especially those involving the carboxyl C-atoms of aromatic L-amino acids.
(e) The enzyme is inhibited irreversibly by diisopropylphosphofluoridate. After hydrolysis of the inactivated enzyme, the only peptide containing the diisopropylphosphoryl (DIP) group has the sequence Met. Gly. Asp. Ser(DIP). Gly. Gly. Pro., although the protein contains a total of 29 serine residues.

(University of Newcastle upon Tyne, General Degree with Honours (Part I), Biochemistry, 1968.)

1.63. Give a brief account of the principle of the light scattering method for molecular weight determination. Two proteins, X and Y, were obtained in a completely pure, homogeneous state from muscle. They were examined by a variety of techniques in order to derive their molecular weights. These included ultracentrifugation, gel-filtration, osmotic pressure, amino acid and N-terminal analysis and also light scattering.

Protein X was examined at a scattering angle of 90° in light of wavelength 4,360 Å and a molecular weight, M, was derived from the Rayleigh ratio, R, using the equation

$$\frac{Kc}{R_{90}} = \frac{1}{M} + Ac \quad . \quad . \quad . \quad . \quad (1)$$

where K and A are constants and c is the concentration of protein. The value for the molecular weight was in good agreement with results obtained by other methods.

The results for protein Y, however, when subjected to the same mathematical treatment, yielded a molecular weight at variance with any of the values obtained by the other methods. The investigator drew a certain conclusion and extended his measurements to include a range of scattering angles. These further results yielded a molecular weight of the order expected.

(a) Explain the significance of the terms K, A and R_{90} in equation (1) showing how the equation is applied in practice to the derivation of molecular weights. (b) How were the results for protein Y used to obtain a molecular weight? (c) Comment on the investigations summarized above. (d) What further molecular parameter of Y could be obtained from the mathematical treatment you describe in (b)?

(University of Manchester, Biological Chemistry Honours Course, Part I, 1967.)

1.64. Describe briefly how the dimensions of the unit cell of a crystal can be determined by X-ray crystallographic methods. Account for the information below in terms of the size and structure of the metallo-protein in question :

(a) A single measurement of the osmotic pressure of a 1 per cent. solution of the protein in 0·2 M sodium chloride, pH 7·2 (the isoelectric pH of the protein), gave a reading of 64 mm of water at 27°.

(b) The copper content of the protein was 0·3 per cent. (Cu = 63·5).

(c) The protein crystallized as well-defined rectangular prisms of square cross-section and density 1·4 g cm^{-1}. When such a crystal was mounted in the path of a monochromatic X-ray beam of wavelength 1·54Å and rotated about the long axis of the crystal, the layer lines corresponded to a spacing of 53·6Å while prominent diffraction maxima in the equator were obtained at angles of 1° 28′, 2° 04′, 2° 57′, 4° 09′, 5° 54′ and 8° 20′. A wide-angle pattern corresponding to a spacing of 1·5Å could be readily obtained.

(d) On exhaustive dialysis against ethylene diamine tetraacetic acid, the copper content of the protein was reduced to 0·02 per cent. and the protein could not be crystallized.

[Avogadro's No. = 6·02 × 10^{23} : 1 Å = 10^{-8} cm.]

(University of Manchester, Biological Chemistry Honours Finals, 1968.)

1.65. Account for the following observations, indicating how you would attempt, where possible, to substantiate your interpretations experimentally :

(a) Sulphatase A of ox-liver exists as a monomer (mol. wt. 107,000) above pH 6·5 and as a tetramer (mol. wt. 411,000) below pH 5·5. At pH 5 and a temperature of 1° the value for $s_{20, w}$ is 13·61S, and is 14·23S at 38°. The $S_{20, w}$ increases as the pH is taken towards the isolectric point of 3·4.

(b) A marked decrease in viscosity is observed after treatment of ovine submaxillary gland glycoprotein with neuraminidase.

(c) The osmotic pressures of mixtures of hyaluronic acid and serum albumin are significantly greater than the sum of the osmotic pressures of solutions containing hyaluronic acid and serum albumin separately at the same concentrations.

(d) Two proteins (A and B) of identical molecular weight were labelled with dimethylaminonaphthalene sulphonyl chloride and depolarisation of fluorescence data were obtained for the products. The relaxation time for A is approximately three times that of B. In the case of B the $1/\rho$ versus T/η plot has a convex curvature.

(University of Manchester, Biological Chemistry Honours Finals, 1968.)

1.66. Show that for proteins which undergo dimerisation the equilibrium constant for the dissociation reaction is given by

$$K_{diss} = \frac{2a^2c}{(1-a)M_m}$$

where : a is the weight fraction of the protein in the monomeric form, c is the concentration of the protein in mg/ml, M_m is the molecular weight of the protein in its monomeric form.

A light scattering experiment on buffered solutions of frog liver glutamate dehydrogenase gave the following results :

$\dfrac{Kc}{R\theta} \times 10^{+6}$	c(mg/ml)
3·50	0·1
3·24	0·2
2·98	0·4
2·83	0·6
2·73	0·8
2·67	1·0
2·48	2·0
2·40	3·0
3·25	4·0
2·31	5·0

When the results were treated by numerical techniques it was found that the enzyme was undergoing dimerisation with a $K_{diss} = 4 \times 10^{-6}$M.

Calculate the concentration (mg/ml) of monomeric glutamate dehydrogenase in a solution of 0·5 mg/ml of the enzyme.

(University of Glasgow, Honours Course Finals, 1970.)

CHAPTER II

ACID-BASE RELATIONSHIPS AND ELECTROLYTE BEHAVIOUR OF AMINO ACIDS AND PROTEINS

Acid-base Equilibria

Brönsted (1923) defined an acid as a compound which donates protons and a base as one which accepts them. Thus

$$\text{acid} \rightleftharpoons \text{base} + H^+ \quad . \quad . \quad . \quad (2.1)$$

Accordingly each acid has its corresponding or *conjugate base*; they are referred to as a *conjugate pair*. Acids and bases can react by *proton exchange*, represented by the general equation

$$\text{acid}_1 + \text{base}_2 \rightleftharpoons \text{base}_1 + \text{acid}_2 . \quad . \quad (2.2)$$

or $$\qquad a_1 + b_2 \quad \rightleftharpoons \quad b_1 + a_2.$$

An *ampholyte* is a compound which displays both acid and base properties, e.g. water and amino acids

$$2\,H_2O \rightleftharpoons H_3O^+ + OH^-$$

$$2\,RCH\overset{+}{N}H_3.COO^- \rightleftharpoons RCH\overset{+}{N}H_3.COOH + RCHNH_2.COO^-.$$

Although, as will be seen from the above equations, the proton is hydrated in aqueous solution to give the hydronium ion, the hydrogen ion is nevertheless frequently referred to as H^+; both methods are used in this chapter. The hydrogen ion activity (H^+) is usually expressed in terms of its negative logarithm, pH

$$\text{pH} = -\log(H^+) \quad . \quad . \quad . \quad (2.3)$$

where (H^+) is the activity. At very great dilution the hydrogen ion activity is equivalent to the *hydrogen ion concentration* $[H^+]$. This point needs, perhaps, some explanation. Only weak electrolytes behave ideally with respect to the colligative properties of their solutions (Chapter I). This is because interaction occurs between charged ions in solution; each ion does not exist as a separate entity but is surrounded by ions of opposite charge. The magnitude of the interaction is determined by the ionic charge and the concentration and charges of all ionic species present in the solution. Debye and Hückel worked out the theory of this departure from ideal behaviour. Consequently physical chemists have introduced the concept of *activity*. The concentration C_i of a given ion multiplied by its *activity coefficient* f_i gives the activity a_i of that ion, the activity exhibiting ideal behaviour. Thus for hydrogen ions

$$a_{H+} = f_{H+}C_{H+} \text{ or } f_{H+}[H^+] \qquad . \qquad . \qquad (2.4)$$

At very great dilution the activity coefficient approaches unity and $a_{H+} = [H^+]$, i.e. $(H^+) = [H^+]$.

Lewis showed that the activity coefficient varies with salt concentration and that the deviation from ideal behaviour of a given kind of ion, provided that the ionic concentration is not too high, depends upon two main factors : (1) the valency of the given ion and (2) the *ionic strength* (I) of the solution. The concept of ionic strength he introduced to enable different electrolytes in similar environments to be compared. It is given by the relationship

$$I = \tfrac{1}{2} \sum m_i z_i^2 = \tfrac{1}{2} \sum c z_i \qquad . \qquad . \qquad (2.5)$$

where m_i is the molarity of the ion, z_i its valency and c its equivalent concentration $(c = m_i z_i)$. Thus the ionic strength is half the sum of the products of molar concentration and the square of the valency for each of the ionic species present in solution.

Example 2.1.—Calculate the ionic strength of 0·5 M-NaCl and 0·5 M-MgCl$_2$ solutions.

$$0·5 \text{ M-NaCl} \quad I = \tfrac{1}{2}[0·5 \times 1^2 + 0·5 \times 1^2]$$
$$= \underline{0·5}$$
$$0·5 \text{ M-MgCl}_2 \quad I = \tfrac{1}{2}[0·5 \times 2^2 + 2(0·5 \times 1^2)]$$
$$= \underline{1·5}$$

In calculating the ionic strength of a weak electrolyte obviously the degree of dissociation must be taken into account to obtain the true ionic concentration.

The activity coefficient varies with salt concentration and for dilute solutions of all electrolytes (e.g. $I < 0·1$ M) the activity coefficient of a given ionic species is represented by this approximate equation

$$pf_i = -\log f_i = \frac{A z_i^2 \sqrt{I}}{1 + \sqrt{I}} \qquad . \qquad . \qquad (2.6)$$

where A is a constant dependent upon the nature and temperature of the solvent (A increases in solvents of low dielectric constant). If the value of A for water is substituted, equation 2.6 for dilute solutions becomes

$$- \log f_i = \frac{0 \cdot 5 z_i^2 \sqrt{I}}{1 + \sqrt{I}} \qquad \cdot \qquad \cdot \qquad \cdot \qquad (2.6a)$$

For very dilute solutions ($I < 0 \cdot 01$ M) equation 2.6a reduces to

$$- \log f_i = 0 \cdot 5 z_i^2 \sqrt{I} \qquad \cdot \qquad \cdot \qquad \cdot \qquad (2.7)$$

In general terms, for a salt instead of a single ionic species, the *mean activity coefficient* is given by

$$- \log f_{\pm} = 0 \cdot 5 z_+ z_- \sqrt{I} \qquad \cdot \qquad \cdot \qquad (2.7a)$$

where z_+ and z_- are the valencies (irrespective of sign) of the cation and anion respectively. The Debye-Hückel treatment of more concentrated solutions leads to a modification of equation 2.6, namely

$$- \log f_i = \frac{A z_i^2 \sqrt{I}}{1 + B d_i \sqrt{I}} \qquad \cdot \qquad \cdot \qquad (2.6b)$$

where B is a constant and d_i is the ionic diameter. Similarly, equation 2.7a now becomes

$$- \log f_{\pm} = 0 \cdot 5 z_+ z_- \sqrt{I} - CI \qquad \cdot \qquad \cdot \qquad (2.7b)$$

where C is a constant dependent upon the electrolyte. Thus an electrolyte has the same activity coefficient in all solutions of the same ionic strength.

Example 2.2.—Find the pH value of $0 \cdot 001$ N hydrochloric acid if the acid is completely dissociated.

Here there is no complication regarding the degree of dissociation and the concentration of H^+ ions is $0 \cdot 001$ g equivalent per litre.

$$\text{Therefore pH} = - \log 0 \cdot 001 = - (\bar{3} \cdot 000)$$
$$= \underline{3 \cdot 0}.$$

Example 2.3.—Find the pH value of a $0 \cdot 001$ N solution of acetic acid. The dissociation constant, K_a, is $1 \cdot 8 \times 10^{-5}$.

Let α be the degree of dissociation and c the concentration. As the acid is monobasic, its molarity will be identical with the normality, i.e. $c = 0 \cdot 001$M.

$$CH_3COOH \rightleftharpoons CH_3COO^- + H^+$$
$$(1 - \alpha)c \qquad \alpha c \qquad \alpha c$$
$$\text{whence } K_a = \frac{\alpha^2 c}{(1 - \alpha)}.$$

Since the degree of dissociation is very small we may equate $1 - \alpha$ to 1, i.e.

$$K_a = \alpha^2 c = 0 \cdot 001 \alpha^2.$$

Therefore

$$\alpha = \sqrt{\frac{1 \cdot 8 \times 10^{-5}}{10^{-3}}} = 0 \cdot 134.$$

Thus the solution contains $\alpha c = 1 \cdot 34 \times 10^{-4}$ g equivalent H^+ per litre. In dilute solution the activity coefficient may be assumed to be unity, therefore

$$\begin{aligned} pH = -\log[H^+] &= -\log (1 \cdot 34 \times 10^{-4}) \\ &= -(\overline{4} \cdot 1271) = 3 \cdot 8729 \\ &= \underline{3 \cdot 87.} \end{aligned}$$

Example 2.3 illustrates the way in which a general expression may be derived to enable the pH of a weak acid to be calculated if the dissociation constant K_a and the concentration c of the acid are known. As $K_a = \alpha^2 c$, $\alpha = \sqrt{K_a / c}$ and hence

$$\begin{aligned} pH = -\log [H^+] &= -\log (\alpha c) \\ &= -\log \sqrt{K_a / c} - \log c \\ &= -\tfrac{1}{2} \log K_a - \tfrac{1}{2} \log c \\ \text{Thus } pH &= \tfrac{1}{2} p K_a - \tfrac{1}{2} \log c \quad . \quad\quad (2.8) \end{aligned}$$

Example 2.4.—What is the pH of a solution containing 176 mg of pyruvic acid per litre? The pK_a of pyruvic acid is $2 \cdot 50$.

Pyruvic acid, $CH_3CO.COOH$, has a molecular weight of 88. The molar concentration of the acid is therefore $\dfrac{176 \times 10^{-3}}{88} = 2 \times 10^{-3}$ M.

Now

$$\begin{aligned} pH &= \tfrac{1}{2} p K_a - \tfrac{1}{2} \log c \\ &= \frac{2 \cdot 50}{2} - \tfrac{1}{2} (\overline{3} \cdot 3010) \\ &= 1 \cdot 25 - \tfrac{1}{2} (-2 \cdot 6990) = 1 \cdot 25 + 1 \cdot 349 \\ &= \underline{2 \cdot 60.} \end{aligned}$$

Ion Product of Water.—The equilibrium constant for the dissociation of water is given by the equation :

$$K = \frac{(H^+)(OH^-)}{(H_2O)} \quad . \quad\quad . \quad\quad . \quad\quad (2.9)$$

and at $25°$ has the value $1 \cdot 8 \times 10^{-16}$.

As water is only very slightly dissociated (H_2O) may be taken as constant (its concentration is $1000/18 = 55 \cdot 5$ moles per litre) and incorporated in the dissociation constant to give $K_w = (H^+)$ (OH^-). The constant K_w is called the *ion product of water*. If the ionic strength of the medium is low the activity coefficients may be taken as unity and K_w equated to the product of the concentrations of the hydrogen and hydroxyl ions. Conductivity measurements indicate that at 25° the concentration of hydrogen ions is 1×10^{-7} gram equivalent per litre and the concentration of hydroxyl ions is, of course, the same. Hence the ion product of water at 25° is :

$$K_w = 10^{-7} \times 10^{-7} = 10^{-14}.$$

Consequently the pH value of pure water is 7 at 25° since

$$\text{pH} = -\log (\text{H}^+) \approx -\log [\text{H}^+] = 7.$$

The ion product, in common with all equilibrium constants, is temperature dependent and at 37°, $K_w = 2 \cdot 45 \times 10^{-14}$, which gives $\text{pH} = \text{pOH} = 6 \cdot 80$.

There are two different ways in which an acid-base pair may react with water

$$A + H_2O \rightleftharpoons H_3O^+ + B$$
$$B + H_2O \rightleftharpoons A + OH^-.$$

The acidic and basic dissociation constants K_a and K_b are, respectively, $K_a = \dfrac{(\text{H}_3\text{O}^+)(\text{B})}{(\text{A})(\text{H}_2\text{O})}$ and $K_b = \dfrac{(\text{A})(\text{OH}^-)}{(\text{B})(\text{H}_2\text{O})}.$

Eliminating $(B) = \dfrac{(\text{A})(\text{OH}^-)}{K_b(\text{H}_2\text{O})}$

$$K_a = \frac{(\text{H}_3\text{O}^+)(\text{OH}^-)(\text{A})}{K_b(\text{H}_2\text{O})^2(\text{A})}$$

$$K_a = \frac{(\text{H}_3\text{O}^+)(\text{OH}^-)}{K_b(\text{H}_2\text{O})^2} \approx \frac{K_w}{K_b}$$

$$K_w = K_a K_b \qquad . \qquad . \qquad . \quad (2.10)$$

The basic dissociation constant K_b is thus related to K_w and K_a by the expression

$$\text{p}K_w = \text{p}K_a + \text{p}K_b$$

and at 25° $\qquad \text{p}K_b = 14 - \text{p}K_a.$

If the pH of a weak base is required and its K_b and concentration are known then the value may be calculated from a general expression derived in the following way. We have seen that

$$K_b = \frac{(\text{A})(\text{OH}^-)}{(\text{B})(\text{H}_2\text{O})}.$$

As the weak base B dissociates only very slightly, to give equal concentrations of the ions A and OH^-, the concentration of B at equilibrium may be equated to the total molar concentration of the base, i.e. c. Further, (H_2O) may be taken as constant and therefore

$$K_b = \frac{(OH^-)^2}{c}, \text{ i.e. } (OH^-) = \sqrt{K_b c}.$$

Hence $\log (OH^-) = \frac{1}{2} \log K_b + \frac{1}{2} \log c$

i.e. $\log K_w - \log (H^+) = \frac{1}{2} \log K_b + \frac{1}{2} \log c$

and $pH = -\log K_w + \frac{1}{2} \log K_b + \frac{1}{2} \log c$

or $pH = pK_w - \frac{1}{2}pK_b + \frac{1}{2} \log c$.   . (2.11)

Alternatively, equation 2.11 can be expressed in terms of pK_a, thus

$$pH = \frac{1}{2}(pK_w + pK_a + \log c) \quad . \quad . (2.11a)$$

Example 2.5.—Calculate the pH of a 0·1 M solution of sodium acetate at 25° given that the dissociation constant of acetic acid is $1·8 \times 10^{-5}$ and the ion product of water is 10^{-14} at this temperature.

Sodium acetate is the salt of a weak acid and a strong base. In aqueous solution sodium acetate is completely dissociated whereas water is only very slightly dissociated. Acetate ions combine with protons from the water to form acetic acid which is present essentially as the undissociated acid. The equilibria occurring may be summarized as

$$Na^+ + CH_3COO^- + H_2O \rightleftharpoons Na^+ + OH^- + CH_3COOH$$

which reveals that for every undissociated molecule of acetic acid formed one hydroxyl ion is liberated, i.e.

$$(CH_3COOH) = (OH^-).$$

Now $K_a = \dfrac{(CH_3COO^-)\,(H^+)}{(CH_3COOH)} = 1·8 \times 10^{-5}$

and $K_w = (H^+)\,(OH^-) = 10^{-14}.$

Since sodium acetate is completely dissociated in solution the concentration of the acetate ion in moles per litre is 0·1. Therefore

$$(H^+) = \frac{K_a\,(CH_3COOH)}{0·1}$$

and also $(H^+) = \dfrac{K_w}{(OH^-)}.$

Hence $(H^+)^2 = \dfrac{K_a K_w}{0·1} = \dfrac{1·8 \times 10^{-5} \times 10^{-14}}{0·1} = 1·8 \times 10^{-18}$

$$(H^+) = 1·342 \times 10^{-9}.$$

Therefore $\quad\quad$ pH = $-\log{(H^+)}$ = $-(\bar{9}\cdot1277)$ = $8\cdot8723$.

Hence the pH of $0\cdot1$ M-sodium acetate is $8\cdot87$.

On the Brönsted theory of acids and bases the acetate ion behaves as a base, with acetic acid as its conjugate acid, e.g.

$$CH_3COO^- + H_2O \rightleftharpoons CH_3COOH + OH^-.$$

Hence, with a molar concentration c of sodium acetate and making the customary assumption concerning the constancy of the concentration of water,

$$K_b = \frac{(CH_3COOH)\,(OH^-)}{(CH_3COO^-)} = \frac{(OH^-)^2}{c}$$

and $\quad\quad$ $(OH^-) = \sqrt{K_b c}$

$$\log{(OH^-)} = \tfrac{1}{2}(\log{K_b} + \log{c}).$$

Substituting for (OH^-) and K_b in terms of K_w, K_a and (H^+),

$$\log{K_w} - \log{(H^+)} = \tfrac{1}{2}(\log{K_w} - \log{K_a} + \log{c})$$

or $\quad\quad$ $-\log{(H^+)} = \tfrac{1}{2}(-\log{K_w} - \log{K_a} + \log{c})$

$$pH = \tfrac{1}{2}(pK_w + pK_a + \log{c}).$$

This derivation thus leads to equation 2.11a. Applying the equation to the data of Example 2.5 we have $pK_w = 14$; $pK_a = -\log(1\cdot8 \times 10^{-5}) = -(\bar{5}\cdot2553) = 4\cdot7447$; $\log c = \log 0\cdot1 = \bar{1}\cdot00 = -1\cdot0$.
Therefore

$$pH = \tfrac{1}{2}(14 + 4\cdot74 - 1\cdot0) = \frac{17\cdot4}{2} = \underline{8\cdot7}.$$

THE HENDERSON-HASSELBALCH EQUATION.—Since

$$acid \rightleftharpoons base + H^+ \text{ (equation 2.1)}$$

$$i.e.\ A \rightleftharpoons B + H^+$$

$$K_a = \frac{(B)(H^+)}{(A)} \text{ and } (H^+) = \frac{K_a(A)}{(B)}.$$

Taking negative logarithms

$$-\log{(H^+)} = -\log{K_a} - \log{\frac{(A)}{(B)}}$$

i.e. $$\mathrm{pH} = \mathrm{p}K_a + \log \frac{(\mathrm{B})}{(\mathrm{A})} \qquad . \qquad . \quad (2.12)$$

This is the *Henderson-Hasselbalch equation.*

Note that this expression employs the activities of B and A, and if concentrations are used then the activity coefficients must be included in the equation

$$\mathrm{pH} = \mathrm{p}K_a + \log \frac{[\mathrm{B}]f_\mathrm{B}}{[\mathrm{A}]f_\mathrm{A}}.$$

Usually $\log f_\mathrm{B}/f_\mathrm{A}$ is reasonably constant for a given system and may be included in the $\mathrm{p}K_a$ term when

$$\mathrm{p}K_a{}' = \mathrm{p}K_a + \log \frac{f_\mathrm{B}}{f_\mathrm{A}} \qquad . \qquad . \quad (2.13)$$

and $$\mathrm{pH} = \mathrm{p}K_a{}' + \log \frac{[\mathrm{B}]}{[\mathrm{A}]} \qquad . \qquad . \quad (2.12a)$$

This equation is of importance for the understanding of the action of buffer systems and indicators for, as will be seen subsequently, these solutions consist of conjugate acid-base pairs. Consequently the pH of such solutions depends on the ratio of base to acid in the pair and its $\mathrm{p}K_a$ value. With a solution of a weak acid in the presence of its salt with a strong base, the salt will be completely dissociated, whereas the weak acid will be only very slightly dissociated ; the concentration of acid in the mixture may therefore be approximated to its molar concentration and the concentration of the conjugate base equated with the concentration of the salt, thus often the Henderson-Hasselbalch equation is rendered in the form

$$\mathrm{pH} = \mathrm{p}K_a{}' + \log \left[\frac{\mathrm{salt}}{\mathrm{acid}} \right] \qquad . \qquad . \quad (2.12b)$$

or, for a weak base in the presence of its salt with a strong acid,

$$\mathrm{pH} = \mathrm{p}K_a{}' + \log \left[\frac{\mathrm{base}}{\mathrm{salt}} \right] \qquad . \qquad . \quad (2.12c)$$

With a proper appreciation of the Brönsted concept of an acid and its conjugate base, equations 2.12b and c are not really necessary.

Example 2.6.—The concentration of H_2CO_3 in blood plasma is about 0·00125 M. By use of the Henderson-Hasselbalch equation, calculate the concentration of $BHCO_3$ in the plasma when the pH is 7·4 and also when it is 7·1. H_2CO_3 in blood has a pK_{a_1}' value of 6·1.

(Glasgow B.Sc. Single Science Course, 1953.)

The acid is H_2CO_3 and its conjugate base the salt $BHCO_3$, thus

$$pH = pK_{a_1}' + \log \frac{[base]}{[acid]}$$

$$= 6\cdot1 \quad + \log \frac{[BHCO_3]}{[H_2CO_3]}.$$

Let x be the concentration of $BHCO_3$ in the plasma, then at pH 7·4

$$7\cdot4 = 6\cdot1 + \log \frac{x}{0\cdot00125}$$

$$\log x = 1\cdot3 + \log 0\cdot00125$$

$$= 1\cdot3 + \bar{3}\cdot0969$$

$$= \bar{2}\cdot3969$$

Therefore $x \approx 0\cdot025$ M.

At pH 7·1 $\log x = 1\cdot0 + \bar{3}\cdot0969$

$$= \bar{2}\cdot0969$$

and $x = 0\cdot0125$ M.

From the Henderson-Hasselbalch equation it will be evident that when the concentration of an acid and its conjugate base are equal, i.e. [A] = [B], the last term of equation 2.12a becomes zero and therefore $pH = pK_a'$. Now in the titration of a weak acid with a strong base [A] = [B] at the mid-point, or half-equivalence point, of the titration and thus the pK_a' corresponds to the pH at this point (Fig. 2.1).

For polyprotic (polybasic) acids, e.g. H_2CO_3, H_3PO_4, the different acid-base pairs are denoted by $A_1 - B_1$, $A_2 - B_2$, etc., and the dissociation constants are referred to as first, second, third (or primary, secondary, tertiary), K_{a_1}, K_{a_2}, K_{a_3}.

Thus $H_2CO_3 \rightleftharpoons H^+ + HCO_3^-$ $K_{a_1} = \dfrac{(H^+)(HCO_3^-)}{(H_2CO_3)}$

$HCO_3^- \rightleftharpoons H^+ + CO_3^=$ $K_{a_2} = \dfrac{(H^+)(CO_3^=)}{(HCO_3^-)}.$

Hence the general form of the Henderson-Hasselbalch equation may be written as

$$pH = pK_{a_1} + \log \frac{(B_1)}{(A_1)}$$

to indicate the appropriate dissociation to which pK_a' refers.

The titration curve of a weak polyprotic acid reveals a series of individual stages, each corresponding to the titration of one of the dissociable protons with one equivalent of alkali, the mid-point of each characterizing the appropriate pK_a' value. This is illustrated for the triprotic acid phosphoric acid in Fig. 2.1, from which

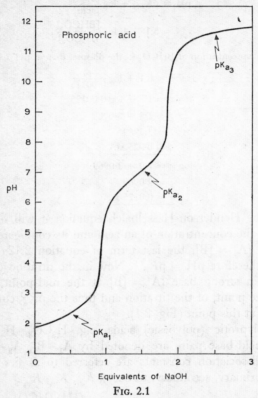

FIG. 2.1

Titration curve for 0·1 M-phosphoric acid with 0·1 N-sodium hydroxide at 20° revealing the three dissociation processes with their corresponding pK_a values given by the half-equivalence point of the three respective stages of the titration.

it will be apparent that the three dissociation processes occur in sequence and, therefore, at a given pH one acid-base pair predominates. pK_{a_1}, pK_{a_2} and pK_{a_3} for phosphoric acid are given by the half-equivalence point of each respective stage.

BUFFER SOLUTIONS.—A buffer solution contains a quantity of an acid and its conjugate base such that their concentrations are scarcely altered by proton exchange with other acid-base pairs present. Thus the solution resists any change in pH value. The pH of the solution is determined by the ratio of base to acid in the buffer pair and its pK_a value.

$$pH = pK_a + \log \frac{(B_{\text{buffer}})}{(A_{\text{buffer}})}.$$

The choice of a buffer pair should be such that the pH to be stabilized falls within a range equal to $pK_a \pm 1$. For values outside this range a different buffer pair should be selected. This point will be readily appreciated when it is realized that the ratio of base to acid varies from 10 at one pH unit above to 0·1 at one unit below the pK_a value.

All weak acids, in the presence of their conjugate bases, constitute buffer solutions. For example, acetic acid, in the presence of its salt sodium acetate, provides a buffer solution the pH of which is governed by the equation

$$pH = pK_a' + \log \frac{[CH_3COO^-]}{[CH_3COOH]}.$$

Since acetic acid is dissociated to a very slight extent whereas sodium acetate is completely ionized, the concentration of acetate ion may be taken equal to the concentration of the salt.

The dissociation constant of acetic acid, K_a, is $1·8 \times 10^{-5}$ at 25° ; therefore $pK_a = -\log (1·8 \times 10^{-5}) = 4·74$, and ideally acetate buffer should be used only to cover the pH range of 3·7 to 5·7. When the salt concentration is equal to that of the acid, as for instance at half neutralization of acetic acid with sodium hydroxide, $pH = pK_a'$ and $[H^+] = K_a'$.

To permit comparison of the efficiency of different buffer solutions Van Slyke (1922) introduced the term *buffer value* which is denoted by β. It is given by the relationship :

$$\beta = \frac{dB}{dpH} \quad . \quad . \quad . \quad . \quad (2.14)$$

where dB is the increment (in gram equivalents per litre) of strong base B added to a buffer solution and dpH the resultant increment in pH. An increment of strong acid is equivalent to a negative increment of base, i.e. $- dB$. β always has a positive value since addition of base increases the pH and addition of acid decreases it, consequently dB and dpH always have the same sign. A solution has a buffer value of unity when one litre will take up 1 gram-equivalent of strong acid or alkali per unit change in pH. Buffer values may be determined directly from a titration curve by reading the values of ΔB and ΔpH for a small increment of B. Ideally, the increment should be infinitesimal, but in actual

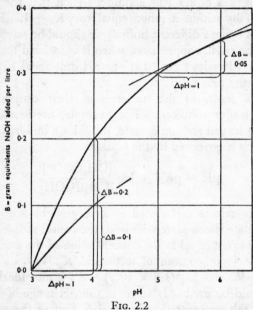

FIG. 2.2

Evaluation of the buffer value β from a titration curve (VAN SLYKE, 1922).

practice measurable increments, provided they are not too large, permit a reasonable evaluation of β. The method is illustrated in Fig. 2.2, and it will be appreciated that values of β may be obtained for any required region of the curve. Alternatively, the tangent to the curve may be drawn at any pH value and its slope evaluated. This, if accurately drawn, gives the exact dB/dpH value and therefore has advantages over the increment method (see Fig. 2.2).

Buffer systems of physiological importance are H_2CO_3 : HCO_3^-, $H_2PO_4^-$: $HPO_4^=$ and H-proteinate : B-proteinate (where B = cation).

INDICATORS.—An indicator is an acid-base pair where the acid and base forms have different colours. A very small quantity of the indicator is added to the solution under investigation so that the concentrations of other acid-base pairs are largely unaffected. The resultant colour of the solution is measured by comparator or colorimeter and the ratio of the concentrations of base and acid forms of the indicator thereby determined. Knowing the pK_a value of the indicator, the pH value of the solution may be calculated from equation 2.12. Indicators, as with buffer solutions, should be used only within the range $pK_a \pm 1$.

Some indicators have only one coloured form, e.g. the acid form of phenolphthalein is colourless and the base form red, whereas others have both forms coloured, e.g. methyl red, which has acid and base forms red and yellow respectively.

The Dipolar Ion Form and Electrolyte Behaviour of Amino Acids

An amino acid ionizes according to the pH of its solution and the following stages may be recognized for an amino acid which has an R group devoid of dissociable groups :

$$
\begin{array}{ccc}
R & R & R \\
| \quad + & | \quad + & | \\
CHNH_3 \rightleftharpoons & CHNH_3 \rightleftharpoons & CHNH_2 \\
| & | & | \\
COOH & COO^- & COO^-
\end{array}
$$

Net charge	$+1$	0	-1
	cation	dipolar ion	anion
	$pH < pI$	isoelectric	$pH > pI$
		and isoionic	
		pI	

The isoelectric species, which bears no net charge, is a dipolar ion. Formerly it was believed that the isoelectric amino acid had the structure $RCHNH_2COOH$, but evidence from various physical and chemical sources, such as spectroscopy, dielectric constant and titration in organic solvents, indicated that the dipolar ion is the correct representation. Amino acids in the crystalline state have high melting points which indicate that they also exist as the doubly charged molecule in the solid state.

D

The dipolar ion possesses two ionizing groups and is therefore characterized by two pK_a values. Consider the simplest amino acid, glycine, which is aminoacetic acid. In Fig. 2.3 the titration curves of acetic acid and glycine are compared. When acetic acid is titrated with sodium hydroxide there is a point of inflexion at half neutralization which corresponds to a pH value of 4·74. This, as already mentioned, is the pK_a value for acetic acid since

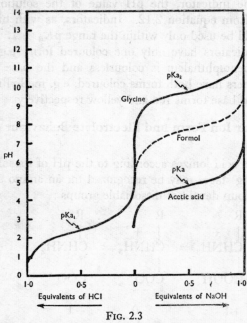

FIG. 2.3

Titration curves of glycine and acetic acid. Note the effect of addition of formaldehyde in the formol titration.

at half neutralization pH = pK_a. It will be noticed that the pH of the solution changes very little during the addition of 0·2 to 0·8 equivalents of alkali ; herein lies the basis of the buffering power of such solutions, to which reference has already been made. The titration curve of glycine with acid and alkali displays two points of inflexion, at pH 2·4 and 9·7. These are referred to as pK_{a1} and pK_{a2} respectively, the former being due to the carboxyl group and the latter to the amino group. Note the convention that pK_a values are numbered from the acid to the alkaline region. It may be noted that a plot of the electrophoretic mobility of a dipolar ion as a function of pH is similar in form to a titration curve.

The fact that pK_{a2} for glycine is due to the amino group may be demonstrated by a procedure, developed by Sörensen, known as the *formol* titration. If formaldehyde is added to an amino acid, either one or two molecules react with the amino group to form methylol derivatives and release a proton. The equilibria set up may be represented as follows:

$$\underset{\underset{COO^-}{|}}{\overset{\overset{R}{|}}{CHNH_3^+}} \underset{+H^+}{\overset{-H^+}{\rightleftharpoons}} \underset{\underset{COO^-}{|}}{\overset{\overset{R}{|}}{CHNH_2}} \underset{-HCHO}{\overset{+HCHO}{\rightleftharpoons}}$$

$$\underset{\underset{COO^-}{|}}{\overset{\overset{R}{|}}{CHNH(CH_2OH)}} \underset{-HCHO}{\overset{+HCHO}{\rightleftharpoons}} \underset{\underset{COO^-}{|}}{\overset{\overset{R}{|}}{CHN(CH_2OH)_2}}$$

Titration of the resultant solution with alkali reveals that the titration curve has shifted appreciably towards the acid side and the stoicheiometric end-point now occurs at approximately pH 9 instead of 12. As pH 9 falls within the range of phenolphthalein, this indicator may be used for the titration. Hence pK_{a2} at 9·7 for glycine must be due to the amino group. The formol titration is a useful method for determining the amino groups of aliphatic amino acids and polypeptides, and for a discussion of the principle of this method the reader is referred to Clark (1952).

All monoaminomonocarboxylic acids which possess no other ionizable group have two pK_a values which do not differ greatly from those of glycine, but with monoaminodicarboxylic and diaminomonocarboxylic acids there are three pK_a values. For instance, aspartic acid has pK_a values for the two carboxyl groups, the α (pK_{a1}) being slightly more acidic than the β group (pK_{a2}). The values are 1·9 and 3·7 respectively, and on account of their close proximity the points of inflexion on the titration curve are not well resolved. pK_{a3} for the amino group of aspartic acid occurs at 9·6. Lysine has $pK_{a1} = 2·2$, $pK_{a2} = 8·9$ and $pK_{a3} = 10·5$ for the carboxyl, α- and ϵ-amino groups respectively, and the

ϵ-amino group is therefore a slightly stronger base than the α group.

Some amino acids possess titratable groups other than the carboxyl and amino groups. The phenolic hydroxyl of tyrosine, thiol of cysteine, guanidino of arginine and the imidazolyl group of histidine behave in this way, the former two as acids and the latter two as bases. Histidine, for example, has $pK_{a1} = 1.8$ for the carboxyl, $pK_{a2} = 6.1$ for the imidazole and $pK_{a3} = 9.2$ for the amino group.

ISOELECTRIC AND ISOIONIC POINTS.—The *isoelectric point* is the pH at which the ionic species bears no net charge and consequently does not migrate in an electric field. The majority of proteins display their minimum solubility at the isoelectric point. The *isoionic point* is defined as the pH at which the number of protons dissociated from acidic groups is equal to the number of protons combined with basic groups. Note particularly that the isoionic point refers to the dissociation and combination of protons only and that the isoelectric and isoionic points are therefore identical only if the dipolar ion combines solely with protons. The isoelectric point is usually denoted by the symbol pI and for an amino acid having two pK_a values is obtained as follows

$$pI = \frac{pK_{a1} + pK_{a2}}{2} \quad . \quad . \quad . \quad (2.15)$$

For amino acids bearing three dissociating groups it is customary to take into account only the two predominating pK_a values, i.e. pK_{a1} and pK_{a2} for monoaminodicarboxylic acids and pK_{a2} and pK_{a3} for diaminomonocarboxylic acids. The following examples illustrate this procedure.

Example 2.7.—Calculate the isoelectric point of (a) glycine, (b) arginine. Glycine has $pK_{a1} = 2.4$ and $pK_{a2} = 9.7$ and arginine has $pK_{a1} = 2.0$, $pK_{a2} = 9.0$ and $pK_{a3} = 12.5$.

(a) *Glycine.* Glycine dissociates in the following manner :

$$\text{Glycine}^+ \rightleftharpoons \text{Glycine}^\pm + \text{H}^+ \qquad K_{a1} = \frac{(\text{Gly}^\pm)(\text{H}^+)}{(\text{Gly}^+)}$$

$$\text{Glycine}^\pm \rightleftharpoons \text{Glycine}^- + \text{H}^+ \qquad K_{a2} = \frac{(\text{Gly}^-)(\text{H}^+)}{(\text{Gly}^\pm)}.$$

At pI $(\text{Gly}^+) = (\text{Gly}^-)$, i.e. $\dfrac{(\text{Gly}^\pm)(\text{H}^+)}{K_{a1}} = \dfrac{K_{a2}(\text{Gly}^\pm)}{(\text{H}^+)}.$

Rearranging and simplifying $(H^+)^2 = K_{a_1} K_{a_2}$

$$(H^+) = \sqrt{K_{a_1} K_{a_2}}$$

$$pH \equiv pI = \frac{pK_{a_1} + pK_{a_2}}{2} = \frac{2 \cdot 4 + 9 \cdot 7}{2} = \underline{6 \cdot 05}.$$

(b) *Arginine.* The three pK_a values refer to the carboxyl, α-amino and guanidino groups respectively. Arginine dissociates in the following manner:

$$\text{Arginine}^{++} \rightleftharpoons \text{Arginine}^{\pm+} + H^+ \qquad K_{a_1} = \frac{(\text{Arg}^{\pm+})(H^+)}{(\text{Arg}^{++})}$$

$$\text{Arginine}^{\pm+} \rightleftharpoons \text{Arginine}^{\pm} \; + H^+ \qquad K_{a_2} = \frac{(\text{Arg}^{\pm})(H^+)}{(\text{Arg}^{\pm+})}$$

$$\text{Arginine}^{\pm} \; \rightleftharpoons \text{Arginine}^{-} \; + H^+ \qquad K_{a_3} = \frac{(\text{Arg}^{-})(H^+)}{(\text{Arg}^{\pm})}.$$

From the above equilibria it will be seen that the arginine species bearing a net charge are Arg^{++}, $\text{Arg}^{\pm+}$ and Arg^-. Therefore at the isoelectric point where there is no net charge the following relationship must hold:

$$2(\text{Arg}^{++}) + (\text{Arg}^{\pm+}) = (\text{Arg}^-)$$

$$\frac{2(H^+)(\text{Arg}^{\pm+})}{K_{a_1}} + \frac{(H^+)(\text{Arg}^{\pm})}{K_{a_2}} = \frac{K_{a_3}(\text{Arg}^{\pm})}{(H^+)}$$

$$\frac{2(H^+)^2(\text{Arg}^{\pm})}{K_{a_1}K_{a_2}} + \frac{(H^+)(\text{Arg}^{\pm})}{K_{a_2}} = \frac{K_{a_3}(\text{Arg}^{\pm})}{(H^+)}$$

$$\frac{2(H^+)^2}{K_{a_1}K_{a_2}} + \frac{(H^+)}{K_{a_2}} = \frac{K_{a_3}}{(H^+)}$$

$$\frac{(H^+)}{K_{a_2}}\left[\frac{2(H^+)}{K_{a_1}} + 1\right] = \frac{K_{a_3}}{(H^+)}$$

$$2pH - \log\left[\frac{2(H^+)}{K_{a_1}} + 1\right] = pK_{a_2} + pK_{a_3}.$$

Now pI must lie between pK_{a_2} and pK_{a_3}, i.e. the pH range where (Arg^-) can exist.

Hence $\qquad\qquad (H^+) \ll 1$ or $\log\left[\dfrac{2(H^+)}{K_{a_1}} + 1\right] \cong 0.$

Therefore $\qquad\qquad pH = \dfrac{pK_{a_2} + pK_{a_3}}{2} = \dfrac{9 \cdot 0 + 12 \cdot 5}{2} = \underline{10 \cdot 75}.$

ELECTROPHORESIS.—An important application of the electrolyte behaviour of amino acids (and of proteins, as will be seen in the following section) is found in the technique of *electrophoresis.* When a solution containing charged ions is placed between electrodes and a direct current is passed, the positively charged cations migrate to the negative electrode (cathode) and the negatively charged anions migrate to the positively charged anode.

The rate of migration of the ions depends upon their size, shape and charge distribution and therefore is influenced by the pH and ionic strength of the medium. At its isoelectric point an ion bears no net charge and does not move towards either electrode. At pH values below the isoelectric point the ion carries a net positive charge and migrates to the cathode, whereas at pH values higher than the isoelectric point migration to the anode occurs on account of the net negative charge. Electrophoresis therefore offers a method for the separation of amino acids or proteins in a mixture by their differential migration in an electric field. By suitable choice of pH and ionic strength effective separation may be achieved. It should be noted that since the ionic strength of the medium influences the migration of an ion it is necessary to define ionic strength as well as pH.

The electrophoresis of small molecules such as amino acids is usually carried out by the technique known as high voltage paper electrophoresis, a form of zone electrophoresis (p. 94) using paper as the support. The migratory behaviour of amino acids in response to variations of pH permits not only the separation of amino acids in a mixture but also affords information concerning the number of dissociable groups in an amino acid molecule and their approximate pK_a values. As previously mentioned, the plot of the electrophoretic mobility of a dipolar ion as a function of pH is similar in form to a titration curve.

Example 2.8.—Devise a procedure for the separation of a mixture of glycine, aspartic acid and lysine. The pK_a values of the amino acids are as follows

	pK_{a1}	pK_{a2}	pK_{a3}
Glycine	2·4	9·7	—
Aspartic acid	1·9	3·7	9·6
Lysine	2·2	8·9	10·5

The isoelectric point of each amino acid is determined as the mean of the two predominating pK_a values. Thus :

Glycine $\quad \dfrac{pK_{a1}+pK_{a2}}{2} = \dfrac{2\cdot4+9\cdot7}{2} = 6\cdot05$

Aspartic acid $\quad \dfrac{pK_{a1}+pK_{a2}}{2} = \dfrac{1\cdot9+3\cdot7}{2} = 2\cdot8$

Lysine $\quad \dfrac{pK_{a2}+pK_{a3}}{2} = \dfrac{8\cdot9+10\cdot5}{2} = 9\cdot7$

Consequently, if electrophoresis were carried out in a buffer of pH 6·05 glycine, carrying no net charge, would remain at the origin. Aspartic acid, being on the alkaline side of its isoelectric point at this pH, would exist as an anion and migrate to the anode while lysine would be present as a cation and so migrate in the opposite direction to the cathode. A separation could thereby be achieved.

(In passing, it is worth noting that in zone electrophoresis, even with ions of such relatively small sizes as amino acids, the rate of migration can be significantly affected by their size. A practical consequence of this effect is the possibility of separating all the amino acids in a mixture by employing paper electrophoresis at pH 2 ; although all amino acids exist as cations at this pH and therefore migrate to the anode, their rates of migration, determined by the sizes of their hydrated cations, usually differ sufficiently to effect a separation.)

ION EXCHANGE CHROMATOGRAPHY.—The ionic nature of amino acids enables them to be separated by means of suitable ion-exchange procedures. Ion-exchange materials are generally insoluble organic macromolecules containing acidic or basic groups (Table 2.1), and the pH-dependent dissociation of these groups determines the number of charges which are available on the material for binding electrostatically an equivalent number of ions of opposite charge from the surrounding medium. Such materials are classed as *cation* or *anion exchangers* according to whether they bind cations or anions from the medium, i.e. whether the ion-exchange material carries, respectively, negative or positive charges.

Cation exchangers may contain weakly acidic groups, such as carboxyl, which are only ionized completely at pH values above 7, or strongly acidic groups, such as sulphonic acid, which are effectively ionized at all pH values. Examples of weakly acidic cation exchangers are methacrylate polymers bearing carboxyl groups, e.g. Amberlite IRC-50, and polysaccharides carrying carboxyl methyl groups, e.g. carboxymethyl cellulose (CM-cellulose)

$$-OCH_2.COOH \rightleftharpoons -OCH_2.COO^- + H^+.$$

Dowex 50 is a strongly acidic resin which exists in the ionized state over the pH range 1-14. Similarly, anion exchangers may be weakly or strongly basic in character. The functional groups of the weakly basic materials include substituted amino groups and the strongly basic quaternary ammonium groups. In use an appropriate cation exchanger, at a pH at which it is completely ionized, is saturated with a solution containing a given cation, e.g. Na^+, which is thus bound to the exchanger.

$$-CH_2.COOH \xrightarrow{\text{suitable pH}} -CH_2.COO^- + H^+$$

$$-CH_2.COO^- + Na^+ \rightarrow -CH_2.COO^- - - -Na^+$$

TABLE 2.1

Characteristics of Ion-Exchange Materials

Resin	Type	Functional group	Behaviour	pH range
Cation Exchangers				
Zeo-Karb 215	Phenolic	$-OH$	Strongly acidic	1-14
Zeo-Karb 225 Dowex 50 Amberlite IR-120	Cross-linked polystyrene	$-SO_3H$	Strongly acidic	1-14
Zeo-Karb 226 Amberlite IRC-50	Methacrylate polymer	$-COOH$	Weakly acidic	6-9
Anion Exchangers				
De-Acidite FF Amberlite IRA-400 Dowex 1 Dowex 2	Cross-linked polystyrene	Quaternary ammonium	Strongly basic	1-14
De-Acidite G-1P	Cross-linked polystyrene	$-N(C_2H_5)_2$	Weakly basic	1-9
De-Acidite H-1P	Cross-linked polystyrene	Mixed $-N(C_2H_5)_3$ $-N^+(C_2H_5)_4$	Weakly basic	1-14
Amberlite IR-45 Dowex 3	Cross-linked polystyrene	Substituted amine	Weakly basic	1-9
Amberlite IR-4B	Phenolic	$-OH$ nuclear amino group	Weakly basic	1-9

This system is then added to a buffer solution containing a different cation, say K^+, still at a pH which maintains the exchange material in its fully ionized form. Exchange of cations then occurs until equilibrium is attained.

$$-CH_2.COO^- ---Na^+ + K^+ \rightleftharpoons -CH_2.COO^- ---K^+ + Na^+$$

Factors such as the relative affinity of the ion-exchanger for the two cations, the temperature and the initial concentrations of the saturated ion-exchanger and the second cation introduced, influence the equilibrium attained. The fact that the ion-exchange process attains an equilibrium means that column operation is usually more effective than batch treatment for, by this means, the exchange may proceed virtually to completion as the cation in the mobile phase percolates through the stationary ion-exchange material. In the example considered, K^+ is progressively removed as the loading buffer solution containing K^+ ions flows down the column ; successive layers of the ion-exchanger are thus exposed to decreasing concentrations of K^+ and this ion is essentially completely removed from the solution. By suitable choice of conditions the cation can be bound to the ion-exchanger in a relatively narrow band ; it will remain bound if washed with a buffer of the same pH, but of lower ionic strength, than the loading buffer.

Elution and recovery of the cation is achieved by passing a suitable buffer solution through the column; this buffer should be either of the same pH, but of greater ionic strength, or of a lower pH such that protons combine with the negative charges of the ion-exchanger, thereby releasing the cation. Alternatively, elution may be effected by a buffer at the same pH and ionic strength provided it contains a cation for which the ion-exchanger has a greater affinity than for the bound cation.

Electrolyte Behaviour of Proteins

Proteins consist of amino acid residues condensed together and their titration behaviour is of considerable interest. Since α-amino and carboxyl groups are used for the formation of peptide bonds, it follows that the polar groups of the amino acid side-chains must provide the major contribution to electrolyte behaviour. The β- and γ-carboxyls of aspartic and glutamic acids respectively, ϵ-amino of lysine, phenolic hydroxyl of tyrosine, guanidino group of arginine, imidazolyl group of histidine and thiol group of cysteine all contribute in this way (Table 2.2). Consequently the actual shape of the titration curve of a protein will depend on the number and type of polar side-chain groups it possesses and on their pK_a values.

From titration data it is possible to make an approximate calculation of the number of side-chain groups of each type present in the protein molecule if the basic assumption is made that the pK_a values of the groups in the side-chain do not differ appreciably from the values characteristic of the free amino acids. Considerable success has been achieved in this way, as confirmed by subsequent hydrolysis and analysis of the protein, although the

TABLE 2.2

pK_a' Values of Acidic and Basic Groups of Proteins

Group	Amino acid	pK_a' (25°)	
		Residue in protein	Free amino acid
α-Carboxyl		3·0–3·2	1·8–2·3
β-Carboxyl	Aspartic acid	3·0–4·7	3·86
γ-Carboxyl	Glutamic acid	~4·5	4·25
Imidazolyl	Histidine	5·6–7·0	6·0
α-Amino		7·6–8·4	8·6–10·7
ε-Amino	Lysine	9·4–10·6	10·53
Thiol	Cysteine	9·1–10·8	8·33
Phenolic hydroxyl	Tyrosine	9·8–10·4	10·07
Guanidino	Arginine	11·6–12·6	12·48
α-Carboxyl	Acetic acid		4·74
	Glycine		2·4
α-Amino	Methylamine		10·7
	Glycine		9·7

pK_a value for a given free amino acid and for the corresponding residue in a protein may differ by more than 1 pH unit, as Table 2.2 reveals. These differences may be ascribed to electrostatic effects arising from other charged groups in the molecule, the possibility of hydrogen bonding and effects of the medium due to the proximity of hydrophobic residues. Some of the ionizing groups may be effectively buried in the native protein structure and not readily accessible to the solvent. The well-known appearance or 'unmasking' of thiol groups as a result of protein denaturation may be attributed to changes in the tertiary structure of the protein exposing these groups to the solvent and to thiol reagents.

Studies of the effect of pH on the activity of enzymes have, in certain cases, been used in attempts to elucidate the amino acid sequences at the active sites. This approach supposes that the primary effect of pH on activity is on the ionization of charged groups at the active site, or on those in close proximity to it, whereas the ionization of groups further removed from the active site will have comparatively little effect. Although the assumptions do not seem fully justified, the analysis of pH-velocity curves has yielded interesting information. For example, the free base of a histidine residue has been implicated in chymotrypsin activity and this is substantiated by other evidence.

Titration curves reveal that proteins possess great buffering capacity ; relatively large quantities of acid and alkali may be added without producing appreciable change in pH. This property is of enormous importance in biological systems where it is found that proteins constitute the main buffering system. For example, about 80 per cent. of the total buffering power of mammalian blood is due to proteins. The buffering capacity of proteins at physiological pH values is mainly due to the imidazolyl and α-amino groups. The well-known Bohr effect with haemoglobin, i.e. the modification of the dissociation curve of oxyhaemoglobin effected by varying the partial pressure of carbon dioxide or by varying pH, has been principally attributed to the imidazolyl groups of the C-terminal histidine residues of the β-chains, together with the α-amino groups of the α-chains. These groups are free in oxyhaemoglobin but in deoxyhaemoglobin their pK_a values are raised, probably by linkage to carboxyl groups.

An important application of the electrolyte behaviour of proteins is found in the technique of electrophoresis which offers a method for the separation of proteins in a mixture by their differential migration in an electric field. By choice of appropriate pH and ionic strength effective separation may be achieved. The classical apparatus for electrophoresis, in which the movement of the proteins is followed by changes of refractive index at the protein boundary, i.e. *moving-boundary electrophoresis*, was invented by Arne Tiselius in 1933. By its use considerable information concerning many proteins has been obtained. This is particularly true of human plasma proteins and it is now customary to classify these on the basis of their mobility in an electrical field. Thus, in decreasing order of mobility, albumin, α-globulin, β-globulin, fibrinogen and γ-globulin may be distinguished.

In the classical method of electrophoresis of solutions care must be taken to avoid convection currents which would disturb the separation. These are produced by the heat of the electric current but can be minimized by working at 4°, the temperature of maximum density of water. However, the effect can be considerably reduced by the use of solid or semi-solid supporting media for the solution which enables separations to be carried out over a wide range of temperature. This modified technique, termed *zone electrophoresis*, employs supports such as paper, ground glass wool, starch grains or starch gel, agar or cellulose acetate and the ions are detected by various properties including colour, biological activity or their capacity for binding dyes. The stabilizing medium not only prevents convection but may also obstruct the free migration of the ions and, in some cases, even act as an ultra-filter for larger molecules, e.g. the high resolution of serum proteins secured in starch gel probably depends on this effect. Zone electrophoresis, particularly that employing filter paper, does not require elaborate apparatus and this factor has brought the technique within the scope of most laboratories, e.g. serum proteins are now examined extensively in clinical practice by means of filter paper electrophoresis and a considerable literature exists on plasma proteins in pathological conditions.

By the use of high voltages separation of peptides, amino acids and carbohydrates (as borate complexes) can be effected.

The Donnan Membrane Equilibrium (Gibbs-Donnan Effect)

The net electric charge of proteins gives rise to their effect in producing an unequal distribution of diffusible ions on either side of a membrane through which the proteins cannot pass.

Donnan, in 1911, propounded the theory of membrane equilibria which accounts for the effect of electrical charges on macromolecules. Donnan equilibria, as they are called, occur whenever a charged molecule is constrained in its movements, as, for example, when a membrane is impermeable to it. The behaviour of charged protein molecules is of great interest in biochemistry ; physiological pH values usually are higher than the pI values of proteins, which therefore exist as anions. These protein anions are accompanied by an equivalent number of sodium ions to preserve electrical neutrality and may thus be regarded as the sodium salts of the proteins. Consider, therefore, the sodium salt of a protein at molar concentration m_1 separated from a solution of sodium chloride of concentration c_2 by a membrane impermeable to protein molecules but permeable to inorganic ions. If the charge on the protein molecule is z^-, there will be a concentration m_1 of protein molecules bearing zm_1 negative charges and an equivalent concentration zm_2 of Na^+ ions. This is illustrated diagrammatically in Fig. 2.4 (a) where the protein is in compartment A. In compartment B is sodium chloride solution at concentration c_2 and hence there is a concentration gradient of Cl^- ions in response to which some diffuse across the membrane into compartment A.

		A		B	
(a) Initial state	Na^+	zm_1	Na^+	c_2	
	Pr^{z-}	m_1	Cl^-	c_2	
(b) Equilibrium state	Na^+	$zm_1 + c_1$	Na^+	$c_2 - c_1$	
	Pr^{z-}	m_1	Cl^-	$c_2 - c_1$	
	Cl^-	c_1			

FIG. 2.4

Initial and equilibrium states in a Donnan equilibrium. Equivalent concentration c_1 of NaCl has been transferred from compartment B to compartment A.

As this would upset the electrical neutrality of the system they are accompanied by an equivalent concentration of Na^+ ions. This, in its turn, sets up a concentration gradient of Na^+ ions which begin to diffuse back from compartment A into B. The two processes continue until equilibrium is set up when the concentration gradient of Cl^- from B to A is balanced by that of the Na^+ from A to B. At equilibrium consider that a concentration c_1 of Cl^- has diffused into compartment A. The equilibrium state will then be that depicted in Fig. 2.4(b).

At constant temperature and pressure the free energy necessary to transfer one mole of Cl^- from compartment B to A in a reversible manner is

$$\Delta G = RT \ln \frac{(Cl^-)_A}{(Cl^-)_B},$$

and similarly for one mole of Na^+

$$\Delta G = RT \ln \frac{(Na^+)_A}{(Na^+)_B}.$$

From thermodynamics we know that the total free energy change is zero at equilibrium (Chapter III).

Therefore $RT \ln \dfrac{(\text{Cl}^-)_A}{(\text{Cl}^-)_B} + RT \ln \dfrac{(\text{Na}^+)_A}{(\text{Na}^+)_B} = 0$

and $$\frac{(\text{Cl}^-)_A}{(\text{Cl}^-)_B} = \frac{(\text{Na}^+)_B}{(\text{Na}^+)_A} \qquad \qquad (2.16)$$

This is a general expression and holds for all univalent cations and anions.

Therefore $$\frac{c_1}{c_2 - c_1} = \frac{c_2 - c_1}{zm_1 + c_1}$$

and $$c_1 = \frac{c_2{}^2}{zm_1 + 2c_2}.$$

Thus the ratio of NaCl in compartment B to that in compartment A is

$$\frac{(\text{NaCl})_B}{(\text{NaCl})_A} = \frac{(\text{Cl}^-)_B}{(\text{Cl}^-)_A} = \frac{c_2 - c_1}{c_1} = 1 + \frac{zm_1}{c_2} \qquad (2.17)$$

which means that the greater the concentration of protein, m_1, relative to that of the salt, c_2, the more uneven will be the final distribution of diffusible ions. However, if the salt concentration c_2 is high, its distribution at equilibrium will approach unity and the Donnan effect will be negligible. Practical use is made of this fact in osmotic pressure measurements (p. 14). Establishment of a membrane equilibrium results in osmotic pressures that are too high being measured and hence introduces error. It can be largely eliminated by working at the isoelectric point when the protein bears no net charge or, alternatively, by employing a concentrated salt solution as solvent for the protein.

For polyvalent cations and anions the general expression for the Donnan distribution becomes

$$\left(\frac{(\text{Anion})_A}{(\text{Anion})_B} \right)^{\frac{1}{z_a}} = \left(\frac{(\text{Cation})_B}{(\text{Cation})_A} \right)^{\frac{1}{z_c}}$$

where z_a and z_c are the valencies of the anions and cations respectively.

The significance of the Donnan effect under physiological conditions is not easily evaluated because of the complicating features of active transport, by which means ions are transported against concentration gradients by processes requiring the expenditure of energy. The static Donnan effect may readily be obscured by such dynamic phenomena. Mammalian blood does, however, afford an example of its importance; the concentration of chloride ions within the erythrocytes is lower than their concentration in the plasma, a finding which is attributable to the Donnan effect. There is evidence also that multiple-charged anions are transported in mitochondria in accordance with a passive Donnan distribution. For equilibrium the relationship

$$\frac{RT}{n} \ln \frac{[A^{n-}]_i}{[A^{n-}]_o} = \frac{RT}{m} \ln \frac{[B^{m-}]_i}{[B^{m-}]_o}$$

holds where the subscripts i and o indicate the concentrations of anions A^{n-} and B^{m-} inside and outside the mitochondrion. As we have already seen, these relationships may be expressed as

$$\left(\frac{[A^{n-}]_i}{[A^{n-}]_o}\right)^{\frac{1}{n}} = \left(\frac{[B^{m-}]_i}{[B^{m-}]_o}\right)^{\frac{1}{m}}$$

so that the appropriate values for each anion should be equal. E. J. Harris has determined the $1/n$th roots of the ratio of the concentration inside and outside the mitochondrion of the anions of pyruvate, malate, phosphoenolpyruvate and citrate and found reasonably good agreement, supporting the existence of a Donnan distribution.

REFERENCES AND SUGGESTED READING

ALBERTY, R. A. (1953). In *The Proteins*, vol. 1A, Chap. 6, ed. H. Neurath & K. Bailey. New York : Academic Press.

BRÖNSTED, J. N. (1923). *Recl. Trav. Chim. Pays-Bas Belg.*, **42**, 718.

BULL, H. B. (1971). *An Introduction to Physical Biochemistry*, 2nd ed. Philadelphia : F. A. Davis.

CLARK, W. M. (1952). *Topics in Physical Chemistry*, 2nd ed., Chap. 16. Baltimore : Williams & Wilkins.

DONNAN, F. G. (1911). *Z. Elektrochem.*, **17**, 572.

EDSALL, J. T. & WYMAN, J. (1958). *Biophysical Chemistry*, vol. 1, Chaps 5 to 9. New York : Academic Press.

HARTLEY, G. S. (1948). State of solution of colloidal electrolytes. *Q. Rev. chem. Soc.*, **2**, 152.

STEIN, W. D. (1962). Behaviour of molecules in solution. In *Comprehensive Biochemistry*, vol. 2, 219, ed. M. Florkin & E. H. Stotz. Amsterdam : Elsevier.

TRAUTMAN, R. (1963). Acid-base properties and electrophoresis of proteins. In *Comprehensive Biochemistry*, vol. 7, 107, ed. M. Florkin & E. H. Stotz. Amsterdam : Elsevier.

VAN SLYKE, D. D. (1922). On the measurement of buffer values and on the relationship of buffer value to the dissociation constant of the buffer and the concentration and reaction of the buffer solution. *J. biol. Chem.*, 52, 525.

PROBLEMS

2.1. Calculate the hydrogen ion concentration of solutions of pH value 6·0 and 9·5.

2.2. What is the pH value of 0·5 per cent. (w/v) hydrochloric acid? (Assume complete dissociation.)

2.3. Calculate the pH value of (a) 0·001 M sulphuric acid and (b) 0·001 M sodium hydroxide, assuming both solutions to be completely ionized. What would be the resultant pH value of the solution if 25 ml of the above sulphuric acid were added to 20 ml of the sodium hydroxide solution?

2.4. 10 ml of a 0·05 M solution of potassium dihydrogen phosphate are added to 10 ml of a solution of disodium hydrogen phosphate. The pH of the resulting solution is 6·725. Calculate the molarity of the disodium hydrogen phosphate if the second dissociation constant for phosphoric acid is $1·4 \times 10^{-7}$.

2.5. Calculate the ionic strength of the following solutions : (a) 0·1 N-NaOH ; (b) 2 N-H₂SO₄ ; (c) 0·025 M-MgSO₄ ; (d) 0·02 M-KH₂PO₄ ; (e) 0·15 M-CaCl₂ ; (f) 0·5 M-Na₂HPO₄ ; (g) 0·075 M-CH₃COONa. For the purpose of these calculations complete dissociation may be assumed in each case although this would not apply in every case. (Consider which of these salts would not be fully dissociated in solution at the given concentrations.) Derive general expressions for the relation between ionic strength and molar concentration for (1) uni-univalent, (2) uni-divalent, (3) di-divalent, (4) uni-trivalent and (5) di-trivalent salts, and use the derivations to check the above calculations.

2.6. Calculate the ionic strength of 0·01 N acetic acid if the dissociation constant of the acid is $1·8 \times 10^{-5}$.

2.7. Calculate the activity coefficients and activities of the ions in aqueous solutions of (a) 5 mM-H₂SO₄ (b) 2 mM-NaCl.

2.8. A solution is 5 mM with respect to MgCl₂ and 2·5 mM with respect to MgSO₄. Determine the ionic strength and estimate the activities of the ions.

2.9. The primary dissociation constant for carbonic acid is $4·31 \times 10^{-7}$. If the pH of a sample of blood is 7·45, determine the ratio of carbonic acid to bicarbonate present.

2.10. Determine the secondary dissociation constant of phosphoric acid if blood of pH 7·4 contains 12·85 mg HPO₄⁼ and 3·21 mg H₂PO₄⁻ per 100 ml plasma respectively.

2.11. The acid dissociation constants of haemoglobin and oxyhaemoglobin are respectively $6·6 \times 10^{-9}$ and $2·4 \times 10^{-7}$. If the normal blood pH value is accepted as 7·4 and both these compounds are present as free acids and as salts, determine which is the more efficient buffering system in the blood.

What is the ratio of free acid to salt for oxyhaemoglobin and haemoglobin in such blood?

2.12. Determine the isoelectric points of the following amino acids at 25°. The values for the pK_a values given refer to this temperature.

(a) α-Alanine : $pK_{a_1} = 2·34$, $pK_{a_2} = 9·69$.

(b) β-Alanine : $pK_{a_1} = 3·60$, $pK_{a_2} = 10·19$.

(c) Asparagine : $pK_{a_1} = 2·02$, $pK_{a_2} = 8·80$.

(d) Cystine : $pK_{a_1} = 1·65$, $pK_{a_2} = 2·26$, $pK_{a_3} = 7·85$, $pK_{a_4} = 9·85$.

(e) Isoleucine : $pK_{a_1} = 2·36$, $pK_{a_2} = 9·68$.

(f) Sarcosine : $pK_{a_1} = 2·23$, $pK_{a_2} = 10·01$ $(CH_2\overset{+}{N}H_3)$.

(g) L-Tyrosine : $pK_{a_1} = 2·20$, $pK_{a_2} = 9·11$, $pK_{a_3} = 10·07$ (OH).

(h) Taurine : $pK_{a_1} = 1·5$ (SO_3H), $pK_{a_2} = 8·74$.

2.13. Determine the isoelectric points of the following dipeptides at 25°. The pK_a values given refer to this temperature.

(a) Glycylglycine : $pK_{a_1} = 3·06$, $pK_{a_2} = 8·13$.

(b) Tyrosyltyrosine : $pK_{a_1} = 3·52$, $pK_{a_2} = 7·69$, $pK_{a_3} = 9·80$, $pK_{a_4} = 10·26$. pK_{a_3} and pK_{a_4} refer to the hydroxyl groups.

(c) Aspartylaspartic acid : $pK_{a_1} = 2·70$, $pK_{a_2} = 3·40$, $pK_{a_3} = 4·70$, $pK_{a_4} = 8·26$. pK_{a_2} and pK_{a_3} refer to the β-carboxyl groups.

(d) Histidylhistidine : $pK_{a_1} = 2·25$, $pK_{a_2} = 5·60$, $pK_{a_3} = 6·80$, $pK_{a_4} = 7·80$. pK_{a_2} and pK_{a_3} refer to the imidazolyl groups.

(e) Glycylalanine : $pK_{a_1} = 3·15$, $pK_{a_2} = 8·25$.

(f) Glycyltyrosine : $pK_{a_1} = 2·98$, $pK_{a_2} = 8·40$, $pK_{a_3} = 10·40$ (OH group).

(g) Aspartylglycine : $pK_{a_1} = 2·10$, $pK_{a_2} = 4·53$ (β-carboxyl), $pK_{a_3} = 9·07$.

2.14. Hastings and Van Slyke obtained the following results for the electrometric titration of 0·1 M-citric acid with sodium hydroxide :

NaOH (moles/litre)	0·0000	0·0197	0·0395	0·0592	0·0790	0·0987	0·1183
pH	2·06	2·51	2·88	3·14	3·42	3·67	3·98
NaOH (moles/litre)	0·1381	0·1578	0·1775	0·1973	0·2170	0·2368	0·2564
pH	4·20	4·46	4·69	4·94	5·18	5·42	5·70
NaOH (moles/litre)	0·2762	0·2782	0·2802	0·2821	0·2861	0·2900	0·2939
pH	6·06	6·07	6·14	6·17	6·32	6·54	6·83

Plot the titration curve and obtain the buffer value for the region of maximum buffering power. Over what range may the solution be used as an efficient buffer system?

(Data from HASTINGS & VAN SLYKE (1922), *J. biol. Chem.*, 53, 269.)

2.15. The following data were obtained in the electrometric titration of 1 mole of potassium dihydrogen phosphate with concentrated sodium hydroxide. The KH_2PO_4 was contained in 1 litre of solution and the volume change due to the addition of the NaOH may be ignored.

Moles NaOH added per mole KH_2PO_4	0·000	0·040	0·125	0·200	0·300	0·475	0·590
Determined pH	4·00	5·50	6·00	6·25	6·45	6·75	7·00

Moles NaOH added per mole KH_2PO_4	0·700	0·800	0·850	0·900	0·940	1·000
Determined pH	7·20	7·40	7·65	7·80	8·00	9·90

Determine the second dissociation constant of phosphoric acid and assess the buffer value at the pH corresponding to pK_{a2}. How does this buffer value compare with those for the KH_2PO_4 : $NaKHPO_4$ system at pH 6·0 and 7·5?

2.16. A plasma sample gave the following analysis : total CO_2 concentration 27 m-moles/litre and bicarbonate concentration 25·7 m-moles/litre. If pK_{a1}' for carbonic acid is 6·11, calculate the pH value of the plasma.

2.17. The pH of plasma samples of arterial and venous blood were 7·45 and 7·38. Analysis for the total CO_2 content of each gave 59·1 and 62·1 volume per cent. respectively. Calculate the partial pressure of CO_2, its concentration in the plasma and the plasma bicarbonate concentration in millimoles per litre for the samples of arterial and venous blood. $pK_{a1}' = 6·11$ for carbonic acid and the solubility factor for CO_2, $K_0 = 0·0301$ when the concentrations are expressed as millimolar and p_{CO_2} in mm Hg. *Note:* [Dissolved CO_2 + H_2CO_3] $= K_0 p_{CO_2}$. All data refer to 38°.

(Consultation of p. 305 may assist in solving this problem.)

2.18. Human red blood cells at pH 7·17 and 37° were in contact with a partial pressure of 25·4 mm CO_2 when oxygenated, and 48·4 mm CO_2 after the oxygen was given up at the same pH. If pK_a' for oxygenated and non-oxygenated red blood cells is 6·070 and 6·018 respectively, calculate the total CO_2 concentration present in the cells. The solubility factor for CO_2 in the cells at 37° is 0·0362 when concentrations are expressed as millimolar and p_{CO_2} in mm Hg.

(Data from DILL, DALY & FORBES (1937), *J. biol. Chem.*, 117, 569.)

2.19. The following data were obtained in experiments on samples of oxygenated and non-oxygenated red blood cells of the ox.

Oxygenated : $p_{CO_2} = 37·3$ mm; pH = 7·22 ; $pK_a' = 6·053$

Non-oxygenated : $p_{CO_2} = 50·7$ mm; pH = 7·22 ; $pK_a' = 6·022$

These figures refer to 37° at which temperature the solubility factor (K_0 for CO_2 in red blood cells is 0·0362 when concentrations are expressed as millimolar and p_{CO_2} in mm Hg. Calculate the bicarbonate concentration in each sample.

(Data from DILL, DALY & FORBES (1937), *J. biol. Chem.*, 117, 569.)

2.20. The following titration data refer to horse haemoglobin (Hb) and oxyhaemoglobin (HbO_2) in the presence of 0·333 M-NaCl :

Haemoglobin		Oxyhaemoglobin	
Acid (−) or base (+) per gram Hb m-equivalent	pH	Acid (−) or base (+) per gram HbO₂ m-equivalent	pH
−0·514	4·280	−0·514	4·280
−0·452	4·415	−0·453	4·410
−0·419	4·525	−0·420	4·525
−0·390	4·610	−0·392	4·618
−0·323	4·842	−0·324	4·860
−0·258	5·160	−0·259	5·188
−0·224	5·320	−0·225	5·430
−0·172	5·690	−0·173	5·800
−0·130	6·072	−0·130	6·055
−0·064	6·541	−0·063	6·430
0·0	6·910	+0·001	6·795
+0·070	7·295	+0·072	7·130
+0·131	7·660	+0·133	7·510
+0·171	7·860	+0·172	7·725
+0·208	8·140	+0·209	8·043
+0·254	8·545	+0·254	8·450
+0·288	8·910	+0·288	8·890
+0·311	9·130	+0·292	8·990
+0·331	9·350	+0·311	9·130
+0·350	9·410	+0·331	9·355
+0·357	9·465	+0·350	9·410
+0·407	9·800	+0·357	9·480
		+0·407	9·800

Plot the titration curves for haemoglobin and oxyhaemoglobin and from these deduce which form of haemoglobin is the stronger acid.

(After GERMAN & WYMAN (1937), *J. biol. Chem.*, 117, 533.)

2.21. In a series of experiments the sodium salt of a protein at equivalent concentration 0·1 was separated from an equal volume of sodium chloride solution by a semi-permeable membrane. If the equivalent concentrations of salt solution were 0·005, 0·010, 0·055, 0·120 and 1·20, calculate the ratios of the distribution of electrolyte at equilibrium in each case.

2.22. The effect of ionic strength on the first dissociation constant of carbonic acid has been investigated and the following data obtained.

Solutions were prepared from purified $NaHCO_3$ and $NaCl$ and were equilibrated at 38° with hydrogen and CO_2 at a tension previously calculated to give the desired pH. Samples of the liquid phase were analysed for total CO_2 content in the Van Slyke manometric gas apparatus. The pH of the solution was measured electrometrically at 38°.

Experiment number	p_{CO_2} mm Hg	Solubility coefficient of CO_2 (α)	Total CO_2 concentration mM	pH	NaCl concentration mM
1	61·6	0·552	12·25	6·977	0·00
7	158·1	0·546	35·10	6·976	24·83
9	152·3	0·544	35·12	6·970	49·65
11	145·0	0·541	34·75	6·970	74·43
13	140·5	0·538	34·86	6·970	99·34
15	136·8	0·532	34·60	6·950	149·00

From this information calculate the concentration of carbonic acid and thence the pK_{a_1}' value for each experiment. Plot the values of pK_{a_1}' so obtained against the square root of the ionic strength (I) and from the resulting graph deduce the relationship between K_{a_1}' and $\sqrt{I}$ and hence the relationship between I and the activity coefficient of the bicarbonate ion.

Note: Where concentrations are expressed as millimolar and p_{CO_2} as mm Hg :

$$pK_{a_1}' = pH - \log [NaHCO_3] + \log p_{CO_2} + \log \frac{\alpha}{760 \times 0.0224}$$

assuming $NaHCO_3$ to be completely dissociated and hence $[NaHCO_3] = [HCO_3^-]$.

(After HASTINGS & SENDROY (1925), *J. biol. Chem.*, 65, 445.)

2.23. A solution of the sodium salt of a non-diffusible substance is separated from an equal volume of a solution of sodium chloride by a semi-permeable membrane. For convenience the compartments will be referred to as A and B respectively. Calculate the concentrations of sodium and chloride ions in compartments A and B at equilibrium when the initial concentrations are as follows :

Initial concentrations (mM)

Non-diffusible substance (compartment A)	Sodium chloride (compartment B)
0·01	1·00
1·00	1·00
1·00	0·50
1·00	0·01

(After CLARK (1952), *Topics in Physical Chemistry*, 2nd ed., p. 153. Baltimore : Williams & Wilkins Co.)

2.24. 40 ml of a 0·000650 M solution of sulphadiazine were titrated with 0·0500 N-NaOH from a micro-burette ; the pH value was measured with a glass electrode, some of the results being as follows :

Titre (ml)	0·139	0·279	0·350
pH value	6·17	6·64	6·90

Calculate the pK_a value for sulphadiazine, neglecting the slight change in total volume during the titration.

(After SPEAKMAN, unpublished data.)

2.25. Describe the experimental methods used in the determination of pH, and discuss the theoretical significance of one pH scale.

10 ml of a solution containing 0·5 g of a protein of molecular weight 76,000 were titrated with 0·1 N hydrochloric acid. From the titration graph the acid groups were shown to be neutralized by the addition of 0·263 ml of the acid, the pH changing from 4·25 to 2·75.

Calculate

(*a*) the amount of acid bound by these groups ;
(*b*) the number of acid groups on each molecule ;
(*c*) an approximate value for the mean dissociation constant of the acid groups.

(Assume that the solutions are ideal, and neglect the small increment in volume in titration.)

(Special Degree of B.Sc., Biological Chemistry, University of Bristol, 1957.)

2.26. Derive an expression which relates the relative concentrations of conjugate acid and base to pH.

A tetracycline antibiotic, containing one tertiary nitrogen atom and two acidic hydroxyl groups, has pK values of 3·45, 7·92 and 9·81. The inhibitory effect of this antibiotic on cultures of *Escherichia coli* can be attributed entirely to the uncharged molecular species. If a concentration of $1·87 \times 10^{-7}$ moles/litre of tetracycline causes a doubling of the mean generation time of a culture at pH 7·00, what concentration of the antibiotic would show the same degree of inhibition at pH 7·80?

(University of Wales, Cardiff, Honours Course Finals, 1966.)

2.27. What factors must be considered in making a buffer solution to stabilize the pH at a given value?

The titration of both groups of the sodium salt of a monoamino monocarboxylic amino acid was found to require 2·34 ml of 2N-HCl. After the addition of 0·59 ml of acid, the pH was 9·69, and after the addition of a total of 1·47 ml of acid, the pH was 2·81. What are the pK'_a values of the two groups? Neglect the small increases in volume on titration. Explain any assumption you make during the calculation.

(University of Newcastle upon Tyne, General Degree with Honours (Part 1), Biochemistry, 1968.)

2.28. Two proteins, A and B, contain the following numbers of ionizable amino acids (residues per mole), each having the corresponding pK'_a value measured in the actual protein:

	pK'_a	A	B
Aspartic and glutamic acid	4·5	16	11
Histidine	6·5	2	9
Cystine	8·0	2	—
Tyrosine	9·95	4	8
Lysine	10·2	14	12
Arginine	>12·0	10	8
α-COOH	3·8	2	1
α-NH₂	7·8	2	1

(*a*) What is the approximate charge state of each of these residues at pH 5·5 and at pH 8·0? At which of these two pH values would you attempt to separate a mixture of A and B by electrophoresis?

(*b*) Equilibrium sedimentation of the mixture of A and B (approximately equal proportions by weight in 0·1 M phosphate buffer pH 7 gave $M_W = M_Z = 31,000$. Would you use gel filtration or ion-exchange chromatography to attempt to separate A and B?

(*c*)　(i) On treatment of protein A with fluoro-dinitro-benzene and subsequent hydrolysis by 6N-HCl, the following derivatives were identified:
imidazole-DNP-histidine
O-DNP-tyrosine
ϵ-DNP-lysine
DNP-valine

(ii) Short-term treatment of protein A with carboxy-peptidase results in the liberation of free glycine.

(iii) Digestion of the protein A with trypsin gives a total of 13 different peptides.
What different conclusions can you draw concerning the primary structure of protein A?

(*d*) The osmotic pressure of a 1 per cent. solution of protein B in 5 M-guanidine hydrochloride was 6·7 mm Hg. What would the *approximate* value of the osmotic pressure be for a 1 per cent. solution of protein A in 5-M guanidine hydrochloride and 0·1 M-mercaptoethanol ?

(e) Equilibrium sedimentation runs were made on protein A in 5 M-guanidine hydrochloride in the absence of reducing agent. The average molecular weights obtained were $M_W = 26,000$ and $M_Z = 37,000$. Can you suggest a qualitative explanation for these results ?

(University of Newcastle upon Tyne, General Degree with Honours (Part I) Biochemistry, 1969.)

2.29.　50 mg of ovalbumin (MW = 54,000) were dissolved in 20 ml of phosphate buffer of pH 7·0. After digesting the protein with pronase, the pH of the solution dropped to pH 6·8. It required 2·50 ml of 0·01 M-NaOH to return the pH to 7·0.

(a) What was the molarity of the phosphate buffer ?

(b) How many peptide bonds per mole of protein were broken ?

(University of Newcastle upon Tyne, Biochemistry Honours Course, Part I, 1970.)

CHAPTER III

THERMODYNAMICS AND BIOCHEMICAL ENERGETICS

BROADLY, the study of chemical reactions may be approached from an analysis of the behaviour of the participating molecules, as in Chemical Kinetics, or from a consideration of the energy transfers that accompany the reactions, as in Chemical Thermodynamics. The latter science deals with the initial and final states of systems and, in general, is not concerned with the speed at which changes occur. Although thermodynamics has contributed to our understanding of processes at the molecular level, the science has developed from laws deduced directly from experience of the behaviour of matter in bulk, and not from the molecular theory of matter.

The First Law of Thermodynamics is concerned with the principle of conservation of energy. It is only possible to convert one form of energy into another form; energy can be neither created nor destroyed.[1] This can be expressed in symbols as follows :

If W is the mechanical work done by the system, Q the quantity of heat absorbed by the system, and ΔU its change in internal energy, then

$$\Delta U = Q - W \qquad . \qquad . \qquad . \qquad (3.1)$$

The quantity U, which is called the *internal* or *intrinsic* energy, includes all forms of energy of the system except those resulting from its position in space. The absolute value of U cannot be determined, but we are concerned only with changes in U such as occur when a chemical reaction takes place. In changing from state 1 to state 2 the change in internal energy is given by

$$\Delta U = U_2 - U_1 \qquad . \qquad . \qquad . \qquad (3.2)$$

In this equation use is made of a notation which finds general

---

[1] Energy changes in biochemical reactions are never, of course, of such magnitude as make the principle of equivalence of mass and energy (Einstein) relevant to these systems.

application in this subject : namely ΔU to denote an *increase* in U of the system, ΔV for an increase in V, and so on. If the change had given rise to a decrease in the value considered, the delta sign would be preceded by a minus. It is important to note that ΔU is independent of the pathway by which the change is carried out ; it depends only upon the initial and final states, and U is known, therefore, as an intrinsic thermodynamic variable. If this were not so it would be possible to create energy from the system and thus violate the First Law. Hess's Law, used in thermochemical calculations, is a restatement of this law as it applies to heats of reaction, although historically Hess's Law preceded it. Q and W, however, are dependent on the pathway of the change.

A special case occurs when no external work is done (i.e. $W = 0$). This is usually covered by saying that the process occurs at *constant volume*, although this does not exclude the possibility that other forms of external work *might* occur. Then :

$$Q_v = \Delta U \qquad . \qquad . \qquad . \qquad (3.3)$$

where the subscript $_v$ indicates constant volume. Thus, at constant volume, ΔU is equal to the observed heat change and, by the convention mentioned, is negative for an exothermic reaction (i.e. energy, as heat, lost to the system) and positive for an endothermic reaction (heat gained by the system).

A reaction involving an increase of volume and occurring at *constant pressure* (as in the case of reactions in open vessels) necessarily implies the performance of work against the surrounding atmosphere. Biochemical reactions usually occur at a constant pressure, that of the atmosphere, and not at constant volume. If the volume increases, a proportion of the intrinsic energy change will be used to perform work against the pressure of the atmosphere and, in consequence, all of the energy liberated will not appear as heat of reaction. It is convenient to make use of another quantity, H, the *heat content* or *enthalpy*, which is related to U by the equation

$$H = U + PV \qquad . \qquad . \qquad . \qquad (3.4)$$

Here P is constant and V, the volume of the system, has, like U, a defined value for any given state ; H, therefore, is also an intrinsic thermodynamic variable. The relation between changes in U and H is readily established. If V_1 is the initial and V_2 the final volume

$$W = PV_2 - PV_1 = P\Delta V.$$

Therefore
$$\Delta U = Q_p - P\Delta V . \qquad . \qquad . \qquad . \qquad (3.5)$$

and
$$Q_p = \Delta U + P\Delta V = \Delta H \qquad . \qquad . \qquad (3.6)$$

At constant pressure then, ΔH is the heat absorbed in the reaction. Values for ΔH may be obtained calorimetrically, but since the bomb calorimeter is a constant volume device, the heats of reaction so measured give changes in internal energy. Values of ΔU, however, may readily be converted to ΔH. If a reaction between gases is represented by

$$A+B \rightarrow C+D+E$$

there is an increase of one in the total number of molecules in the system when it occurs at any temperature T. Since $PV = nRT$, where n is the number of moles, in this case $n = 1$ and the work done due to the increase in volume $P\Delta V$ is therefore RT and $\Delta H = \Delta U + RT$. Volume changes when extra molecules of solids or liquids arise are negligible compared with those for gases. Hence, in the example given, if A were a solid, C a liquid and B, D and E all gases, the work done due to the increase in volume would also be RT. Some examples of complete combustion are :

1. Solid glucose to liquid water and gaseous carbon dioxide. (Note that the subscripts $_S$, $_L$ and $_G$ refer to the state.)

$$C_6H_{12}O_{6(S)} + 6O_{2(G)} \rightarrow 6CO_{2(G)} + 6H_2O_{(L)}$$

$\Delta H = -673,000$ calories[1] or -673 kilocalories, i.e. an exothermic reaction.

2. Solid maltose to liquid water and gaseous carbon dioxide.

$$C_{12}H_{22}O_{11(S)} + 12O_{2(G)} \rightarrow 12CO_{2(G)} + 11H_2O_{(L)}$$

$\Delta H = -1,350$ kilocalories.

3. Solid urea to liquid water and gaseous carbon dioxide and nitrogen.

$$CO(NH_2)_{2(S)} + 1{\cdot}5O_{2(G)} \rightarrow CO_{2(G)} + N_{2(G)} + 2H_2O_{(L)}$$

$\Delta H = -152$ kilocalories.

[1] Although the calorie is not an SI unit and will eventually be abandoned in favour of the joule (1 calorie = 4·185 joules), it is retained in this present edition.

Example 3.1.—The combustion of solid urea to liquid water and gaseous carbon dioxide and nitrogen in a bomb calorimeter at 25° liberates 152·3 kilocalories. Determine the heat content change for the reaction.

As will be seen from case 3 above, the increase in volume attendant on this reaction is $2 - 1·5 = 0·5$ mole of gas.

Hence
$$\Delta H = \Delta U + 0·5RT$$
$$= -152300 + 0·5 \times 1·987 \times 298$$
$$= -152300 + 296 = -152,004 \text{ calories}$$
$$= -152·0 \text{ kilocalories.}$$

REACTION CALORIMETRY.—The direct calorimetric measurement of the enthalpy changes associated with chemical reactions, known as *reaction calorimetry*, has become of considerable importance for the collection of thermodynamic data for biochemical reactions. Pioneer investigators in the field of biochemical calorimetry have included Calvet and Prat in France and Benzinger, Kitzinger and Sturtevant in the U.S.A., who designed and built microcalorimeters capable of measuring the minute quantities of heat associated with enzymic reactions and, for example, the endogenous metabolism of micro-organisms. Benzinger has been responsible for the technique termed *heat-burst calorimetry* which measures the heat pulse associated with a reaction, e.g. $\Delta H°$ has been measured for the hydrolysis of ATP (Fig. 3.1). The development within recent years of suitable commercial apparatus has greatly facilitated and extended work of this type (see, for example, Brown, 1969).

Instruments designed to measure the enthalpy changes resulting from the mixing of two reactants are of two general types, batch calorimeters and flow calorimeters. The batch instruments are the simplest type and provided the volumes are relatively large, e.g. about 100 ml, the technical problems associated with initiation of reaction and mixing of reactants do not arise. However, biochemical reactants are rarely available in such generous quantities and while some batch microcalorimeters have been successfully designed, the development of the continuous flow calorimeter by Stoesser and Gill (1967) was a landmark. This instrument permits two reactant solutions to be thermally equilibrated during passage through separate platinum tubes and then combined at a mixing Y-junction, a thermopile measuring the heat evolved in the reaction. Sturtevant and his colleagues have recently carried out flow calorimetric work with various biochemical systems such as the oxidation of ferro- to ferri-cytochrome c, the enthalpy of binding of NAD^+ to glyceraldehyde 3-phosphate

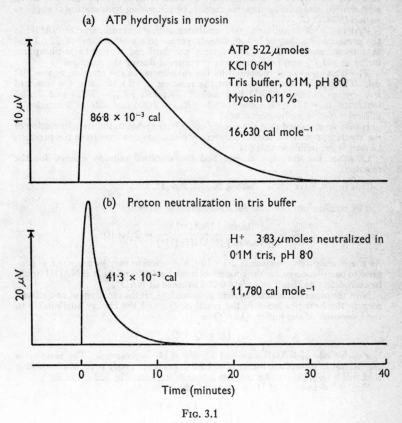

(a)   ATP hydrolysis in myosin

ATP 5·22 μmoles
KCl 0·6M
Tris buffer, 0·1M, pH 8·0
Myosin 0·11%

86·8 × 10⁻³ cal

16,630 cal mole⁻¹

10 μV

(b)   Proton neutralization in tris buffer

H⁺ 3·83 μmoles neutralized in 0·1M tris, pH 8·0

41·3 × 10⁻³ cal

11,780 cal mole⁻¹

20 μV

0   10   20   30   40

Time (minutes)

FIG. 3.1

Hydrolysis of ATP measured by heat-burst calorimetry
(a) The heat change of enzymic ATP hydrolysis including the subsequent neutralization of one proton per ATP molecule.
(b) Proton neutralization in the same buffer.
The difference between (a) and (b) is the heat of reaction of ATP hydrolysis proper at pH 8·0, i.e. −4800 cal mole⁻¹.

(After KITZINGER & BENZINGER (1955). *Z. Naturforsch.*, **10B**, 375 and (1965) *Fractions*, **2**, No. 2.)

dehydrogenase and the enthalpy change when the S-peptide of ribonuclease-S recombines with the S-protein to regenerate the active enzyme.

Example 3.2.—The temperature change for the reaction

$$\text{Pyruvate} + \text{NADH} + \text{H}^+ \rightleftharpoons \text{lactate} + \text{NAD}^+$$

catalysed by lactate dehydrogenase has been determined by the use of an isothermally jacketed calorimeter capable of measuring temperature changes to within $0 \cdot 0001°$ C.

Pyruvate ($0 \cdot 981$ millimole) was incubated with $0 \cdot 206$ millimole of NADH in the presence of $285 \mu g$ of lactate dehydrogenase in a total volume of 25 ml and at a temperature of $25°$. The system was buffered with $0 \cdot 15$ M-phosphate buffer at pH $7 \cdot 3$ and no change in pH occurred during the reaction.

The heat equivalent measured by the calorimeter in the reaction was $-2 \cdot 02$ cal. The equilibrium constant for the reaction is $2 \cdot 5 \times 10^{11}$ and the standard free energy change $-15 \cdot 55$ kcal mole^{-1}. At $25°$, $pK_{a1} = 2 \cdot 124$ and ΔH of ionization is -1773 cal mole^{-1} while $pK_{a2} = 7 \cdot 206$ and ΔH of ionization is 822 cal mole^{-1} for phosphoric acid.

It may be assumed that the energy contribution due to the transformation of the reactant, pyruvate, having a different ionization constant from the product, lactate, is negligible at this pH value.

Calculate the enthalpy change and the standard entropy change for the reaction.

(Data from S. KATZ (1965). *Biochim. Biophys. Acta*, **17**, 226.)

The equilibrium constant for the reaction is given by

$$K = \frac{[\text{lactate}] \, [\text{NAD}^+]}{[\text{pyruvate}] \, [\text{NADH}] \, [\text{H}^+]} = 2 \cdot 5 \times 10^{11}.$$

As the equilibrium constant is $2 \cdot 5 \times 10^{11}$, the reaction may be assumed to progress to completion under the conditions used, i.e. $0 \cdot 206$ m-mole of NADH would be completely oxidized and yield $0 \cdot 206$ m-mole of NAD$^+$.

Now the observed heat equivalent produced, Q_e, is the resultant of two effects, namely the heat produced in the reaction, Q_r, and the energy attributable to the ionization of the buffer, Q_h. Thus

$$Q_e = Q_r + Q_h.$$

For each mole of NADH oxidized 1 mole of H$^+$ is removed. The reaction is carried out in phosphate buffer of pH $7 \cdot 3$; at this pH only pK_{a2} of phosphoric acid will be significant and this has a heat of ionization of 822 cal mole^{-1}.

Now $0 \cdot 206$ m-mole of H$^+$ is removed, therefore

$$Q_h = 822 \times 0 \cdot 206 \times 10^{-3} = 0 \cdot 169 \text{ cal.}$$

Hence
$$Q_r = -2 \cdot 02 - 0 \cdot 169 = -2 \cdot 19 \text{ cal.}$$

The enthalpy change for the reaction, ΔH_r, is thus given by

$$\Delta H_r = \frac{2 \cdot 19 \times 1000}{0 \cdot 206} \text{ cal mole}^{-1} = \underline{-10 \cdot 63 \text{ kcal mole}^{-1}.}$$

The standard enthalpy charge $\Delta H°$ is related to ΔH_r by the equation

$$\Delta H° = \Delta H_r + \Sigma L \text{ reactants} - \Sigma L \text{ products}$$

where L is the relative partial molal enthalpy.

It may be assumed, however, that the contribution of any of the relative partial molal enthalpy terms will be small and that their cumulative effect will be relatively insignificant since they tend to cancel. Hence

$$\Delta H° \approx \Delta H_r = -10 \cdot 63 \text{ kcal mole}^{-1}.$$

Since $\Delta G° = \Delta H° - T\Delta S°$

$-15.55 = -10.63 - 298\Delta S°$.

Whence $\Delta S° = \dfrac{4.92 \times 1000}{298} = 16.54$ cal degree^{-1} mole^{-1}.

The Second Law.—Kelvin stated this law as follows: 'It is impossible to take heat from a system and convert it into work without other simultaneous changes occurring in the system or in its environment.' Lewis has given an alternative statement of the law in the words: 'Every process that occurs spontaneously is capable of doing work; to reverse any such process requires the expenditure of work from outside.'

Although in particular instances, such as the oxidation of hydrocarbons in the internal-combustion engine, the ability of a chemical reaction to perform work can readily be appreciated, for most of the reactions encountered in the laboratory difficulties are presented. Reference should be made to a textbook of chemical thermodynamics (that of Butler (1951) is particularly recommended on this point) as to how the conversion of energy of reaction into work can be visualized and evaluated. Meanwhile, the contribution of the Second Law towards the problem of assessing chemical affinity—the 'driving force' of reactions—may be noted. A great impetus was given to thermochemical research in the last century by the belief that the heat of a reaction measured its affinity: the greater the amount of energy liberated, the more affinity the reactants possessed for each other. This view was rendered untenable by the discovery of endothermic reactions which could proceed, and so exhibit 'affinity', although they absorbed heat in the process. When, however, attention was focused on the work that a reaction was able to perform rather than on the heat that it liberated, such difficulties did not arise, for the Second Law states that if a reaction is inherently incapable of work it will not be spontaneous; that is, will not proceed of its own accord and will exhibit no chemical affinity. Negative values for the work done by a spontaneous reaction, in contrast to negative values for heat liberated, will not be encountered. The extent to which a reaction can proceed before it reaches equilibrium is hence related to the energy which is capable of conversion into work when the reaction takes place reversibly, i.e. the maximum work. There is another intrinsic thermodynamic variable known as the *free energy* (and defined, as seen later, by equation 3.7, namely $G = H - TS$, where G is the symbol denoting free energy), the change in which, during a reaction occurring at constant temperature and pressure, is equal to this maximum work.

Another concept arising from the Second Law is that of *entropy*, denoted by S. If at a fixed temperature T an amount of heat Q is absorbed reversibly by a system, its entropy rises by Q/T. Entropy is an intrinsic thermodynamic function most readily appreciated by considering the changes that occur when entropy alters. Thus, if the transfer of heat were reversible, that is, if the heat flowed from a system that was at a higher temperature than its surroundings only by an infinitesimally small amount, then the loss of entropy by the system would equal the gain by the surroundings. But, in actual fact, the system will always be at a definitely higher temperature; perfectly reversible systems are not encountered in Nature, and Q/T for the system will be less than for the surroundings. For system and surroundings taken together there will have been an overall increase in S. Now, heat flows from a higher to a lower temperature spontaneously, and as a consequence of the levelling out of the temperature difference, there is a destruction of an ordered arrangement. Thus, if the transfer of heat occurred between two gases, there would be originally a segregation of fast-moving molecules, at the higher temperature, from molecules that move more slowly on the average. This distinction is abolished when the temperatures are equated, and as a result the energy becomes distributed in a more random manner. Fast-moving molecules of high energy are no longer segregated from those with lower energies. Spontaneous changes occur, in general, with an increase in disorder, leading to a final state characterized by a higher probability than the initial state, and this is on a parallel with an increase in S. For example, in solution the random coil form of DNA has a higher entropy than the helical form, principally because the random structure is more probable than the highly organized helical structure. Another view which may assist the visualizing of entropy is to regard the intrinsic energy as made up of the energy of chemical linkage that is liberated as free energy in reactions, plus the random energy of vibration and rotation of atoms and molecules. Certain molecules, in particular those that are complicated, have a high capacity for taking up this type of random disordered energy and they have a high entropy. The biological activity of many proteins has been shown to depend upon their conformation, i.e. the three dimensional configuration they display. The oxygenation of haemoglobin is accompanied by a conformational change and the activity of certain enzymes is altered by combination at sites other than their active centres with compounds which function as allosteric effectors (p. 224) and produce changes in conformation of the enzyme molecules to which they are attached. All changes of conformation, if of sufficient magnitude, are accompanied by attendant entropy changes, thus emphasizing the importance of being able to measure such entropy changes in biochemical reactions.

The relation between free energy, heat content and entropy is

$$G = H - TS \qquad . \qquad . \qquad . \qquad (3.7)$$

Thus, if a system is changed from one state to another at the same temperature, the change in free energy associated with this change is

$$\Delta G = \Delta H - T\Delta S \qquad . \qquad . \qquad . \qquad (3.8)$$

Since G is an intrinsic property, ΔG is independent of the path of the reaction. It may be determined by any of three important methods (a) e.m.f. measurements (b) equilibrium data, and (c) from purely thermal data by means of the Third Law of thermodynamics.

Because ΔG is a measure of the capacity of a system for doing work it enables one to decide whether or not a reaction may occur spontaneously. Spontaneous reaction can occur only if there is a *decrease* in free energy (such reactions are termed *exergonic*); if there is an increase in free energy, then work must be put into the system to bring about the change. This is an *endergonic* reaction. By convention, a decrease in free energy is denoted by a negative sign and an increase by a positive sign. Hence if the free energy change for a reaction is -5000 calories the reaction may occur spontaneously and there is a decrease in G of 5000 calories. It does not automatically follow, however, that because a given reaction has a high negative value for ΔG the reaction takes place at a measurable rate. ΔG measures only the difference in free energy between the initial and final states of the reaction and does not give any information about the rate of reaction. To bring about appreciable reaction a catalyst may be necessary.

The free energy change of a reversible reaction is related to the equilibrium constant of the reaction (see p. 131) by the equation

$$\Delta G° = -RT \ln K \qquad . \qquad . \qquad (3.9)$$

where $\Delta G°$ is the *standard free energy change* and K the equilibrium constant. $\Delta G°$ signifies that the free energy change refers to the reaction when all the reactants and products are in their standard states. The standard state is a convenient reference condition in which the activities are arbitrarily defined as unity for pure liquids or solids, gases at 1 atmosphere, and compounds in solution at approximately 1 M concentration[1] at a given temperature, usually 25°. It follows, therefore, that $\Delta G°$ is a constant for any given reaction. Values of $\Delta G°$ are additive : i.e. if such values are known for two reactions, that of a third may be calculated by methods the same as those applied to heats of reaction when the Law of Hess is used. The standard free energy $\Delta G°$ must not be confused with the free energy ΔG. They are related for the reaction $A + B \rightleftharpoons C + D$ by the equation

$$\Delta G = \Delta G° + RT \ln \frac{[C][D]}{[A][B]} \qquad . \qquad . \qquad (3.10)$$

where [A] and [B] are the initial concentrations of the reactants and [C] and [D] the final concentrations of the products. It is ΔG and not $\Delta G°$ which determines whether or not spontaneous reaction may occur. This distinction has not always been observed in the biochemical literature. $\Delta G°$ is, however, the value always tabulated for a given reaction because it is a defined quantity, whereas ΔG can have any value depending on the conditions as implied by equation 3.10. Unlike ΔG, to a very good first approximation ΔH is not dependent on the state or concentration of the reactants and hence may be equated with $\Delta H°$ (see Example 3.2).

---

[1] Strictly speaking the standard state for a solute is one of unit *activity*, the activity scale being so chosen that the ratio of activity to concentration (either molar or molal) tends to unity as the concentration approaches zero. Hence a solution of unit activity will not, in general, be exactly 1 molar (or molal). In thermodynamics the concentration is usually taken as molal, since it is then independent of temperature. For biochemical work in aqueous media the use of molar concentrations does not introduce appreciable error.

Special mention must be made of the value taken for the concentration of water when standard free energy values are being calculated from equilibrium data. As a pure liquid its activity in the standard state may be taken as unity, but the actual molal concentration is $1000/18 = 55\cdot55$ in dilute aqueous solutions. Hence care must be exercised to assign the same value to water as has been used in defining $\Delta G°$; otherwise a difference of 2475 calories will arise at 37°. In calculating equilibrium constants, the activity of water, either in the pure state or in dilute solution, is usually taken as unity.

When a reversible system is at equilibrium, the free energy change is zero, i.e. $\Delta G = 0$, and therefore

$$\Delta H = T\Delta S.$$

The entropy of the system and its surroundings is at a maximum and the system is in its most probable state. Sometimes an endergonic reaction, which will not of itself proceed because of the increase in free energy involved, can be made to do so by coupling it with an exergonic reaction. For this to occur there must be an intermediate common to both reactions. Consider the following three reactions :

1. $A + B \rightleftharpoons C + D$ $\Delta G°_1$ small positive (endergonic),

2. $D + L \rightleftharpoons M + N$ $\Delta G°_2$ large negative (exergonic),

3. $A + B + L \rightleftharpoons C + M + N$ $\Delta G°_3$ moderately negative (exergonic).

Compound D, formed in reaction 1, is a reactant in reaction 2, and the overall reaction then becomes 3, which is moderately exergonic ($\Delta G°_1 + \Delta G°_2$) and hence compound C will be formed from A and B. This may be regarded as an example of the principle of Le Chatelier whereby a reaction is made to proceed in the direction of completion by removal of one of the products. In this case D is removed by coupling it to a second reaction.

In living organisms many examples of coupled reactions occur. Of particular importance are reactions involving the so-called 'energy-rich' or 'high energy' phosphate compounds. The hydrolysis of most phosphate esters results in a free energy change of

E

approximately -2 kilocalories per mole, whereas the hydrolysis of compounds such as adenosine triphosphate (ATP), adenosine diphosphate (ADP), creatine phosphate and arginine phosphate gives a much greater free energy change of approximately -7 to -12 kilocalories per mole. Substances which liberate this greater amount of energy on hydrolysis are termed 'high-energy' compounds. It was subsequently discovered that certain other compounds of metabolic importance, such as the thiol esters of coenzyme A, also displayed high free energies of hydrolysis and so could be classified as high energy compounds. At the present time the following principal classes of high energy compounds may be recognized:

1. Phosphoric acid anhydrides, e.g. ADP, ATP.
2. Phosphoric-carboxylic acid anhydrides, e.g. acetyl phosphate.
3. Phosphoguanidines, e.g. creatine phosphate and arginine phosphate.
4. Enol phosphates, e.g. phospho-enol pyruvate.
5. Thiol esters, e.g. esters of coenzyme A such as succinyl-CoA.

Table 3.1 records the free energies of hydrolysis of some of these high energy compounds. Their importance lies in the fact that free energy liberated in metabolic processes is transformed into the chemical energy characteristic of compounds containing these bonds, and which manifests itself as a high free energy of hydrolysis when the phosphate group is subsequently split off. High-energy compounds thus serve as 'storehouses' of energy for the organism. It is currently believed that all the chemical energy derived from the oxidation of foodstuffs, other than that dissipated as heat, must first be converted to high-energy phosphate compounds before it can be utilized for the performance of mechanical work. Similarly the radiant energy utilized in photosynthesis is converted to potential chemical energy in the form of ATP by reactions occurring in the chloroplasts. ADP and inorganic phosphate can thus be converted to ATP by the energy derived from a suitable metabolic reaction or from sunlight ; the free energy thereby conferred can be transferred subsequently to other compounds or utilized for the performance of work. In muscle, for instance, the protein myosin acts as an enzyme specific for the hydrolysis of the terminal phosphate of ATP (adenosine triphosphatase activity) ; the energy liberated is used for the contraction process and mechanical work is done. As there is only a limited quantity of ATP present in muscle, creatine phosphate serves as a reservoir of high energy phosphate groups and regenerates ATP from ADP, thus :

(1) ATP $\xrightarrow{\text{myosin}}$ ADP + inorganic phosphate + free energy

(2) ADP + creatine phosphate → ATP + creatine.

TABLE 3.1

Free Energies of Hydrolysis of some Phosphoric Acid and Coenzyme A Derivatives

Compound	$-\Delta G^{\circ\prime}$ kcal mole^{-1}	pH	Temp. °C
(a) *Low-energy Phosphate Compounds*			
Glycerol 1-phosphate . . .	2·2	8·5	38
Glucose 6-phosphate . . .	3·0	8·5	38
Fructose 6-phosphate . . .	3·5	8·5	38
Glucose 1-phosphate . . .	4·75	8·5	38
(b) *High-energy Compounds*			
Adenosine triphosphate^{4-} (terminal group)	8·6	7·0	20
Adenosine diphosphate^{3-} (terminal group)	7·8	7·0	—
Enol-phosphopyruvate^{3-}	13·3	7·0	20
Creatine phosphate^{2-} . . .	10·4	7·7	20
Arginine phosphate^{-} . . .	11·8	7·7	20
Acetyl phosphate^{2-} . . .	10·5	7·0	25
Acetyl Coenzyme A . . .	8·2	7·0	—
Succinyl Coenzyme A . . .	7·8	7·0	—

Note : $\Delta G^{\circ\prime}$ is the standard free energy change for the reaction at the specified pH ; ΔG° refers to unit hydrogen ion activity, i.e. pH 0.

Data from A. W. D. AVISON & J. D. HAWKINS (1951), *Quart. Rev. Chem. Soc.*, V, 171 ; H. GUTFREUND (1954), *Ann. Rep. Chem. Soc.*, 51, 235 ; K. BURTON (1961), *Biochemists' Handbook*, p. 94, ed. C. Long. London : E. & F. N. Spon Ltd.)

The transference of a phosphate group and free energy is achieved in this manner. Creatine is subsequently rephosphorylated under resting conditions, when the demand for ATP for contraction has decreased, by a reversal of reaction (2) above. Creatine phosphate thus functions as a storehouse of energy which can be drawn upon during muscle contraction.

In many biosynthetic reactions instead of splitting to yield ADP and inorganic phosphate, ATP undergoes pyrophosphate cleavage to give inorganic pyrophosphate (PP$_i$) and adenosine monophosphate (AMP), thus

$$ATP \longrightarrow AMP + PP_i.$$

The activation of fatty acids and amino acids involves reactions of this type, e.g.

$$R.COOH + ATP + CoASH \rightleftharpoons R.CO.SCoA + AMP + PP_i$$
$$R.CHNH_2COOH + ATP \rightleftharpoons R.CHNH_2.CO - AMP + PP_i.$$

The carboxyl group of the fatty acid forms a thioester linkage with the thiol group of coenzyme A while the carboxyl of the amino acid is bound in an anhydride linkage with the 5′-phosphate group of AMP, inorganic pyrophosphate being eliminated in both cases. Unlike ATP and ADP, AMP is not a high energy compound ; it can be rephosphorylated to ATP only after conversion to ADP by the action of the enzyme adenylate kinase, which catalyses the reaction

$$ATP + AMP \rightleftharpoons 2ADP.$$

Thus, at any given time, a living cell contains not only ATP and ADP but also AMP.

The generation of high-energy phosphate compounds occurs by three major mechanisms, substrate phosphorylation, oxidative or respiratory chain phosphorylation and photophosphorylation, characteristic of anaerobic, aerobic and photosynthetic tissues respectively. Glycolysis is the major metabolic sequence for substrate phosphorylation and the tricarboxylic acid cycle for oxidative phosphorylation (p. 356). These processes occur in a series of stages releasing graded amounts of energy, thus ensuring that the overall liberation of free energy in the oxidation of glucose or pyruvic acid occurs gradually, not suddenly nor in great amount. The organism is enabled to 'harness' the available energy in the form of energy-rich phosphate compounds.

A word of caution is necessary here about the use of the term 'energy-rich phosphate bond' which formerly enjoyed considerable vogue in biochemical circles and led to certain misunderstandings. It was found convenient to indicate a high energy compound by a 'squiggle' notation which denotes that the hydrolysis of a given phosphate group is associated with a high free energy of hydrolysis. Thus ADP and ATP were written as A-P∼P and

A-P~P~P respectively, a formulation which did much to foster the use of the expression 'energy-rich phosphate bond'. The term is obviously misleading, however, for it refers not to bond energy as defined by chemists but to the free energy resulting from hydrolysis of the phosphate compound. Bond energy in its chemical (and correct) sense is the energy which must be absorbed to break a bond of a gaseous molecule with the production of neutral gaseous atoms or radicals. The concept that energy is located in a high-energy phosphate bond, and is released when the bond dissociates, gained currency in some quarters. Gillespie, Maw & Vernon (1953) have emphasized that the notion of energy-rich phosphate bonds is irreconcilable with physicochemical principles. Thus, bond energies are always *positive*, i.e. energy must be utilized to break the bond, and consequently the ideas of a bond dissociation process yielding energy and of energy being stored in the bond are physically meaningless. They point out that even if the correct bond energies were known for the particular bonds concerned, they would not generally bear any simple relationship to the free energies of hydrolysis since other quantities are involved, such as the energies of formation of new bonds, the total entropy change and heats of ionization and solvation of reactants and products. These difficulties are surmounted if it is remembered that free energies of hydrolysis are the relevant quantities concerned when considering high- and low-energy phosphate compounds and that the free energy of hydrolysis is a property of the molecule as a whole.

The large free energy of hydrolysis associated with these phosphate compounds has been attributed to three factors. The first is that the hydrolysis products have greater thermodynamic stability than the parent compound because of increased possibilities of resonance. Secondly, these compounds contain negatively charged acid groups in close proximity, and on hydrolysis some of the energy released arises from the electrostatic repulsion of these groups. Thirdly, the energy changes involved in the ionization or neutralization of the groups produced by hydrolysis may contribute to the high free energy of hydrolysis.

From the values for $\Delta G^{o\prime}$ of hydrolysis recorded in Table 3.1 it will be apparent that while the concept of high- and low-energy compounds has been a fruitful one for the understanding of biochemical energetics, there is not in fact a sharp line of demarcation between these groups. Indeed, ATP itself occupies a position that is intermediate between compounds with high and those with low $\Delta G^{\circ\prime}$ values. Lehninger, following the earlier suggestions of Dixon, considers the free energy of hydrolysis of ATP as forming the mid-point of a thermodynamic scale of phosphorylated compounds. If this scale is arranged in descending order of $\Delta G^{\circ\prime}$ then those compounds highest in the scale undergo more complete hydrolysis at equilibrium than do those low in the scale, i.e. their position in the scale indicates the ease with which they release their phosphate groups. This thermodynamic scale is thus a quantitative measure of the affinity of the compound for its phosphate group. Lehninger has introduced the term *phosphate group transfer potential* as an arbitrary means of expressing the relative affinity of a compound for its phosphate group ; it is defined numerically as $- \Delta G^{\circ\prime}$ in kcal.

The unique role of ATP in metabolism is explicable in terms of its bridging position between phosphate compounds with a high phosphate group potential and those having a low potential, so that it may function in the transfer of phosphate groups from the former to the latter compounds. Thus ADP accepts phosphate groups in specific enzymic reactions from phosphate compounds of high potential and the ATP so formed donates its terminal group enzymically to suitable acceptor molecules, such as glucose, and thereby raises their energy content ; these are the low potential compounds. Position in the scale consequently denotes the direction of enzymic transfer of phosphate at equilibrium when equimolar reactants are taken. The central role of ATP and ADP in metabolism is therefore emphasized.

We have already noted that a living cell contains AMP as well as ADP and ATP. The total adenylate pool (AMP + ADP + ATP) can be considered to be essentially constant, at least over short time periods, and consequently the energy status of the cell at any particular moment will depend on the relative concentrations of the three components. If all the adenylate pool is in the form of ATP the cell will be in its maximum energetic state, whereas if it is all in the form of AMP it will be devoid of high energy components and so be in its lowest energetic state. The total energy stored in the adenylate system is clearly proportional to the average number of anhydride-bound phosphate groups per adenosine moiety. This number varies between zero for AMP and two for ATP.

To enable the energetic state of a system to be expressed quantitatively Atkinson has introduced the concept of *energy charge*, which he defines as half of the number of anhydride-bound phosphate groups per adenosine. (The half was introduced because an index which ranges from zero to one was preferred to one which ranges from zero to two.) The energy charge is thus defined by the expression

$$\text{Energy charge} = \tfrac{1}{2}\left(\frac{[\text{ADP}] + 2[\text{ATP}]}{[\text{AMP}] + [\text{ADP}] + [\text{ATP}]}\right)$$

for any given set of concentrations of AMP, ADP and ATP. The maximum energy charge, i.e. when all the adenylate is present as ATP, is therefore equal to 1·0 and the lowest, when all is present as AMP, is 0·0. Atkinson has pointed out that the energy charge is conceptually analogous to the charge of an electrochemical storage or accumulator cell, an analogy which can be helpful in visualizing metabolic relationships. On the assumption that adenylate kinase is present in excess, catalysing the reaction

$$2\text{ADP} \rightleftharpoons \text{AMP} + \text{ATP},$$

so that the concentrations of the three adenylates are near equilibrium values at all times, their relative concentrations will vary with the energy charge of the system as shown in Fig. 3.2. Consideration of this diagram reveals that, for any given point on the charge scale, a metabolic process that utilizes ATP will move the system to the left and the concentrations of the individual adenylates will increase or decrease as indicated by their respective curves. Conversely, a process that regenerates ATP will move the system to the right. The function of the regulatory interactions in energy metabolism is to maintain the processes involved in ATP generation and utilization in balance, and the concept of the energy charge has proved extremely useful in such studies.

It has been discovered that many regulatory enzymes in both anabolic and catabolic pathways are sensitive to ATP, ADP or AMP which serve as either positive or negative effectors for these allosteric proteins (see p. 224). Atkinson has suggested that it is the energy charge of the ATP-ADP-AMP system which regulates the pathways that produce and utilize high-energy compounds.

The relationship between metabolic sequences generating and utilizing ATP as a function of energy charge is illustrated in Fig. 3.3. At the point of intersection of the curves there is a metabolic steady state in which the rate of ATP production is equal to the rate of ATP utilization. Should the energy charge rise or fall from the value characteristic of the steady state, regulatory processes will come into operation to restore the balance. Thus if the energy charge increases above the steady state value the ATP-

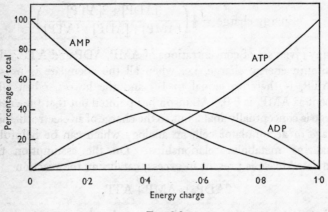

FIG. 3.2

Relative concentrations of AMP, ADP and ATP as functions of the energy charge (half of the number of anhydride-bound phosphate groups per adenosine moiety), assuming equilibration of the adenylate kinase reaction and the equilibrium constant to be 0·8, i.e. $K = \dfrac{[ATP]\,[AMP]}{[ADP]^2} = 0·8$.

(After ATKINSON (1968).)

generating systems are decelerated due to the inhibition of their regulatory enzymes in response to the changed concentrations of ATP, ADP and AMP ; likewise the ATP-utilizing systems are accelerated. Should the energy charge decrease then the converse applies. In this manner the adenine nucleotide system operates to ensure that the energy-producing and utilizing processes of a cell are maintained in a steady state for optimum cellular economy.

Living organisms are not one hundred per cent. efficient in their utilization of the free energy made available by catabolic reactions, although they are considerably more efficient than man-made machines which rely on combustion of a fuel as the source of energy. We may, therefore, distinguish between the free energy made available by catabolic reactions and the biologically useful energy which the organism can derive from this. The biologically useful energy will be in the form of high-energy phosphate, usually as ATP, and so the efficiency is directly related to the facility with which ATP synthesis can be coupled with catabolic reactions.

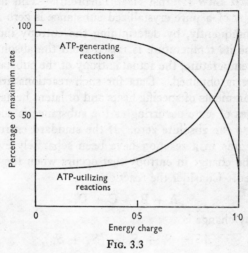

Fig. 3.3

Relationship of ATP-generating and ATP-utilizing reactions to the energy charge of the adenylate pool. At the point of intersection of the curves there is a metabolic steady state when the rate of production of ATP is equal to its rate of utilization.

An interesting example of biological efficiency has been studied by Bauchop and Elsden (1960) with yeast and *Zymomonas mobilis*. These micro-organisms both ferment glucose to ethanol and carbon dioxide, essentially in accordance with the equation

$$C_6H_{12}O_6 \longrightarrow 2C_2H_5OH + 2CO_2,$$

but by quite different enzymic mechanisms. The free energy change of the reactions is the same, however, because it depends solely on the initial and final states and not on the reaction pathway. Molar growth yield studies (see p. 319) revealed that one mole of glucose as an energy source supported the growth of twice as much yeast as *Z. mobilis*, indicating that the energy yield was twice as great for yeast. This can be correlated with the fact that glucose is fermented via the Embden-Meyerhof glycolytic sequence in yeast and by the Entner-Doudoroff route in *Z. mobilis*, with yields per mole of glucose of two and one moles of ATP respectively. It affords, therefore, an interesting example of two organisms carrying out the same overall reaction but differing in the amount of biologically useful energy which they can derive from an identical change in free energy.

The Third Law (Nernst Heat Theorem).—This is stated as, 'The entropy of a pure crystallized substance is zero at absolute zero.' Consequently, by determining the entropy increase of a substance as its temperature is raised from the absolute zero to any given temperature, the actual entropy of the substance at that temperature is obtained. Data for such reactions are obtained from measurements of specific heats and of latent heats associated with changes of state occurring as the substance is cooled from, say, 25° C to the absolute zero. If the standard entropies of all the compounds in a reaction have been separately evaluated in this way, the change in entropy that occurs when they react can be calculated. Consider the reaction

$$A + B \rightarrow C + D.$$

The entropy change is

$$\Delta S = S_C + S_D - (S_A + S_B),$$

and, knowing ΔS, the free energy change for the reaction can be obtained from equation 3.8.

Parks and Huffman (1932) have compiled tables of entropies of compounds, and their studies reveal a relationship between chemical constitution and entropy. This is of value for the empirical calculation of entropies of unknown substances. The following examples illustrate the use of these entropy 'rules'. All the data refer to 25°.

> For each H atom in the molecule S increases by 11·3 calories per °C per mole.
>
> For each C atom in the molecule S decreases by 13·4 cal degree^{-1} mole^{-1}.

The entropy contribution of oxygen is dependent on the type of linkage. Thus :

> For each terminal O atom, as in $-OH$ of $-COOH$, S increases by 0·9 cal degree^{-1} mole^{-1}.

For each secondary O atom, as in $-OH$ of $-CHOH$, S decreases by 4·6 cal degree^{-1} mole^{-1}.

For each carbonyl O atom, as in $> C = O$ of $-COOH$, S increases by 24·4 cal degree^{-1} mole^{-1}.

Consequently the type of oxygen bond in a molecule plays an important role in determining the entropy of a compound, and the large value of the carbonyl oxygen may well be a predominant factor in the entropy value.

Example 3.3.—Use the Parks and Huffman rules to calculate the entropy of pyruvic acid at 25° C.

$$CH_3-C-C \begin{smallmatrix} O \\ \nearrow \\ \searrow OH \end{smallmatrix}$$

4 H atoms	$4 \times 11·3 = + 45·2$
3 C atoms	$- 3 \times 13·4 = - 40·2$
1 terminal O atom	$0·9 = + 0·9$
2 carbonyl O atoms	$2 \times 24·4 = + 48·8$

Total $= S°_{298}$(pyruvic acid) $= + \underline{54·7}$ cal deg^{-1} mole^{-1}

Although the application of these empirical rules has been of great value when the entropy of a compound was required but thermal data were not available, it will be appreciated that thermodynamic quantities obtained more directly are greatly to be preferred. The very fact that modifications in structure profoundly affect entropy is an indication of the uncertainty that may be introduced if this quantity is calculated for an 'unknown' compound by the application of empirical rules.

Biochemical reactions generally occur in solution and under such conditions hydrogen bonding and the solvation of substrates are important factors in determining entropy changes. Charged ions and polar substrates in a polar solvent such as water attract the water molecules and become solvated, i.e. solvent molecules are oriented about the solute and, since their freedom for rotation and translation is restricted, the entropy is low. This effect is related to the magnitude of the charge and, additionally, anions usually exert a greater solvating effect than cations. As the entropy of the solvent is assumed to be normal in such cases, negative entropy values are assigned to small ions.

Ionization produces an appreciable decrease in entropy since the charged species so formed partially immobilize solvent molecules by solvation, e.g. the entropy of ionization of acetic acid is $-22 \cdot 1$ cal degree^{-1} mole^{-1}. In the case of the acidic ionization of dipolar ions such as amino acids, the entropy change is smaller (about -8 cal degree^{-1} mole^{-1}) since the dipolar ion is itself solvated. When oppositely charged ions react to form neutral entities the reverse phenomenon occurs, i.e. solvation decreases and the entropy change is positive and large. Other examples include the reaction of an ion with a neutral molecule to produce another ion of identical charge, or reactions where charges are neither neutralized nor produced ; in these cases the entropy change is usually small and may be either positive or negative.

Protein denaturation is an example of reactions which are accompanied by very large increases of entropy (certain aspects of this process are discussed in Chapter V, p. 171). Native proteins consist of long chains of amino acid residues held in compact highly orientated configurations such as the a-helix. The unfolding and breakdown of these structures in denaturation involve large increases in the rotational degrees of freedom of the polypeptide chains and thus confer very large positive entropy changes which, for different proteins under various conditions, may have values as high as 100 to 320 cal degree^{-1} mole^{-1}.

REFERENCES AND SUGGESTED READING

ATKINSON, D. E. (1968). In *The Metabolic Roles of Citrate*, p. 23, ed. T. W. Goodwin. London and New York : Academic Press.

BAUCHOP, T. & ELSDEN, S. R. (1960). The growth of micro-organisms in relation to their energy supply. *J. gen. Microbiol.*, 23, 457.

BENZINGER, T. H. (1969). Ultrasensitive reaction calorimetry. In *A Laboratory Manual of Analytical Methods of Protein Chemistry*, vol. 5, 93, ed. P. Alexander & H. P. Lundgren. Oxford : Pergamon Press.

BROWN, H. D., editor (1969). *Biochemical Calorimetry*. New York : Academic Press.

BURTON, K. & KREBS, H. A. (1953). The free-energy changes associated with the individual steps of the tricarboxylic acid cycle, glycolysis and alcoholic fermentation and with the hydrolysis of the pyrophosphate groups of adenosinetriphosphate. *Biochem. J.*, 54, 94.

BUTLER, J. A. V. (1951). *Chemical Thermodynamics*, 4th ed. London : Macmillan and Co.

FLORKIN, M. & MASON, H. S., editors (1960). *Comparative Biochemistry*, vol. I, Chaps 2 and 4 ; vol. II, Chaps 1, 2, 3 and 4. New York : Academic Press.

GEORGE, P. & RUTMAN, R. J. (1960). The 'high energy phosphate bond' concept. *Prog. Biophys. biophys. Chem.*, 10, 1.

GILLESPIE, R. J., MAW, G. A. & VERNON, C. A. (1953). The concept of phosphate bond-energy. *Nature*, **171**, 1147.

GUTFREUND, H. (1951). The nature of entropy and its role in biochemical processes. *Adv. Enzymol.*, **11**, 1.

INGRAHAM, L. L. & PARDEE, A. B. (1967). Free energy and entropy in metabolism. In *Metabolic Pathways*, **3**, 1, ed. D. M. Greenberg. New York: Academic Press.

KREBS, H. A. & KORNBERG, H. L. (1957). *Energy Transformations in Living Matter*. Berlin: Springer-Verlag.

PARKS, G. S. & HUFFMAN, H. M. (1932). *The Free Energies of Some Organic Compounds*. New York: Chemical Catalogue Company.

SPANNER, D. C. (1964). *Introduction to Thermodynamics*. New York: Academic Press.

STOESSER, P. R. & GILL, S. J. (1967). Precision flow-microcalorimeter. *Rev. scient. Instrum.*, **38**, 422.

WILKIE, D. R. (1960). Thermodynamics and the interpretation of biological heat measurements. *Prog. Biophys. biophys. Chem.*, **10**, 260.

PROBLEMS

Note.—A study of Example 4.1 may assist the solution of some of these problems. Further problems of a similar nature will be found in Chapter IV.

3.1. The change in internal energy when solid palmitic acid is combusted in a bomb calorimeter at 20° is $- 2384 \cdot 1$ kcal. The equation for the reaction is:

$$C_{16}H_{32}O_{2(S)} + 23O_{2(G)} \rightarrow 16CO_{2(G)} + 16H_2O_{(L)} .$$

Determine the heat content change for the reaction.

3.2. The heat of combustion of stearic acid (solid) at 15° and 1 atmosphere pressure is $- 2698 \cdot 0$ kcal. What is the change in internal energy associated with the reaction?

3.3. The heat of combustion (ΔH) of liquid ethanol to gaseous carbon dioxide and liquid water is $- 325 \cdot 70$ kcal per mole at 15°. Calculate the internal energy change for the reaction.

The heats of formation of gaseous carbon dioxide and liquid water are $- 94 \cdot 45$ and $- 68 \cdot 37$ kcal respectively at 15°. Determine the heat of formation of ethanol from its elements at this temperature.

3.4. At 25° and 1 atmosphere pressure the standard free energy of formation of solid α-D-glucose is $- 215,800$ calories. The standard free energies of formation of gaseous carbon dioxide and liquid water are $- 94,450$ and $- 56,690$ calories respectively under similar conditions. Find the standard free energy of combustion of glucose to carbon dioxide and water at this temperature and pressure.

3.5. The standard free energy of formation of aqueous divalent succinate ion is $- 165,090$ calories at 25° and 1 atmosphere. If the standard free energy of the reaction

$$\text{Fumarate}^- \text{ (aq.)} + H_2 \rightarrow \text{succinate}^= \text{ (aq.)}$$

is $- 20,470$ calories under the same conditions, calculate the standard free energy of formation of aqueous divalent fumarate ion at 25°.

3.6. The standard free energies of formation of aqueous solutions of monovalent L-cysteine anion and divalent L-cystine anion are $- 70,270$ and $- 137,190$ calories respectively. What is the standard free energy change for the reduction of cystine to cysteine? The values refer to 25° and 1 atmosphere pressure.

3.7. The most general method for obtaining heats of reaction is by the determination of heats of combustion of the reactants and products. Recent attention has been focused on the direct calorimetric determination of the heat of reaction where this is possible. The following data illustrate both experimental methods.

Huffman and Fox determined the heats of combustion of solid α-D-glucose and β-D-glucose to gaseous carbon dioxide and liquid water at 25° and 1 atmosphere. The values obtained were, respectively, − 669·58 and − 671·08 kcal per mole. Determine the heat of reaction for the mutarotation process.

Sturtevant has made direct calorimetric measurements of the heats of mutarotation in solution and the heats of solution of the two forms of glucose. He obtained the following results at 298° K:

α-D-glucose (S) → α-D-glucose (aq.) ΔH =   10,716 joules per mole.

α-D-glucose (aq.) → β-D-glucose (aq.) ΔH = − 1,162 joules per mole.

β-D-glucose (aq.) → β-D-glucose (S) ΔH = − 4,680 joules per mole.

Calculate the heat of mutarotation of solid α-D-glucose to β-D-glucose. Compare this value with that obtained by the former method and express the difference as a percentage of the value obtained by the direct measurement, and also as a percentage of the average heat of combustion of the two glucose isomers. Comment on the accuracy of heats of reaction obtained by the heat of combustion method.

(Data from HUFFMAN & FOX (1938), *J. Amer. chem. Soc.*, **60**, 1400, and STURTEVANT (1941) *J. phys. Chem.*, **45**, 127.)

3.8. The heats of combustion of glucose, pyruvic acid and ethanol at 18° and 1 atmosphere are, respectively, − 674·00, − 279·10 and − 326·66 kcal per mole. What quantity of heat is evolved in the formation of 1 mole of (*a*) pyruvic acid (*b*) ethanol by the fermentation of glucose at 18° and 1 atmosphere? Neglect heats of dilution, etc.

3.9. The heat of combustion of solid anhydrous citric acid is − 475·0 kcal and that of solid malic acid − 320·1 kcal at 15° and 1 atmosphere pressure. Calculate the heat of reaction for the conversion of citric acid to malic acid at 15° and 1 atmosphere. The heats of formation of liquid water and gaseous carbon dioxide at the same temperature and pressure are − 68·37 and − 94·45 kcal respectively.

3.10. Calculate the heat of reaction of the formate hydrogenlyase system of *Escherichia coli*:

$$H_2 + HCO_3^- \rightleftharpoons HCOO^- + H_2O$$

from the following thermal data, all of which applies to 298° K.

HCOOH (L) ΔH = − 99,750 cal.

CO_2(G) ΔH = − 94,240 cal.

H_2O (L) ΔH = − 68,310 cal.

CO_2 (G) $\rightleftharpoons CO_2$ (sat. aq.) + 4844 cal.

$H_2CO_3 \rightleftharpoons H^+ + HCO_3^-$ − 2075 cal.

HCOOH (L) $\rightleftharpoons$ HCOOH (aq.) + 100 cal.

HCOOH (aq.) $\rightleftharpoons H^+ + HCOO^-$ + 13 cal.

Note.—To get from CO_2 (sat. aq.) to H_2CO_3 Woods writes CO_2 (sat. aq.) + $H_2O = H_2CO_3$ and finds ΔH by substituting values of ΔH for H_2O and CO_2 (sat. aq.). This presumably implies that no calories are absorbed or evolved when CO_2 reacts with H_2O or, alternatively, that the heat change in the formation of H_2CO_3 has been included in the CO_2(G) $\rightleftharpoons CO_2$ (sat. aq.) equation.

(Data quoted by WOODS (1936), *Biochem. J.*, **30**, 515.)

3.11. From the entropy data of Parks and Huffman given on p. 124, calculate the entropy at 25° of the following compounds of biochemical interest : (a) lactic acid, (b) malic acid, (c) α-oxoglutaric acid, (d) oxaloacetic acid, (e) acetylmethylcarbinol.

3.12. The oxidation of phosphoglyceraldehyde to phosphoglyceric acid involves a free energy change of − 29,000 cal. Under biological conditions this reaction is achieved via the formation of phosphoglycerylphosphate and requires the participation of nicotinamide adenine dinucleotide (NAD) as coenzyme, and also inorganic phosphate.

The reduction of NAD, viz.

$$NAD^+ + 2H \rightleftharpoons NADH + H^+,$$

has $\Delta G = 13,000$ cal and the overall reaction

Phosphoglyceraldehyde + phosphate + $NAD^+ \rightarrow$

phosphoglycerylphosphate + NADH + H^+

has $\Delta G = -1000$ cal.

From these data, what do you deduce with regard to the nature of phosphoglycerylphosphate? State how phosphoglycerylphosphate is converted biologically into phosphoglyceric acid and indicate the value of this mechanism to the organism.

What would happen if phosphate were replaced by arsenate in the oxidation of phosphoglyceraldehyde?

The oxidation of lactic acid to pyruvic acid has $\Delta G = -9000$ cal. Is this compatible with the triosephosphate dehydrogenase and lactic dehydrogenase systems being 'coenzyme-linked'?

(Glasgow Honours Course Finals, 1951.)

3.13. Summarize the methods by which standard free energy changes for reactions can be obtained. The enzyme phosphoglucomutase catalyses the interconversion of glucose 1-phosphate and glucose 6-phosphate. At 38° C the equilibrium mixture contains 5·4 per cent. of the 1-phosphate. Calculate the equilibrium constant and the free energy change of the reaction.

(Leeds Honours Course Finals, 1955.)

3.14. Hexokinase was incubated at pH 6 with a solution containing glucose 6-phosphate (G6P) and adenosine diphosphate (ADP). When equilibrium had been attained the following concentrations (millimolar) were determined : G6P, 7·0 ; ADP, 4·0 ; glucose and adenosine triphosphate (ATP) each 0·26.

Calculate the free energy change ($\Delta G'$) for the hexokinase reaction at pH 6 assuming activity coefficients of unity.

Use the following information to calculate the free energy change for the hydrolysis of ATP to ADP at pH 7 : (a) for the hexokinase reaction $\Delta G'$ is 25 per cent. greater at pH 7 than at pH 6, (b) for the hydrolysis of G6P to glucose at pH 7, $\Delta G' = -3·1$ kcal. (All experiments were performed at the standard temperature of 25° C when $RT \log_e x$ can be expressed as $1364 \log_{10} x$ calories.)

(Leeds Honours Course Finals, 1963.)

CHAPTER IV

EQUILIBRIA

CONSIDER the reversible chemical reaction

$$A + B \rightleftharpoons C + D.$$

The rate of the forward reaction is $k_1(A)(B)$ and the rate of the reverse reaction is $k_{-1}(C)(D)$, where the activities of the reactants are represented by (A), (B), (C) and (D) and k_1 and k_{-1} are the respective rate constants. At equilibrium the rates of the forward and reverse reactions will be equal so that, employing equilibrium activities (denoted by subscript e),

$$k_1(A)_e(B)_e = k_{-1}(C)_e(D)_e \quad . \quad . \quad . \quad (4.1)$$

and

$$\frac{(C)_e(D)_e}{(A)_e(B)_e} = \frac{k_1}{k_{-1}} = K \quad . \quad . \quad . \quad (4.2)$$

where K is the *thermodynamic equilibrium constant* of the reaction. Note the convention that the activities of the products of the reaction, i.e. right-hand side of the chemical equation, are put in the numerator. This equation represents the Law of Mass Action, first enunciated by Guldberg and Waage in 1864. If concentrations are used in place of activities a *concentration equilibrium constant* K_c is obtained which is equal to the thermodynamic constant only at infinite dilution (see p. 135).

From the above relation it follows that, irrespective of the initial activities of A, B, C and D, at equilibrium the activities of all molecular species are related to one another in such a way that equation 4.2 holds. Consequently, if K is known, the position of equilibrium may be calculated from any initial activities of the reacting substances.

The example quoted is a simple case where one molecule of each substance reacts, but frequently more complicated reactions are encountered. The general case of a reaction such as :

$$aA + bB + cC + \cdots \rightleftharpoons pP + qQ + rR + \cdots$$

where a, b, etc., represent the respective numbers of molecules of

A, B, etc., reacting, leads to the following formulation of the equilibrium constant

$$K = \frac{(P)_e^p (Q)_e^q (R)_e^r \cdots}{(A)_e^a (B)_e^b (C)_e^c \cdots} \qquad . \qquad . \qquad . \qquad (4.3)$$

It will be noted that the equilibrium activity of each substance is raised to the power of the number of its molecules participating in the reaction.

The Law of Mass Action may also be derived thermodynamically and the relationship of the equilibrium constant to the free energy change expressed as :

$$\Delta G = -RT \ln K + RT \ln \frac{(P)^p (Q)^q (R)^r \cdots}{(A)^a (B)^b (C)^c \cdots} \qquad . \qquad (4.4)$$

This may be contracted by the use of algebraic notation to

$$\Delta G = -RT \ln K + RT \sum a \ln (A) \qquad . \qquad (4.5)$$

where the second right-hand term represents the sum of the logarithms of the activity of each reacting substance multiplied by the number of molecules of the substance, with correct sign, i.e. all numerator activities are positive and all denominator activities negative. If all the substances are in their standard states of unit activity, the last term becomes $RT \ln 1$ which is zero and hence equation 4.5 reduces to

$$\Delta G^\circ = -RT \ln K \qquad . \qquad . \qquad . \qquad (4.6)$$

where ΔG° is the *standard free energy change* for the reaction. It follows, therefore, that

$$\Delta G = \Delta G^\circ + RT \sum a \ln (A) \qquad . \qquad . \qquad (4.7)$$

Consequently, the standard free energy change of a reaction may be calculated from the equilibrium constant by means of equation 4.6 and then used in equation 4.7 to obtain the free energy change for the reaction with any arbitrary concentrations. Once obtained, standard free energies have the valuable property of being additive like heats of reaction and so enable the value for a particular reaction to be calculated from those previously established. Table 4.1 records the relationship between the standard free energy change and the equilibrium constant for temperatures of 25° and 37°.

In many cases when it may prove difficult or even impossible to measure an equilibrium constant directly, recourse to calculation by indirect methods is necessary. Equation 4.6 is the basis of all such calculations. For example, $\Delta G°$ for several partial reactions, the summation of which is the reaction in question, may be added together to give $\Delta G°$ for this reaction and hence K. $\Delta G°$ for the partial reactions may be obtained either by equilibrium or by e.m.f. measurements. Sometimes the values of the standard free

TABLE 4.1

Relationship between Standard Free Energy Change and Equilibrium Constant

$$\Delta G° = -RT \ln K = 2.303 \, RT \log K$$

K	$\log K$	$\Delta G°$ cal mole^{-1}	
		25°	37°
0·0001	−4	5452	5676
0·001	−3	4089	4257
0·01	−2	2726	2838
0·1	−1	1363	1419
1·0	0	0	0
10·0	1	−1363	−1419
100·0	2	−2726	−2838
1000·0	3	−4089	−4257
10000·0	4	−5452	−5676

energy of formation of compounds from their elements can be found in suitable reference books (Landolt-Börnstein; Parks and Huffman) and $\Delta G°$ for the total reaction obtained by summation. It is important to note that the value of the standard free energy of formation (usually denoted by $\Delta G_f°$) employed must refer to the state which obtains in the reaction. For instance, if the compound is reacting in solution $\Delta G_f°$ must refer to the solution of appropriate activity and not simply to $\Delta G_f°$ of the solid compound. Since $\Delta G_f°$ of a saturated solution is equal to that of the solid compound, this value must be added to the standard free energy of dilution of a saturated solution to a solution of the given activity. Furthermore, should the compound be present in solution as an ion, the standard free energy of ionization must also be added. For a substance dissolved in water the standard state is that at which the concentration of the solute is approximately 1 molar (see p. 114), and the following relationship holds for the transference of a substance from the standard to another state by alteration of either concentration, pressure or temperature

$$\Delta G = \Delta G^\circ + RT \ln a,$$

where a is the activity.

Example 4.1.—Calculate the equilibrium constant for the formation of the dipeptide glycylglycine from two molecules of glycine at 37·5° C. The standard free energies of formation from their elements to the standard state in aqueous solution (1 M activity, except for water, which is 1 mole fraction) are, at 37·5° C, respectively: glycine, $- 87,710$ calories; glycylglycine, $- 115,630$ calories; water, $- 56,200$ calories.

The reaction is
$$2 \text{ glycine} \rightleftharpoons \text{glycylglycine} + H_2O.$$

Hence $\quad \Delta G^\circ = \Delta G_f^\circ \text{ (glycylglycine)} + \Delta G_f^\circ(H_2O) - 2 \Delta G_f^\circ \text{ (glycine)}$
$$= -115630 - 56200 - 2(- 87710)$$
$$= -171830 + 175420$$
$$= 3590 \text{ calories.}$$

Now $\quad \Delta G^\circ = -RT \ln K$
$$\therefore 3590 = -2\cdot303 \times 1\cdot987 \times 310\cdot5 \log K$$
$$\log K = -2\cdot526 = \overline{3}\cdot474$$
$$\therefore \underline{K = 0\cdot00298.}$$

Consider now the reversible reaction
$$a A + b B \rightleftharpoons c C + d D.$$

The change in free energy for this reaction is expressed by equation 4.5, which becomes

$$\Delta G = -RT \ln K + RT \ln \frac{(C)^c (D)^d}{(A)^a (B)^b}.$$

At thermodynamic equilibrium in a reversible reaction ΔG is zero and since $\Delta G = \Delta H - T \Delta S$ it follows that, at equilibrium, the capacity of the system to do work is at a minimum and the entropy is at a maximum.

Example 4.2.—The following data were obtained in an experiment to determine the equilibrium constant of the reaction

$$\textit{iso}\text{-propanol} + NAD^+ \rightleftharpoons \text{acetone} + NADH + H^+,$$

which is carried out by crystalline yeast alcohol dehydrogenase.

pH	Acetone $(10^2 \times$ M$)$	Iso-propanol $(10^2 \times$ M$)$	NADH $(10^5 \times$ M$)$	NAD$^+$ $(10^5 \times$ M$)$
(a)	8·78	19·5	3·92	4·74	5·41
(b)	7·28	1·54	8·49	4·51	6·00
(c)	7·18	1·51	8·33	3·43	5·72

The values given are for the concentrations of the compounds at equilibrium in experiments carried out at 25° C.

Determine whether or not the equilibrium constant is influenced by pH and also the standard free energy change for the reaction. The activity coefficients of all the reactants may be taken as unity.

(After BURTON & WILSON (1953), *Biochem. J.*, **54**, 86.)

The activities are equal to the concentrations of the reactants since all activity coefficients are unity and the equilibrium constant is therefore given by the equation

$$K = \frac{\text{(acetone)(NADH)(H}^+)}{\text{(iso-propanol)(NAD}^+)} = \frac{\text{(acetone)(NADH)}}{\text{(iso-propanol)(NAD}^+)} \times \frac{1}{\text{antilog pH}}$$

since $(\text{H}^+) = \dfrac{1}{\text{antilog pH}}$.

Hence in experiment (a) at pH 8·78

$$K = \frac{19 \cdot 5 \times 10^{-2} \times 4 \cdot 74 \times 10^{-5}}{3 \cdot 92 \times 10^{-2} \times 5 \cdot 41 \times 10^{-5}} \times \frac{1}{\text{antilog } 8 \cdot 78}$$

$$= \frac{19 \cdot 5 \times 4 \cdot 74}{3 \cdot 92 \times 5 \cdot 41 \times 6 \cdot 026 \times 10^8} = 7 \cdot 23 \times 10^{-9} \text{ M.}$$

In experiment (b) at pH 7·28

$$K = \frac{1 \cdot 54 \times 4 \cdot 51}{8 \cdot 49 \times 6 \cdot 00} \times \frac{1}{\text{antilog } 7 \cdot 28} = 7 \cdot 16 \times 10^{-9} \text{ M.}$$

and in experiment (c) at pH 7·18

$$K = \frac{1 \cdot 51 \times 3 \cdot 43}{8 \cdot 33 \times 5 \cdot 72} \times \frac{1}{\text{antilog } 7 \cdot 18} = 7 \cdot 20 \times 10^{-9} \text{ M.}$$

From these results it is obvious that the value of K has not been appreciably affected by changing the pH value from 8·78 to 7·18. The average value of the equilibrium constant from the above three values is $7 \cdot 196 \times 10^{-9}$ M, and this may now be used to calculate the standard free energy change of the reaction.

$$\Delta G° = -RT \ln K = -2 \cdot 303 \, RT \log K$$

$$= -2 \cdot 303 \times 1 \cdot 987 \times 298 \log 7 \cdot 196 \times 10^{-9}$$

$$= -2 \cdot 303 \times 1 \cdot 987 \times 298 \, (\overline{9} \cdot 8571 = -8 \cdot 1429)$$

$$= \underline{11,105 \text{ calories.}}$$

Effect of Temperature on the Equilibrium Constant

Van't Hoff deduced the following relationship for the variation of the equilibrium constant with absolute temperature

$$\frac{d \ln K}{dT} = \frac{\Delta H}{RT^2} \quad . \quad . \quad . \quad (4.8)$$

where ΔH is the heat of reaction at constant pressure. This is known as the *van't Hoff equation* or *isochore*. Integrating equation 4.8

$$\ln K = \frac{\Delta H}{RT} + \text{constant} \quad . \quad . \quad . \quad (4.9)$$

Thus if $\log K$ is plotted against $1/T$ a straight line of slope $-\Delta H/2 \cdot 303 \ R$ will be obtained enabling ΔH for the reaction to be evaluated. If values of the equilibrium constant are determined at two temperatures (K_1 and K_2 at temperatures T_1 and T_2 respectively) then equation 4.9 may be written in the form

$$\ln \frac{K_2}{K_1} = \frac{-\Delta H}{R}\left(\frac{1}{T_2} - \frac{1}{T_1}\right) \quad . \quad . \quad (4.10)$$

$$\ln \frac{K_2}{K_1} = \frac{-\Delta H}{R}\left(\frac{T_2 - T_1}{T_1 T_2}\right) \quad . \quad . \quad (4.10a)$$

As we have already seen (p. 114), to a good first approximation ΔH may be equated with $\Delta H°$. Consequently, with measurements of the equilibrium constant of a reaction at two, but preferably more, temperatures it is possible to determine $\Delta H \approx \Delta H°$ and simultaneously $\Delta G°$ and $\Delta S°$ from the relationships $\Delta G° = -RT \ln K$ and $\Delta G° = \Delta H° - T\Delta S°$.

Activities and Concentrations

In order that a distinction may be made between equilibrium constants calculated from the equilibrium activities of the reactants and those calculated from equilibrium concentrations they are referred to as the thermodynamic and concentration equilibrium constants and often denoted in the literature by K_a and K_c respectively. For the reaction

$$A + B \rightleftharpoons C + D$$

they are related in the following manner

$$K_a = \frac{(C)(D)}{(A)(B)} = \frac{f_C[C]f_D[D]}{f_A[A]f_B[B]}$$

$$\text{and} \quad K_c = \frac{[C][D]}{[A][B]}$$

$$\text{whence} \quad K_a = K_c \frac{f_C f_D}{f_A f_B}.$$

Thus K_a can be obtained from K_c provided that the activity coefficients of all reactants are known. If the concentrations are molal and not molar then the appropriate molal activity coefficients must be employed, i.e. γ_A, γ_B, etc.

Apparent Equilibrium Constants

Quite often in biochemical work the equilibrium constant of a reaction involving hydrogen ions is calculated without inclusion of the hydrogen ion activity, e.g. in reactions involving NAD, such as :

$$C_2H_5OH + NAD^+ \rightleftharpoons CH_3CHO + NADH + H^+$$

the equilibrium constant at a given temperature may be written as :

$$K_{app} = \frac{(CH_3CHO)(NADH)}{(C_2H_5OH)(NAD^+)}$$

and the hydrogen ion term is omitted from the numerator. The constant obtained in this manner is termed an *apparent equilibrium constant*, denoted by K_{app}, to distinguish it from the thermodynamic equilibrium constant calculated by inclusion of the hydrogen ion activity. Examples of this usage will be found in Problems 4.11 and 4.12. K_{app} thus defines the position of equilibrium at a particular pH value, that of the hydrogen ion activity omitted from the numerator of the equation. Clearly then the apparent equilibrium constant varies with the pH, in contrast to the thermodynamic constant which is independent of pH. It is necessary therefore always to specify the pH at which an apparent equilibrium constant was determined, as well as the temperature.

Since the standard free energy change is calculated from the equilibrium constant by means of equation 4.6 it will be obvious that the value of $\Delta G°$ depends on whether an apparent or thermodynamic equilibrium constant is used. Thus

$$\Delta G° = RT \ln K$$

and

$$\Delta G°' = -RT \ln K_{app}$$

so that $\Delta G^{o\prime}$ is dependent on both pH and temperature, which must always be specified, whereas ΔG^o is dependent on temperature. ΔG^o must refer to the standard state of all reactants so that, for reactions involving hydrogen ions, unit activity of $(H^+) = 1$, i.e. pH 0. As few, if any, biochemical reactions can occur at this pH, $\Delta G^{o\prime}$ for a specified biological pH is usually employed. For much biochemical work $\Delta G^{o\prime}$ is specified as pH 7·0 and 25°.

The relationship between K and K_{app} for the NAD reaction cited is given by

$$K = K_{app} (H^+).$$

If logarithms are taken and the equation is multiplied throughout by $-RT$ then

$$-RT \ln K = -RT \ln K_{app} - RT \ln (H^+)$$

whence $\qquad \Delta G = \Delta G^{o\prime} - 2\cdot303\ RT \log (H^+)$

and $\qquad \Delta G = \Delta G^{o\prime} + 2\cdot303\ RT\ \text{pH} \qquad . \qquad . \qquad . \qquad (4.11)$

Thus $\Delta G^{o\prime}$ differs from ΔG^o by the term $2\cdot303\ RT\ \text{pH}$. $\Delta G^{o\prime}$ may be related to the standard electrode potential, E'_o, at a stated pH other than zero (see p. 344) by the equation

$$\Delta G^{o\prime} = 2E'_o F - 2RT \ln (H^+)$$

where F is the faraday.

Coupled Reactions

It is frequently the case that two reversible reactions have an intermediate common to both and, should they occur simultaneously in the same isothermal system, they are then able to give rise to *coupled reactions*. This situation is one which is often encountered in living cells. Suppose the reactions are

$$1.\ A + B \rightleftharpoons C + D$$

$$2.\ D + L \rightleftharpoons M + N$$

where D, a product of the first reaction, becomes a reactant in the second. The overall coupled reaction (3), the sum of reactions 1 and 2, becomes

$$3.\ A + B + L \rightleftharpoons C + M + N.$$

Now the equilibrium constants for reactions 1 and 2 are given by

$$K_1 = \frac{[C][D]}{[A][B]} \quad \text{and} \quad K_2 = \frac{[M][N]}{[D][L]}$$

where the concentrations are those of the systems at equilibrium. Solving both equations for [D] we have

$$[D] = \frac{K_1[A][B]}{[C]} \quad \text{and} \quad [D] = \frac{[M][N]}{K_2[L]}$$

and therefore

$$K_1 K_2 = \frac{[C][M][N]}{[A][B][L]}.$$

But this expression gives the equilibrium constant K_3 for the coupled reaction 3, thus

$$K_3 = K_1 K_2.$$

In other words, the equilibrium constant for the coupled net reaction is equal to the product of the equilibrium constants for the two component reactions. The existence of coupling does not therefore influence the equilibrium constants for the individual component reactions but the equilibrium constant for the overall reaction is determined by their values.

We have already noted how the equilibrium constant is related to the standard free energy change of the reaction. Consequently, for the coupled systems discussed, we may write

$$\Delta G_1^\circ = -RT \ln K_1$$
$$\Delta G_2^\circ = -RT \ln K_2$$
$$\Delta G_3^\circ = -RT \ln K_3.$$

Further, since $K_3 = K_1 K_2$

if we take logarithms, and multiply throughout by $-RT$

$$-RT \ln K_3 = -RT \ln K_1 - RT \ln K_2$$

whence $\qquad \Delta G_3^\circ = \Delta G_1^\circ + \Delta G_2^\circ.$

This relationship emphasizes the point made in Chapter III (p. 114) that values of ΔG° are additive and, knowing ΔG° for two reactions, that of a third related reaction can be calculated. In this connexion it was pointed out that an endergonic reaction, which would not of itself proceed on account of the increase in free energy involved, could be made to do so by coupling it to a suitable exergonic reaction. This is a device commonly associated with many of the reactions involved in biosynthetic pathways.

Example 4.3.—Extracts of *Escherichia coli* contain the enzymes aspartate ammonia lyase (aspartase) and fumarate hydratase which catalyse the undernoted reactions with the free energy changes (pH 7·4, 37°) recorded

$$\text{fumarate}^= + NH_4^+ \rightleftharpoons \text{aspartate}^= \quad \Delta G^{o\prime} = -3720 \text{ cal mole}^{-1}$$
$$\text{fumarate}^= + H_2O \rightleftharpoons \text{malate}^= \quad \Delta G^{o\prime} = -700 \text{ cal mole}^{-1}$$

What is the free energy change and the equilibrium constant for the formation of aspartate from malate and ammonium ion at pH 7·4 and 37° ? If the extract is incubated with 0·1 M-malate and 1·0 M-NH_4^+ what will be the equilibrium concentration of aspartate ?

The required reaction, namely

$$\text{malate}^= + NH_4^+ \rightleftharpoons \text{aspartate}^= + H_2O$$

is a coupled reaction and the sum of the two enzymic reactions, if the fumarate hydratase reaction is written in the reverse order. For the reverse direction the free energy change will be equal in magnitude but opposite in sign, i.e. +700 cal mole^{-1}
Thus we may write

$$\text{fumarate}^= + NH_4^+ \rightleftharpoons \text{aspartate}^= \quad \Delta G^{o\prime} = -3720 \text{ cal mole}^{-1}$$
$$\text{malate}^= \rightleftharpoons \text{fumarate}^= + H_2O \quad \Delta G^{o\prime} = +700 \text{ cal mole}^{-1}$$

Sum : $\text{malate}^= + NH_4^+ \rightleftharpoons \text{aspartate}^= + H_2O \quad \Delta G^{o\prime} = -3020 \text{ cal mole}^{-1}$

The coupled reaction is thus exergonic and will proceed from left to right, resulting in the net synthesis of aspartate. The equilibrium constant for the reaction at 37° will be given by :

$$\Delta G^{o\prime} = -RT \ln K$$

Thus
$$-3020 = -1\cdot987 \times 310 \times 2\cdot303 \log K$$
$$3020 = 1419 \log K$$
$$2\cdot129 = \log K$$
$$K = 134\cdot6.$$

If the reaction is started with malate at 0·1 M and NH_4^+ at 1 M and we let the equilibrium concentration of aspartate be xM, then since

$$\frac{[\text{aspartate}^=]}{[\text{malate}^=][NH_4^+]} = 134\cdot6$$

$$\frac{x}{(0\cdot1 - x)(1\cdot0 - x)} = 134\cdot6$$

$$134\cdot6x^2 - 149\cdot06x + 13\cdot46 = 0.$$

We can now solve this quadratic equation using the appropriate formula (Appendix 5)

$$x = \frac{149\cdot06 \pm \sqrt{((-149\cdot06)^2 - 4 \times 134\cdot6 \times 13\cdot46)}}{2 \times 134\cdot6}$$

$$= \frac{149 \cdot 06 \pm \sqrt{(22230 - 7246)}}{2 \times 134 \cdot 6} = \frac{149 \cdot 06 \pm \sqrt{14984}}{2 \times 134 \cdot 6}$$

$$= \frac{149 \cdot 06 \pm 122 \cdot 4}{2 \times 134 \cdot 6} = \frac{271 \cdot 46}{269 \cdot 2} \text{ or } \frac{26 \cdot 66}{269 \cdot 2}$$

Therefore $x = 1 \cdot 001$ or $0 \cdot 099$.

Since the reaction was started with $0 \cdot 1$ M-malate, however, clearly only the second of these values is a feasible solution to the problem posed. It will be seen therefore that the reaction has proceeded 99 per cent. to completion, i.e. virtually all of the malate has been converted to aspartate.

An extension of the concept of energetic coupling is to be found in various reactions where the hydrolysis of ATP (or of some alternative high energy compound) provides the exergonic driving force for an endergonic reaction, but as part of a single, more complex reaction, rather than as two individual component reactions. An example of this is furnished by the phosphorylation of glucose. To enter metabolic pathways glucose must first be converted to a phosphorylated derivative and this is normally achieved by the enzyme hexokinase which, in the presence of ATP and magnesium ions, catalyses the reaction

glucose $+$ ATP $\rightleftharpoons$ glucose 6-phosphate $+$ ADP.

For thermodynamic purposes this reaction may, however, be regarded as the sum of two reactions which, at pH 7 and 25°, have the undernoted free energy changes

$$\text{ATP} + \text{H}_2\text{O} \rightleftharpoons \text{ADP} + \text{H}_3\text{PO}_4$$
$$\Delta G^{\circ\prime} = -7700 \text{ cal mole}^{-1}$$

glucose $+$ $\text{H}_3\text{PO}_4 \rightleftharpoons$ glucose 6-phosphate $+$ H_2O
$$\Delta G^{\circ\prime} = +3140 \text{ cal mole}^{-1}$$

Sum: glucose $+$ ATP $\rightleftharpoons$ glucose 6-phosphate $+$ ADP
$$\Delta G^{\circ\prime} = -4560 \text{ cal mole}^{-1}$$

Thus the endergonic phosphorylation of glucose is achieved by coupling with the highly exergonic hydrolysis of ATP to ADP and inorganic phosphate, giving an overall exergonic reaction. It should be noted, however, that there is no experimental evidence to suggest that two separate reactions occur ; the mechanism involves one single, exergonic enzyme-catalysed reaction. This pattern is commonly encountered in metabolism where the hydrolysis of ATP furnishes the energy for endergonic synthesis.

REFERENCES

JOHNSON, M. J. (1960). Enzymic equilibria and thermodynamics. In *The Enzymes*, **3**, 407. Ed. P. D. Boyer, H. Lardy & K. Myrbäck.

LANDOLT-BÖRNSTEIN (1923-36). *Physikalisch-chemische Tabellen*, 5th ed. Berlin: Springer-Verlag.

PARKS, G. S, & HUFFMAN, H. M. (1932). *The Free Energies of Some Organic Compounds*. New York : Chemical Catalogue Company.

PROBLEMS

Note.—In all the problems in this chapter it may be assumed, unless stated to the contrary, that the activity coefficients are unity.

4.1. Calculate the equilibrium constant of the hydrolysis of DL-alanylglycine by a suitable peptidase at 37·5° C.

ΔG_f° values for the compounds involved, in their standard state in aqueous solution are, at 37·5° C, as follows : alanylglycine, $-$ 114,680 cal; alanine, $-$ 87,300 cal; glycine, $-$ 87,710 cal; water, $-$ 56,200 cal.

4.2. Calculate the standard free energy of formation of the hippurate ion if the equilibrium constant for the formation of hippurate from benzoate and glycine is 0·0142 at 37·5° C.

ΔG° values at 37·5° C are : benzoate ion, $-$ 49,710 cal; glycine, $-$ 87,710 cal for the standard states in aqueous solution, and for water, $-$ 56,200 cal.

4.3. Determine the standard free energy of formation of benzoylglycyl-glycine in aqueous solution at 25° C from the following data. The equilibrium constant of the formation of benzoylglycylglycine from benzoate ion and glycylglycine is 0·1564 at 25° C and, at the same temperature, the standard free energy change for the condensation of two glycine molecules to yield glycyl-glycine is 3530 calories. Standard free energies of formation of benzoate ion and glycine at 25° C are, respectively, $-$ 51,175 cal and $-$ 89,140 cal for the standard states in aqueous solution. ΔG_f° for water at 25° C is $-$ 56,690 cal.

4.4. Calculate the equilibrium constants at 38° C for the following reactions :

(a) oxaloacetate$^-$ + H_2O $\rightleftharpoons$ pyruvate$^-$ + HCO_3^-.

(b) fumarate$^=$ + $2H_2O$ $\rightleftharpoons$ lactate$^-$ + HCO_3^-.

(c) fumarate$^=$ + H_2O $\rightleftharpoons$ malate$^=$.

The values for the standard free energy of formation of these compounds at 38° C are, according to Borsook, as follows : pyruvate$^-$, $-$ 106,460 cal; lactate$^-$, $-$ 117,960 cal; HCO_3^- $-$ 139,200 cal; malate$^=$, $-$ 199,430 cal; H_2O, $-$ 56,200 cal; fumarate$^=$, $-$ 142,525 cal; oxaloacetate$^=$, $-$ 184,210 cal. These values apply to the standard state in aqueous solution.

(Data from EVANS, VENNESLAND & SLOTIN (1943), *J. biol. Chem.*, **147**, 771.)

4.5. Calculate the equilibrium constant at 25° C of the 'aldehyde mutase' reaction, which may be expressed as :

$$2CH_3CHO + H_2O \rightleftharpoons CH_3COOH + CH_3CH_2OH.$$

The standard free energies of formation of the respective reactants at 25° C are : acetaldehyde aq., $-$33,000 cal; acetic acid aq., $-$96,210 cal; ethanol aq., $-$43,850 cal and water (liquid), $-$56,560 cal.

What concentration of acetaldehyde will be in equilibrium with 0·1 M-acetic acid and 0·1 M-ethanol?

The assumption has been made above that the acid is not appreciably dissociated ; the reaction is usually carried out in a bicarbonate-CO_2 buffer system and may be represented as :

$$2CH_3CHO \text{ aq.} + HCO_3^- \rightleftharpoons CH_3COO^- + CH_3CH_2OH \text{ aq.} + CO_2(g).$$

This formulation takes account of the ionization and neutralization of the acid. ΔG_f° at 25° C for HCO_3^- is $-140,000$ cal; for CH_3COO^- is $-89,720$ cal; and for gaseous CO_2 is $-94,100$ cal.

Calculate the equilibrium constant of this reaction and the concentration of acetaldehyde in equilibrium with 1 M-acetate and ethanol if the bicarbonate concentration is 0·03 M and the partial pressure of CO_2 is 0·05 atmosphere.

(After DIXON (1939), *Ergeb. Enzymforsch.*, 8, 217.)

Note.—This reaction is now known to consist of two coupled reactions, alcohol dehydrogenase and aldehyde dehydrogenase, viz.

$$CH_3CHO + NADH_2 \rightleftharpoons CH_3CH_2OH + NAD$$
$$CH_3CHO + H_2O + NAD \rightleftharpoons CH_3COOH + NADH_2$$

$$\overline{2CH_3CHO + H_2O \rightleftharpoons CH_3CH_2OH + CH_3COOH}$$

(RACKER (1949), *J. biol. Chem.*, 177, 883.)

4.6. The enzyme phosphoglucomutase catalyses the interconversion of glucose 1-phosphate and glucose 6-phosphate. At 38° C, the equilibrium mixture contains 5·4 per cent. of the 1-phosphate. Calculate the equilibrium constant and the standard free energy change of the reaction.

4.7. The triosephosphate isomerase of muscle catalyses the reversible conversion of glyceraldehyde 3-phosphate to dihydroxyacetone phosphate. At 37° C the standard free energy change of the reaction is -2000 calories. Calculate the equilibrium concentrations of the two reactants.

4.8. The aspartate ammonia-lyase (aspartase) enzyme of *Escherichia coli* catalyses the reversible reaction :

$$\text{Fumarate}^- + NH_3 \rightleftharpoons \text{L-aspartate}^-.$$

Starting with fumarate and ammonia concentrations of 0·1 M, 28 per cent. of the original ammonia remained at equilibrium when the reaction was carried out at 37° C. The activity coefficients for the compounds involved are $f_{fum}.0·7$, $f_{asp}.0·7$ and $f_{NH_3}0·22$. Determine the standard free energy change for the reaction.

(Data from BORSOOK & HUFFMAN (1933), *J. biol. Chem.*, 99, 663.)

4.9. The following data were obtained in an experiment to investigate the effect of temperature on the equilibrium constant of the fumarate hydratase (fumarase) system which catalyses the reaction

$$\text{Fumarate}^- + H_2O \rightleftharpoons \text{Malate}^-.$$

Temperature °C.	15	20	25	30	35	40	45	50
Equilibrium constant, K	4·786	4·467	4·074	3·631	3·311	3·090	2·754	2·399

Determine graphically the heat change for the hydration of fumarate and calculate also the standard free energy change of the reaction at 25° C and 38° C.

(After SCOTT & POWELL (1948), *J. Amer. chem. Soc.*, 70, 1104.)

4.10. The malate dehydrogenase enzyme from horse heart catalyses the reversible reaction

$$\text{L-malate}^- + NAD^+ \rightleftharpoons \text{oxaloacetate}^- + NADH + H^+.$$

The equilibrium constant of this reaction was determined by following the change in extinction at 340 nm in the initial 10 minutes ; this enabled the concentration of reduced NAD to be obtained and that of oxaloacetate was assumed to be the same. The concentrations of L-malate and NAD^+ were obtained from the initial concentrations and those of the NADH formed. These equilibrium figures are recorded below for experiments carried out at three different pH values and at 25°.

	pH	NADH $(=oxaloacetate)$ $(10^5 \times M)$	NAD^+ $(10^5 \times M)$	L-Malate $(10^3 \times M)$
(a)	8·81	2·82	32·4	5·27
(b)	8·83	3·27	41·3	5·14
(c)	7·55	0·79	43·7	5·19

Calculate the equilibrium constant for each experiment and comment on the effect of pH on its value.

The activity coefficients may be assumed to be unity in every case. Determine the standard free energy change of the reaction.

(After BURTON & WILSON (1953), *Biochem. J.*, **54**, 86.)

4.11. The enzymatic synthesis of citric acid has been studied spectrophotometrically by coupling the condensation reaction (1) with the malate dehydrogenase system (2)

(1) Acetyl-CoA + oxaloacetate$^-$ + H_2O $\rightleftharpoons$ citrate^{3-} + CoA + H^+.

(2) L-malate$^-$ + NAD^+ $\rightleftharpoons$ oxaloacetate$^-$ + NADH + H^+.

(3) Net reaction : L-malate$^-$ + acetyl-CoA + NAD^+ + H_2O $\rightleftharpoons$ citrate^{3-} + CoA + NADH + $2H^+$.

In this way the reduction of NAD may be followed and the equilibrium constant of reaction (3) determined. This has been done by measuring the ratio of NAD to NADH at varying ratios of malate to citrate with constant initial concentrations of acetyl-CoA and constant concentrations of condensing enzyme (enzyme (1)) and malate dehydrogenase (enzyme (2)). The following data were obtained at pH 7·2 and 22° :

Equilibrium concentrations ($\times 10^3$M)					
Citrate	CoA	NADH	L-malate	Acetyl-CoA	NAD^+
16·7	0·028	0·028	0·139	0·057	0·132
16·7	0·057	0·057	1·80	0·028	0·103
16·7	0·065	0·065	3·44	0·020	0·095
116·7	0·039	0·039	3·46	0·046	0·121

Calculate the apparent equilibrium constant of reaction (3) from this information. The concentration of water may be taken as 55·5 M. The apparent equilibrium constant of reaction (2) at pH 7·2 and 22° has been found to be $2·33 \times 10^{-5}$. From this deduce the apparent equilibrium constant of the citrate synthesis reaction (reaction 1) at pH 7·2 and 22° and use this value to calculate the standard free energy change of the reaction.

(Data from STERN, OCHOA & LYNEN (1952), *J. biol. Chem.*, **198**, 313.)

4.12. The equilibrium constant of the 'malic' enzyme of wheat germ, which catalyses the reaction :

$$Pyruvate^- + CO_2 + NADPH \rightleftharpoons L\text{-}malate^- + NADP^+$$

has been investigated. The change in extinction due to NADP was followed spectrophotometrically at 340 nm at pH 7·3 and 22° C. The gas phase employed was nitrogen containing CO_2 to various concentrations and concentrations of L-malate, pyruvate and NADP in the solutions were determined enzymatically. The values of (H_2CO_3) below give the total concentration in solution of free CO_2 and may be used as such in the calculation of the equilibrium constants. The following data were obtained :

Initial concentration of reactants $10^3 \times$ moles/litre				Concentration of reactants at equilibrium, $10^3 \times$ moles/litre		
NADP+	L-malate	Pyruvate	(H_2CO_3)	NADPH	NADP+	L-malate
0·0530	0·657	32·6	1·8	0·0195	0·0335	0·637
0·0530	0·986	16·3	1·8	0·0343	0·0187	0·952
0·1060	0·657	32·6	1·8	0·0371	0·0689	0·620
0·0530	0·657	16·3	3·6	0·0178	0·0352	0·639
0·0530	1·32	16·3	7·2	0·0170	0·0360	1·303

From these data determine the apparent equilibrium constants of the individual reactions and obtain an average value for them.

Note.—Since the concentrations of pyruvate and CO_2 are large relative to those of NADP and L-malate, their variation in a given experiment may be neglected. Hence they may be assumed to be constant in the calculation of the individual equilibrium constants.

The equilibrium constant of the malate dehydrogenase reaction

$$\text{malate}^- + \text{NAD}^+ \rightleftharpoons \text{oxaloacetate}^- + \text{NADH}$$

has been determined and a value of $2·33 \times 10^{-5}$ obtained at pH 7·2 and 22° C. Assuming that the value of the equilibrium constants would be independent of the nicotinamide nucleotide participating in the reaction, calculate the apparent equilibrium constant for the carboxylation of pyruvate, i.e.

$$\text{pyruvate}^- + \text{CO}_2 \rightleftharpoons \text{oxaloacetate}^-.$$

(Data from HARARY, KOREY & OCHOA (1953), *J. biol. Chem.*, 203, 595.)

Note.—The authors of this work calculate the apparent equilibrium constants in accordance with the equations given above which omit the hydrogen ion term associated with the reduced form of the nicotinamide nucleotide (compare with Problem 4.11). As an additional exercise, calculate the equilibrium constants for the above reactions, taking into account the hydrogen ion term.

4.13. The synthetic action of intestinal alkaline phosphatase in the formation of glycerophosphate from glycerol and inorganic phosphate has been studied at 38° under appropriate conditions. The reaction mixtures were allowed to react for varying times and the final concentrations of glycerol and of ester formed were determined. It was found that practically all of the ester formed was of the α form. Data for four experiments carried out at pH 8·5 and for one at pH 5·8 are given below :

pH	Time hours	Initial molar concentration			Final molar concentration		
		Glycerol	Phosphate	H_2O	Glycerol	Total ester	α-ester per cent.
	168	7·10	0·405	27·0	6·99	0·110	98
	144	1·70	0·480	48·8	1·677	0·0226	99
8·5	700	11·19	0·0863	10·3	11·14	0·0552	87
	800	11·13	0·0863	10·3	11·13	0·0583	84
5·8	48	7·08	0·408	26·9	7·02	0·0610	99

Determine the equilibrium constant for the formation of total ester and the standard free energy change of the reaction at both pH values. The standard state of water should be taken as 55·5 M.

(After MEYERHOF & GREEN (1949), *J. biol. Chem.*, 178, 655.)

4.14. The synthesis of formic acid from hydrogen and carbon dioxide by a reversal of the formate hydrogenlyase enzyme of *Escherichia coli* has been investigated and the equilibrium constant of the reaction :

$$H_2 + HCO_3^- \rightleftharpoons H.COO^- + H_2O$$

determined at two different temperatures. Experiments were carried out in Warburg manometers at pH 7·4 using formate concentrations covering the range 0·1 − 0·0125 M, a bicarbonate concentration of 0·025 M and a gas phase of 5 per cent. CO_2 in hydrogen. Initial rates of H_2 uptake or evolution were measured in order that the concentration of bicarbonate should not decrease appreciably during the experiment.

Figures obtained for the gas exchanges are given in the following table (controls have been subtracted from all the figures given).

Formate concentration M	Hydrogen uptake or evolution μl/h	
	25·1° C	37·7° C
0·10	+200	+296
0·05	+ 72	+140
0·025	− 25	+ 25
0·0125	− 93	− 46

The partial pressure of hydrogen at equilibrium was 0·916 and 0·884 atmospheres respectively for 25·1° and 37·7° (these values are corrected for atmospheric pressure and water vapour pressure).

(a) Determine graphically the concentration of formic acid present at equilibrium for the two temperatures and then use these values to calculate the equilibrium constants. The activity of hydrogen may be taken as its partial pressure and the activity of water as its mole fraction (0·988 under the experimental conditions employed). The activity coefficients of HCO_3^- and H.COO— may be taken as both equal to 0·762.

(b) Calculate the free energy change of the reaction.
(c) Calculate the heat of reaction.

(After WOODS (1936), *Biochem. J.*, 30, 515.)

4.15. The equilibrium constants of the aconitate hydratase (aconitase) system have been determined at 25° and 38°, when the following equilibrium concentrations of tricarboxylic acids were obtained at pH 7·4.

Temperature	Equilibrium concentrations (per cent.)		
	Iso-*citric acid*	cis-*aconitic acid*	*citric acid*
25	6·20	2·90	90·90
38	6·60	4·30	89·10

Determine the equilibrium constants from these data.

(After KREBS (1953), *Biochem. J.*, 54, 78.)

4.16. The aspartate : 2-oxoglutarate aminotransferase system carries out the reaction

L-glutamate + oxaloacetate $\rightleftharpoons$ 2-oxoglutarate + L-aspartate

Krebs has determined the equilibrium concentrations of all reactants at 25°
and pH 7·4 in a series of experiments in which the reaction was allowed to
continue for 30, 40 and 50 minutes. Data from one such experiment are given
below. Use this information to comment on the attainment of equilibrium in
the time intervals investigated.

Period incubation (min)	Equilibrium concentrations (μl substrate/ml solution)			
	Oxaloacetate	Glutamate	2-Oxoglutarate	Aspartate
30	111	121	298	307
40	106	120	289	298
50	108	123	297	305

(After KREBS (1953), *Biochem. J.*, 54, 82.)

4.17.   The equilibrium constant of the alanine : 2-oxoglutarate aminotransferase system has been determined by Krebs and has a value of 1·50 at 25° and
pH 7·4.

i.e. L-glutamate + pyruvate $\rightleftharpoons$ 2-oxoglutarate + L-alanine ($K = 1\cdot50$)

If equal initial concentrations of L-glutamate and pyruvate of 430 μl substrate
per ml are reacted under these conditions determine the equilibrium concentrations of all reactants.

(After KREBS (1953), *Biochem. J.*, 54, 82.)

4.18.   The determination of the equilibrium constant of the beef heart
transhydrogenase reaction, which is

$$NAD^+ + NADPH \rightleftharpoons NADH + NADP^+$$

has been carried out and the following experimental data obtained when the
reaction was allowed to continue for 40 minutes at 37° C.

pH	Time (min)	Concentration ($\mu moles/ml$)			
		$NADP^+$	$NADH$	$NADPH$	NAD^+
6·5	0			0·88	1·58
	40	0·66	0·60	0·31	0·99
6·5	0	1·33	1·18		
	40	0·83	0·56	0·45	0·52
6·0	0			0·96	1·57
	40	0·63	0·55	0·25	1·04
6·0	0	1·35	0·74		
	40	0·90	0·36	0·42	0·48
7·5	0			0·92	1·08
	40	0·56	0·60	0·42	0·58
7·5	0	0·98	1·08		
	40	0·68	0·63	0·53	0·50

Compare the equilibrium constants for the reaction in both directions at the
various pH values and comment on the consistency of the values obtained.

(After KAPLAN, COLOWICK & NEUFELD (1953), *J. biol. Chem.*, 205, 1.)

4.19. (a) The calcium ions in human serum combine with the protein(s) in a way which has been expressed by the following equation :

$$Ca\ Prot \rightleftharpoons Ca^{2+} + Prot^-$$

The following analytical data are given by McLean and Hastings :

Concentrations expressed as m-moles/kg H_2O

Total Protein	Total Calcium	Free Ca^{2+}
9·69	3·07	1·35
7·10	2·27	1·15
4·70	1·49	0·90
2·31	0·74	0·55

Use the mass law equation to calculate the dissociation constant, K, for each of the four sets of data.

(b) Does the agreement of the calculated dissociation constants in part (a) establish the correctness of the postulated reaction? What restrictions do these data place on any alternative formulation of the reaction?

(c) Suppose some calcium chloride is added to serum and the following values obtained soon afterwards :

Total Protein . . .	9·36 m-moles/kg H_2O	
Total Calcium . . .	4·47	,,
Free Ca^{2+} . . .	2·51	,,

Does it appear that equilibrium has been reached? If not, what will be the final concentration of each species?

(d) If this final equilibrium mixture in part (c) is diluted with an equal volume of isotonic sodium chloride, will the resulting mixture be in equilibrium? If not, in which direction will the reaction proceed?

(e) The neutral salt calcium citrate, Ca_3Cit_2, has a primary dissociation which appears to go to completion :

$$Ca_3Cit_2 \rightleftharpoons Ca^{2+} + 2CaCit^-$$

The secondary dissociation

$$CaCit^- \rightleftharpoons Ca^{2+} + Cit^{3-}$$

is found to have a dissociation constant of $6·03 \times 10^{-4}$.

If 500 mg of anhydrous sodium citrate, Na_3Cit, are dissolved in 1 litre of the equilibrium mixture of part (c), what will be the final concentration of free calcium ions and calcium protein complex in the resulting solution? Neglect volume changes caused by the addition of the solid salt.

(Data from McLean & Hastings (1935), *J. biol. Chem.*, **108**, 285, and Hastings, McLean, Eichelberger, Hall & DaCosta (1934), *J. biol. Chem.*, **107**, 351. Harvard Medical Sciences 201 ab.)

4.20. Crystalline yeast alcohol dehydrogenase catalyses the reaction :

$$iso\text{propanol} + NAD^+ \rightleftharpoons \text{acetone} + NADH + H^+$$

The equilibrium constant at 25° was found to be $7·19 \times 10^{-9}$ M (Burton and Wilson).

Calculate the standard free energy change for the reaction at 25°.

From heat of combustion and ancillary data $\Delta G° = 5·89$ kcal for the reaction :

$$iso\text{propanol(aq.)} \rightleftharpoons \text{acetone(aq.)} + H_2(g)$$

Determine the standard free energy change for the hydrogenation of NAD^+, viz.

$$NAD^+ + H_2 \rightleftharpoons NADH + H^+$$

and also compute the standard electrode potential at pH 7 and 25° for this reaction.

(Glasgow Honours Course Finals, 1956.)

F

4.21. The following data were obtained by Smith and Gunsalus (1957) in a series of experiments to determine the equilibrium constant of the isocitrate glyoxylate lyase (isocitratase) enzyme of *Pseudomonas aeruginosa* which catalyses the reaction :

$$\text{Glyoxylate} + \text{succinate} \rightleftharpoons \text{isocitrate}$$

The values are given in millimoles per litre, temperature 27°, pH 7·6.

Additions, 0 min		*Found at equilibrium, 30 min*		
Succinic acid	*Glyoxylic acid*	*Succinic acid*	*Glyoxylic acid*	*Isocitric acid*
6·67	3·53	5·97	2·97	0·73
3·33	3·53	3·08	2·90	0·39
5·00	5·00	4·55	4·52	0·45
Isocitric acid				
4·57		3·94	4·07	0·67
2·68		2·40	2·40	0·26

Calculate an average value for the equilibrium constant and also determine the standard free energy change for isocitrate synthesis.

Compare the effect of concentration of reactants on isocitrate synthesis by considering the percentage of the substrate present as isocitrate when the initial concentrations of the reactants are : (a) 1 mM, (b) 1 M. How do you account for your findings ?

(Glasgow Honours Course Finals, 1958. After SMITH & GUNSALUS (1957), *J. biol. Chem.* **229**, 305.)

4.22. The equilibrium constant for the enzymic hydrolysis of a phosphate ester at 17° C was 32, and at 37° C was 50. Calculate the heat of reaction and free energy change at 37° C. What can you deduce from the fact that these quantities differ ?

$$(4 \cdot 575 \log_{10} x = R \log_e x).$$

(Leeds Honours Course Finals, 1952.)

4.23. The equilibrium constant of the galactokinase reaction:

$$(\alpha + \beta)\text{-D-galactose} + \text{ATP} \rightarrow \alpha\text{-D-galactose 1-phosphate} + \text{ADP}$$

was measured from the following data at pH 7·0 and 25° with 25 mM $MgCl_2$ (an excess of magnesium at the concentrations of nucleotide used).

Initial reactants	*Concentrations (mM) of products at equilibrium*			
	Galactose	*Galactose 1-phosphate*	*ATP*	*ADP*
Galactose and ATP	1·2	11·6	1·06	2·78
α-D-Galactose 1-phosphate and ADP	0·37	6·1	1·51	2·49

Under the same conditions of pH and temperature, and in the presence of magnesium, the equilibrium constant of the phosphatase reaction:

$(\alpha + \beta)$-D-glucose 6-phosphate + water→
$(\alpha + \beta)$-D-glucose + orthophosphate

was measured with purified intestinal alkaline phosphatase. The results were as follows:

Initial reactants	Concentrations (mM) of products at equilibrium			
	Ortho-phosphate	Glucose	Glucose 6-phosphate	Mg^{2+}
Glucose and orthophosphate	⌠248 ⌡246	214 239	0·200 0·225	2 5
Glucose, ortho-phosphate and glucose 6-phosphate . .	167	253	0·156	2

The equilibrium constant of the phosphoglucomutase reaction:

α-D-glucose 1-phosphate→$(\alpha + \beta)$-D-glucose 6-phosphate

was also measured under the same conditions using purified muscle phosphoglucomutase. The results obtained were as follows:

Initial reactants	Concentrations (mM) of products at equilibrium	
	$(\alpha + \beta)$-D-glucose 6-phosphate	α-D-glucose 1-phosphate
Glucose 6-phosphate	13·6	0·77
Glucose 1-phosphate	12·6	0·90

Use these data to compute the free energy of hydrolysis of ATP at pH 7·0 and 25° in the presence of excess Mg^{2+}, i.e.

ATP + water→ADP + orthophosphate.

Note :—It may be assumed that the free energy of hydrolysis of α-D-galactose 1-phosphate and of α-D-glucose 1-phosphate is independent of the orientation of the hydroxyl group at position 4 of the pyranohexose.

(Data from Atkinson, Johnson & Morton (1959), *Nature*, **184**, 1925.) (Glasgow Honours Course Finals, 1963.)

4.24. Explain the term 'free-energy'. What relevance has this concept to an understanding of biochemical reactions? Glucose 1-phosphate (0·01 M) was incubated at 38° C with the enzyme phosphoglucomutase. At equilibrium the concentration of glucose 1-phosphate had fallen to 0·00054 M. Calculate the standard free energy change for the reaction:

Glucose 1-phosphate ⇌ Glucose 6-phosphate.

[Absolute zero = −273° C ; gas constant = 1·99 cal per mole per degree; $\log_e x = 2·3 \log_{10} x$.]

(University of Wales, Cardiff, Honours Course Finals, 1964.)

4.25. The standard free energy changes of the reactions

$$\text{Glycerol} + \text{HPO}_4^{2-} \rightleftharpoons \text{Glyceraldehyde 3-phosphate}^{2-} + \text{H}_2\text{O} + \text{H}_2$$

and

$$\text{L-Glycerol 1-phosphate}^{2-} \rightleftharpoons \text{Dihydroxyacetone phosphate}^{2-} + \text{H}_2$$

are $+ 14\cdot72$ kcal and $+ 10\cdot24$ kcal respectively. If at pH 7 and 25° C the equilibrium constant of the reaction

$$\text{Dihydroxyacetone phosphate}^{2-} \rightleftharpoons \text{Glyceraldehyde 3-phosphate}^{2-}$$

is $0\cdot045$, what will be the equilibrium constants for the hydrolysis of (a) L-glycerol 1-phosphate^{2-}, and (b) the racemic equilibrium mixture DL-glycerol 1-phosphate^{2-}, (the gas constant $R = 1\cdot987$ cal per degree)?

(University of Wales, Cardiff, Honours Course Finals, 1966.)

CHAPTER V

REACTION KINETICS

THE study of the rate behaviour or kinetics of chemical and enzymic reactions furnishes a powerful method for the analysis of the properties and mechanisms of such processes, and is therefore an important approach for the investigator. The velocity of a chemical reaction is dependent upon a variety of factors such as the concentrations of the reacting molecules, the temperature and pressure, the presence or absence of catalysts, and the pH of the medium. The reaction velocity may be measured by following the rate of disappearance of a reactant or the rate of appearance of a product per unit time, although the latter technique is often preferable since experimentally it is usually easier to measure the appearance of small amounts of a newly formed product than a small decrease in concentration of a reactant. To be of value the time interval must be short and the measurement made in the initial stages of the reaction, i.e. the *initial rate*, otherwise considerable error may arise due to the alteration in reaction rate as a function of time, the result of continuously altering concentrations of reactants, inhibition by products, reversibility of reaction, or other factors. Usually the rate of a reaction is a function of the concentration of one or more of the substances present. This is conveniently expressed in terms of the *order of the reaction*.

The overall order of reaction is the experimentally determined number of atoms or molecules the concentrations of which determine the observed kinetics, and it is equal to the sum of the exponents of the concentration terms in the rate equation. Thus, for a reaction in which it is found experimentally that the rate of increase of concentration of a product Y (denoted by v_Y) is given by

$$v_Y \propto [A]^a [B]^b \ldots$$

then the overall order of reaction is equal to $(a+b+\ldots) = n$ and the reaction is described as being of order a with respect to A, of order b with respect to B, and so on. The foregoing equation can be written in the form

$$v_Y = k[A]^a [B]^b \ldots$$

where k is a constant, the *rate constant* or *specific reaction rate*.
The difference between the order of the reaction and the *mole-cularity* must be emphasized. The molecularity is the number of
atoms or molecules that take part in the actual molecular mechan-ism ; they are not necessarily the same. For instance, many first-order reactions involve more than one molecule in their mechanism
and therefore are not unimolecular. (Note that in the following
discussion concentrations are considered to be equivalent to the
activities.)

Zero order reactions occur when the rate is entirely independ-ent of the concentration of reacting substance, i.e.

$$v = k \qquad . \qquad . \qquad . \qquad . \quad (5.1)$$

Here n is zero. This may happen in photochemical reactions and
under certain circumstances in surface catalysis and in enzymic
reactions (see p. 190). A zero-order reaction may be recognized
by plotting the velocity of reaction as a function of the concentra-tion of reactant ; the velocity will be found to have a constant
value independent of concentration.

First order reactions take place at a rate proportional to the
concentration of only one reacting substance. Suppose a molecule
A breaks down to give a product or products

$$A \rightarrow X, \text{ etc.}$$

The velocity v may be measured by the rate of disappearance of A
or the appearance of X and is proportional to the concentration
of A. Thus

$$v = \frac{-d[A]}{dt} = \frac{d[X]}{dt} = k[A] \qquad . \qquad . \quad (5.2)$$

This may be written in the form $\frac{-d[A]}{[A]} = kdt$.

Integrating the expression and converting to $\log_{10}$

$$\log \frac{A_0}{A} = \frac{kt}{2 \cdot 303} \qquad . \qquad . \qquad . \quad (5.3)$$

where A_0 is the initial concentration and A the concentration
after time t. The rate constant k has the dimensions of reciprocal

time and is usually expressed as sec^{-1} or min^{-1}. Notice particularly that k is independent of the concentration units employed. For first order kinetics the plot of log A_0/A (or more generally log C, where C is the concentration of reactant) against t gives a straight line of slope $-k/2\cdot303$. An alternative notation is sometimes helpful. If the initial concentration of A, i.e. A_0, is represented by a, and after time t a concentration x of product has been formed, then the concentration of A remaining at that time will be $a-x$. Hence equation 5·3 may be written

$$\log \frac{a}{a-x} = \frac{kt}{2\cdot303} \qquad . \qquad . \qquad . \qquad (5.3a)$$

Many first order biochemical reactions are apparent or pseudo unimolecular reactions although the mechanism is more complex. For example, the rate of reaction may be equal to $k(A)(B)$ where substance A disappears in the reaction but substance B does not. B, therefore, is a catalyst, the concentration of which remains constant throughout the reaction. Many enzyme reactions come within this category. Since (B) is constant, the velocity becomes

$$V = k'(A), \text{ where } k' = k(B)$$

i.e. first order although the mechanism is bimolecular. The Michaelis-Menten treatment of enzyme kinetics (Chapter VI) reveals that at low substrate concentrations the reaction is first order with respect to substrate but, as the enzyme becomes saturated with substrate, this changes to zero order (see Fig. 6.1).

Some important first order reactions, which are of considerable interest in biochemistry, are radioactive decay processes (see Chapter XI). The wide application of radioactive elements as 'tracers' in biological experiments necessitates a thorough understanding of the decay processes by biochemists. The rate of decay is given by

$$-\frac{dn}{dt} = kn \qquad . \qquad . \qquad . \qquad . \qquad (5.4)$$

where n is the number of radioactive atoms and k is the rate constant, in this case usually called the *decay constant*.

Integrating, $\qquad n = n_0 e^{-kt}$

$$kt = \ln \frac{n_0}{n} = 2\cdot303 \log \frac{n_0}{n} . \qquad . \qquad . \qquad (5.5)$$

where n_0 is the number of atoms at time $t = 0$ and n the number of atoms after time t. (Note the resemblance to the expression for the exponential growth of bacteria (p. 315) where $n = n_0 e^{kt}$. Here the power kt is positive since the numbers are *increasing* with time.)

The *half-life period* for the decay process is the time after which the concentration of decomposing substance has decreased to half its original concentration, i.e. $n = \dfrac{n_0}{2}$. Hence the half-life period $t_{\frac{1}{2}}$ is given by the expression

$$t_{\frac{1}{2}} = \frac{\ln 2}{k} = \frac{2 \cdot 303 \log 2}{k} = \frac{0 \cdot 693}{k} \quad . \quad . \quad (5.6)$$

Compare this with the mean generation time (m.g.t.) of a bacterial culture (p. 315). The m.g.t. $= \dfrac{0 \cdot 693}{k}$, where k is the growth constant. The half-life period (or m.g.t.) can be obtained graphically by plotting the logarithm of the number of radioactive atoms (or number of bacteria) against the time and reading off the time interval corresponding to a change of $0 \cdot 301$ (i.e. log 2) in the ordinate. This is shown in Fig. 5.1.

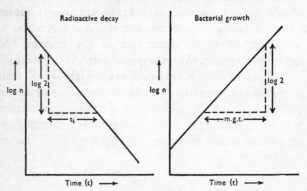

FIG. 5.1

Graphical determination of half-life period in a radioactive decay process and the mean generation time in bacterial growth.

Example 5.1.—A sample of radioactive iodine ^{130}I, giving 316 counts per minute in a Geiger–Müller tube, was assayed at intervals of time. From the following data obtained, determine the half-life period of this isotope.

Time (hours)	5	10	15	18·5	25
Counts/min	240	182	138	115	79

This problem may be solved either by calculation or graphically. Firstly, by calculation, we have the expression

$$2 \cdot 303 \log \frac{n_0}{n} = kt \qquad . \qquad . \qquad . \text{(equation 5.5)}$$

Here we can substitute for n_0 and n the given values for the counts and thus determine the decay constant k. At zero time $n_0 = 316$ and after 5 hours $n = 240$. Hence

$$k = \frac{2 \cdot 303}{5} \log \frac{316}{240} = \frac{2 \cdot 303 \times 0 \cdot 1195}{5}.$$

Furthermore, the half life $t_{\frac{1}{2}}$ is given by equation 5.6

$$t_{\frac{1}{2}} = \frac{0 \cdot 693}{k} = \frac{0 \cdot 693 \times 5}{2 \cdot 303 \times 0 \cdot 1195} = 12 \cdot 6 \text{ hours.}$$

Secondly, the graphical solution demands that the logarithm of the radioactive count be plotted against the time.

Time (hours)	0	5	10	15	18·5	25
Counts/min	316	240	182	138	115	79
Log (counts/min)	2·4997	2·3802	2·2601	2·1399	2·0607	1·8976

From the resulting graph, the time interval corresponding to a decrease of log 2 (0·3) is read off and this gives the half-life period, which is 12·5 hours by this method. See Fig. 5.2.

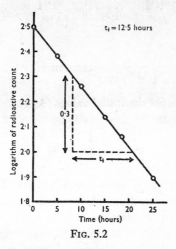

FIG. 5.2

Second order reactions occur at rates dependent on two concentration terms, i.e. $n = 2$. Thus reactions of the type

$$A + B \rightarrow X, Y, \text{ etc.}$$

are of second order and, in fact, this example illustrates the *simplest* type of second order reaction (which is also bimolecular). For this reaction the rate equation is

$$\frac{-d[A]}{dt} = \frac{-d[B]}{dt} = k[A][B] \qquad . \qquad . \quad (5.7)$$

Integration of this expression between limits gives

$$\frac{1}{[A_0] - [B_0]} \log \frac{[B_0][A]}{[A_0][B]} = \frac{kt}{2 \cdot 303} \qquad . \qquad . \quad (5.8)$$

where $[A_0]$ and $[B_0]$ are the initial concentrations of A and B at time $t = 0$ and $[A]$ and $[B]$ their concentrations after time t. The graphical evaluation of k is achieved by plotting log $[B_0][A]/[A_0][B]$ versus t when the slope is equal to $k([A_0] - [B_0])/2 \cdot 303$. For the special case where A and B are the same substance, i.e.

$$2A \rightarrow X, Y \text{ etc.}$$

or where $[A_0] = [B_0]$, the rate of reaction is

$$\frac{-d[A]}{dt} = k[A]^2 \text{ or } \frac{-d[A]}{[A]^2} = kdt \qquad . \qquad . \quad (5.9)$$

Integration gives

$$\frac{1}{[A]} = kt + \text{constant} \qquad . \qquad . \qquad . \quad (5.10)$$

and between limits for the initial values (A_0 at $t = 0$) and those at any time (A, t) we have

$$\frac{1}{[A]} - \frac{1}{[A_0]} = kt \qquad . \qquad . \qquad . \quad (5.11)$$

It should be noted that if the reaction between A and B is carried out with equal concentrations of A and B the rate equation (5.7) assumes the form of equation 5.10.

Using the alternative notation of representing the initial concentrations of A and B by a and b, and assuming that after time t an amount x has reacted, then the concentrations of A and B at time t will be respectively $(a - x)$ and $(b - x)$. The rate of reaction is thus

$$\frac{dx}{dt} = k(a - x)(b - x) \qquad . \qquad . \qquad . \quad (5.12)$$

Integrating and using the information that $x = 0$ at $t = 0$

$$\ln \frac{a(b-x)}{b(a-x)} = (b-a)kt \qquad . \qquad . \quad (5.13)$$

or
$$t = \frac{2 \cdot 303}{k(a-b)} \log \frac{b(a-x)}{a(b-x)} \qquad . \qquad . \quad (5.14)$$

A plot of $\log \dfrac{b(a-x)}{a(b-x)}$ versus t is thus linear for a second order reaction and the slope is $k(a-b)/2 \cdot 303$.

Again, for the special case where $a = b$, or where A and B are the same substance, the expression becomes

$$\frac{dx}{dt} = k(a-x)^2$$

and
$$\frac{x}{a(a-x)} = kt \qquad . \qquad . \qquad . \quad (5.15)$$

Under these particular circumstances the half-life of a second order reaction may be readily obtained from equation 5.15, for at $t_{\frac{1}{2}}$, $x = a/2$. Thus

$$t_{\frac{1}{2}} = \frac{1}{ka} \qquad . \qquad . \qquad . \qquad . \quad (5.16)$$

In contrast to a first order reaction, the half-life is therefore inversely proportional to the initial concentration of reactant.

It will be noted that the rate constant k for a second order reaction is not independent of the units in which the concentrations are expressed, but has the dimensions $L^3 M^{-1} t^{-1}$. It is often expressed in litres per mole per second.

From the foregoing discussion it will be apparent that the order of a reaction may be determined by appropriate graphical methods.

Higher Orders of Reaction.—Reactions of third or higher order can also occur, as, for example, in the denaturation of certain proteins. They can be treated in an analogous manner to first and second order reactions as already discussed, but the expressions obtained are naturally more complex.

A method of wide applicability for the study of reaction

velocities is the comparison of reciprocals of the times required to effect a given change. Provided that the initial concentration of reactant is kept constant, then irrespective of the order of the reaction

$$k = \frac{C}{t} \qquad \qquad \text{(5.17)}$$

where C is a term involving the concentration changes and, by definition, is kept constant. Accordingly under different experimental conditions

$$\frac{k_1}{k_2} = \frac{1}{t_1} \Big/ \frac{1}{t_2} = \frac{t_2}{t_1} \qquad \qquad \text{(5.18)}$$

where k_1 and k_2 are the rate constants under two different conditions and t_1 and t_2 are the times required to effect the same amount of chemical change. In this method no knowledge of the order of the reaction is required, but it is assumed that the order does not change from one set of conditions to the other.

Example 5.2.—The hydrolysis of sucrose to a mixture of glucose and fructose (termed the 'inversion' of sucrose) may be followed by means of a polarimeter. As sucrose is dextro-rotatory and the resulting mixture of glucose and fructose is laevo-rotatory, measurement of the angle of rotation of plane polarized light at various time intervals enables the reaction to be followed. The data given below were obtained by Lewis for the inversion of sucrose in 0·9 N-HCl at 25°.

Time (min)	0·0	7·18	18·00	27·05
Rotation (°)	+24·09	+21·405	+17·735	+15·00
Time (min)	36·80	56·07	101·70	∞
Rotation (°)	+12·40	+7·80	+0·30	−10·74

Determine the order of the reaction and obtain a value for the velocity constant.

The rotation at zero time, $\alpha_0 = 24·09°$, corresponds to pure sucrose and that at infinite time, $\alpha_\infty = -10·74°$, corresponds to the complete hydrolysis to glucose and fructose. Consequently the initial concentration of sucrose is proportional to $\alpha_0 - \alpha_\infty = +34·83°$ and the concentration of sucrose at any time t, when the rotation is α, is proportional to $\alpha - \alpha_\infty$. The reaction involved is:

$$C_{12}H_{22}O_{11} + H_2O \rightarrow C_6H_{12}O_6 + C_6H_{12}O_6$$

and we might expect, therefore, the reaction either to be of first or of second order. This can be decided by testing the experimental values in the respective equations for these orders or, alternatively, by plotting (a) log C against t for first order and (b) $\frac{1}{C}$ against t for second order to discover which gives a linear plot.

In either case the velocity constant may be obtained from the slope of the line. Here we shall illustrate this latter method.

Time (min)	Rotation (α)	Sucrose concn. C ($\alpha - \alpha_\infty$)	log C	$\frac{1}{C}$
0	$+24 \cdot 09$ (α_0)	$+34 \cdot 83$	$1 \cdot 5420$	$0 \cdot 0287$
$7 \cdot 18$	$21 \cdot 405$	$32 \cdot 145$	$1 \cdot 5072$	$0 \cdot 0311$
$18 \cdot 00$	$17 \cdot 735$	$28 \cdot 475$	$1 \cdot 4555$	$0 \cdot 0350$
$27 \cdot 05$	$15 \cdot 00$	$25 \cdot 74$	$1 \cdot 4106$	$0 \cdot 0389$
$36 \cdot 80$	$12 \cdot 40$	$23 \cdot 14$	$1 \cdot 3643$	$0 \cdot 0432$
$56 \cdot 07$	$7 \cdot 80$	$18 \cdot 54$	$1 \cdot 2681$	$0 \cdot 0539$
$101 \cdot 70$	$0 \cdot 30$	$11 \cdot 04$	$1 \cdot 0429$	$0 \cdot 0981$
∞	$-10 \cdot 74$ (α_∞)	—	—	—

From these figures the graphs are drawn (Fig. 5.3) and it will be seen that the reaction fits the first order equation. From the slope of the line the velocity constant is obtained and has a value of $0 \cdot 01137$ min^{-1}.

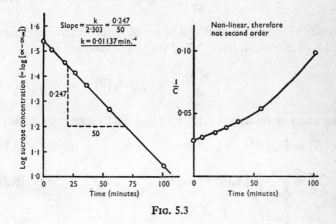

FIG. 5.3

Consecutive Reactions.—In many cases the reactions which occur are more complex than those so far considered. For example, it is quite common for the product of one reaction to participate immediately in another reaction, thus

$$A \longrightarrow B \longrightarrow C.$$

Such a sequence is referred to as *consecutive reactions*. Clearly, if the rate of conversion of B to C is very much greater than the rate of conversion of A to B, then the latter will be the rate-limiting step in the appearance of C and the kinetic treatment will be simplified. However, in many consecutive reactions this situation does not hold and we must now consider examples of this type.

If k_1 and k_2 are the appropriate rate constants then

$$A \xrightarrow{k_1} B \xrightarrow{k_2} C$$

and the following equations apply.

The rate of disappearance of A is given by

$$-\frac{d[A]}{dt} = k_1[A] \qquad \cdot \qquad \cdot \qquad \cdot \qquad (5.19)$$

and the rate of appearance of C by

$$\frac{d[C]}{dt} = k_2[B] \qquad \cdot \qquad \cdot \qquad \cdot \qquad (5.20)$$

The rate of formation of B will be equal to the difference between its rate of formation from A and its rate of conversion to C, i.e.

$$\frac{d[B]}{dt} = k_1[A] - k_2[B] \qquad \cdot \qquad \cdot \qquad (5.21)$$

Integration of equations 5.19, 5.20 and 5.21 yields respectively

$$[A] = [A_0]e^{-k_1 t} \qquad \cdot \qquad \cdot \qquad \cdot \qquad \cdot \qquad \cdot \qquad (5.19a)$$

$$[B] = \frac{k_1}{k_2 - k_1}[A_0](e^{-k_1 t} - e^{-k_2 t}) \qquad \cdot \qquad \cdot \qquad \cdot \qquad (5.21a)$$

$$[C] = [A_0]\left(1 - \frac{k_2}{k_2 - k_1}e^{-k_1 t} + \frac{k_1}{k_2 - k_1}e^{-k_2 t}\right) \qquad \cdot \qquad (5.20a)$$

where $[A_0]$ is the concentration of A at time zero and $[A]$, $[B]$ and $[C]$ are respectively the concentrations of A, B and C at time t.

In Fig. 5.4 are recorded the plots of equations 5.19a, 5.20a and 5.21a for consecutive reactions where $[A_0] = 1$, $k_1 = 0.1$ and $k_2 = 0.05$. Since $[B]$ must be zero at $t = 0$ and $t = \infty$, there must be a time when it has a maximum value. The concentration of C is zero at $t = 0$ and increases slowly in the initial stages, most rapidly when $[B]$ is at a maximum, and thereafter at a gradually decreasing rate until, finally, $[C]$ approaches the value $[A_0]$. The concentration of A declines steadily throughout, as shown in Fig. 5.4.

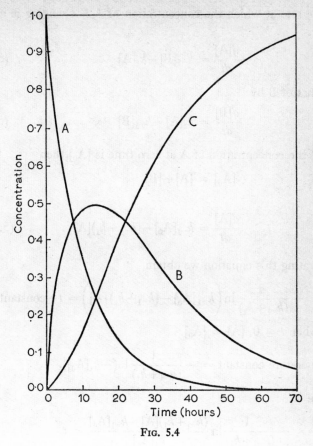

FIG. 5.4

Concentration progress curves for A, B and C in the consecutive first order

reactions A $\xrightarrow{k_1}$ B $\xrightarrow{k_2}$ C when $k_1 = 0\cdot1$ and $k_2 = 0\cdot05$.

Reversible Reactions.—When a reaction is reversible the kinetics of the process become more complicated. Consider a reaction in which both the forward and reverse reactions are of first order. The process may be represented as

$$A \underset{k_{-1}}{\overset{k}{\rightleftharpoons}} B$$

where k_1 and k_{-1} are the respective rate constants.

The rate at which the concentration of A is changing is given by

$$\frac{d[A]}{dt} = k_{-1}[B] - k_1[A] \qquad \cdot \qquad \cdot \qquad \cdot \quad (5.22)$$

and that of B by

$$\frac{d[B]}{dt} = k_1[A] - k_{-1}[B] \qquad \cdot \qquad \cdot \qquad \cdot \quad (5.23)$$

and if the concentration of A at zero time is $[A_0]$ then

$$[A_0] = [A] + [B]$$

and

$$\frac{d[A]}{dt} = k_{-1}[A_0] - (k_{-1} + k_1)[A] \; . \qquad \cdot \quad (5.22a)$$

Integrating this equation we obtain

$$-\frac{1}{(k_{-1} + k_1)} \ln \left(k_{-1}[A_0] - (k_{-1} + k_1)[A] \right) = t + \text{constant.}$$

But when $t = 0$, $[A] = [A_0]$

and therefore constant $= -\dfrac{1}{(k_{-1} + k_1)} \ln(-k_1[A_0])$.

Hence

$$-\frac{1}{(k_{-1} + k_1)} \ln \frac{(k_{-1} + k_1)[A] - k_{-1}[A_0]}{k_1[A_0]} = t$$

and $\quad [A] = \dfrac{k_1[A_0]e^{-(k_{-1} + k_1)t} + k_{-1}[A_0]}{k_{-1} + k_1}$

i.e. $\quad [A] = [A_0]\left[\dfrac{k_{-1} + k_1 e^{-(k_{-1} + k_1)t}}{k_{-1} + k_1}\right] \qquad \cdot \qquad \cdot \qquad \cdot \quad (5.24)$

Therefore

$$[B] = [A_0] - [A_0]\left[\frac{k_{-1} + k_1 e^{-(k_{-1} + k_1)t}}{k_{-1} + k_1}\right]$$

i.e. $\quad [B] = [A_0]\left[\dfrac{k_1 - k_1 e^{-(k_{-1} + k_1)t}}{k_{-1} + k_1}\right] \qquad \cdot \qquad \cdot \qquad \cdot \quad (5.25)$

Equations 5.24 and 5.25 thus give the concentrations of A and B respectively at time t. When $t = \infty$, these equations become

$$[A] = [A_0]\frac{k_{-1}}{k_{-1}+k_1} \qquad \cdot \qquad \cdot \qquad \cdot \qquad (5.26)$$

$$[B] = [A_0]\frac{k_1}{k_{-1}+k_1} \qquad \cdot \qquad \cdot \qquad \cdot \qquad (5.27)$$

Reversible Consecutive Reactions.—Consider now the case of consecutive reactions which are also reversible, of the type

$$\begin{array}{cc} k_{+1} & k_{+2} \\ A \rightleftharpoons B \rightleftharpoons C \\ k_{-1} & k_{-2} \end{array}$$

The following rate equations apply

$$\frac{d[A]}{dt} = k_{-1}[B] - k_{+1}[A] \qquad \cdot \qquad \cdot \qquad \cdot \qquad (5.28)$$

$$\frac{d[B]}{dt} = k_{+1}[A] + k_{-2}[C] - (k_{+2}+k_{-1})[B] \qquad \cdot \qquad (5.29)$$

$$\frac{d[C]}{dt} = k_{+2}[B] - k_{-2}[C] \qquad \cdot \qquad \cdot \qquad \cdot \qquad (5.30)$$

and $\qquad [A_0] = [A] + [B] + [C]$

where $[A_0]$ is the concentration of A at zero time.

Whence

$$\frac{d^2[A]}{dt^2} + \frac{d[A]}{dt}(k_{+1}+k_{+2}+k_{-1}+k_{-2}) + [A](k_{+1}k_{+2}+k_{-1}k_{-2}+k_{+1}k_{-2})$$
$$= k_{-1}k_{-2}[A_0]$$

$$\frac{d^2[B]}{dt^2} + \frac{d[B]}{dt}(k_{+1}+k_{+2}+k_{-1}+k_{-2}) + [B](k_{+1}k_{+2}+k_{-1}k_{-2}+k_{+1}k_{-2})$$
$$= k_{+1}k_{-2}[A_0]$$

and

$$\frac{d^2[C]}{dt^2} + \frac{d[C]}{dt}(k_{+1}+k_{+2}+k_{-1}+k_{-2}) + [C](k_{+1}k_{+2}+k_{-1}k_{-2}+k_{+1}k_{-2})$$
$$= k_{+1}k_{+2}[A_0]$$

Hence[1]

$$[A] = c_1 e^{m_1 t} + c_2 e^{m_2 t} + \frac{k_{-1}k_{-2}[A_0]}{k_{+1}k_{+2} + k_{-1}k_{-2} + k_{+1}k_{-2}} \quad . \quad . \quad (5.31)$$

$$[B] = c_3 e^{m_1 t} + c_4 e^{m_2 t} + \frac{k_{+1}k_{-2}[A_0]}{k_{+1}k_{+2} + k_{-1}k_{-2} + k_{+1}k_{-2}} \quad . \quad . \quad (5.32)$$

$$[C] = c_5 e^{m_1 t} + c_6 e^{m_2 t} + \frac{k_{+1}k_{+2}[A_0]}{k_{+1}k_{+2} + k_{-1}k_{-2} + k_{+1}k_{-2}} \quad . \quad . \quad (5.33)$$

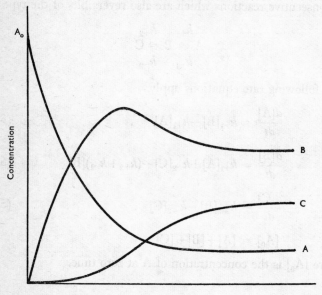

FIG. 5.5

One possible example of concentration progress curves for A, B and C in the reversible consecutive first order reactions $A \underset{k_{-1}}{\overset{k_{+1}}{\rightleftharpoons}} B \underset{k_{-2}}{\overset{k_{+2}}{\rightleftharpoons}} C.$

where $m_1 + m_2 = -(k_{+1} + k_{-1} + k_{+2} + k_{-2})$

and $m_1 m_2 = k_{+1}k_{+2} + k_{-1}k_{-2} + k_{+1}k_{-2}$

$c_1 m_1 + c_2 m_2 = -k_{+1}[A_0]$

[1] For the solution of differential equations see, for example, E. L. Ince (1956) *Integration of Ordinary Differential Equations*, 7th ed. Edinburgh: Oliver & Boyd.

$$c_1 + c_2 = [A_0] \left[\frac{k_{+1}k_{+2} + k_{-2}k_{+1}}{k_{+1}k_{+2} + k_{-1}k_{-2} + k_{+1}k_{-2}} \right]$$

$$c_3 m_1 + c_4 m_2 = k_{+1}[A_0]$$

$$c_3 + c_4 = -[A_0] \left[\frac{k_{+1}k_{-2}}{k_{+1}k_{+2} + k_{-1}k_{-2} + k_{+1}k_{-2}} \right]$$

$$c_5 m_1 + c_6 m_2 = 0$$

$$c_5 + c_6 = -[A_0] \left[\frac{k_{+1}k_{+2}}{k_{+1}k_{+2} + k_{-1}k_{-2} + k_{+1}k_{-2}} \right].$$

The relationships between [A], [B] and [C] may be expressed graphically and the shapes of the resulting concentration progress curves for A, B and C are dependent on the relative magnitudes of the four velocity constants; one example is shown in Fig. 5.5.

Effect of Temperature on Reaction Rates

The velocity of a chemical reaction is generally increased by an increase in temperature. The *temperature coefficient* of a reaction, commonly expressed as the Q_{10} value, is the ratio of the velocity at temperature $t + 10°$ to that at temperature $t°$. Q_{10} usually has a value in the region of 2. Enzymic reactions differ from most chemical ones because the temperature increment of velocity usually breaks down between 45 and 50° due to the thermal destruction or denaturation of the enzyme protein. Consequently the Q_{10} of an enzyme for a range above 40° will be generally much less than 2.

An increase of temperature may produce an increase in the overall velocity of an enzymic reaction in various ways. For example, the rate of combination of the enzyme with its substrate may be promoted, i.e. an increase in k_{+1} (see p. 186), or the rate of breakdown of the enzyme-substrate complex to the products of the reaction may be stimulated, i.e. an increase in k_{+2}. The affinity of the enzyme for a coenzyme or activator may be increased and the temperature might have an effect on the ionization and pH functions of some or all of the components of the reaction. Despite the apparent complexity of temperature effects they are nevertheless amenable to experimental analysis if suitable conditions are chosen. For full details of the investigation of temperature effects the reader is referred to Dixon and Webb (1964).

Arrhenius, in 1889, proposed the following relationship between reaction velocity and temperature (compare this with the van't Hoff equation, p. 135).

$$\frac{d \ln k}{dT} = \frac{E}{RT^2} \qquad . \qquad . \qquad . \quad (5.34)$$

where k is the rate constant, R the gas constant ($1 \cdot 987$ calories per degree per mole), T the absolute temperature and E is a constant known as the *activation energy*. Integration of the expression gives

$$\ln k = -\frac{E}{RT} + \text{constant (a)}$$

and

$$\log k = -\frac{E}{2 \cdot 303 RT} + \text{constant (b)} \qquad . \quad (5.35)$$

The value of the activation energy may be obtained from the plot of $\log k$ against $1/T$. The slope of the line is $\dfrac{-E}{2 \cdot 303 R}$ from which E can be evaluated. The determination of the energy of activation of enzymic reactions is usually carried out in this way by plotting the logarithm of the reaction velocity against $\dfrac{1}{T}$ and assuming that the velocity is directly proportional to the rate constant. It is important to note that activation energies obtained from enzyme velocities of reaction and the Arrhenius equation are meaningful only if the velocities are those secured when the enzyme is saturated with substrate, i.e. when $V = k_{+2}[E_0]$ (p. 187). These aspects of the effect of temperature on enzyme reactions are treated fully by Dixon and Webb (1964). The Arrhenius equation sometimes applies to complex biological systems taken as a whole, for instance bacterial growth, but the significance of such observations is doubtful and should not be held to imply relation to rate-determining or 'master' reactions in growth.

The Collision Theory of Reaction Rates

On the collision theory of chemical reactions only a certain proportion of the molecules which collide with one another react. Not every collision results in reaction, and only those molecules possessing sufficient energy will do so. The activation energy E is the 'extra' energy which is required for reaction to occur. The magnitude of this energy is dependent on the type of reaction occurring and therefore the fraction of activated molecules will vary from one reaction to another.

The probability that any molecule will have sufficient energy to react, i.e. energy of E per mole, at a temperature T, is given by the frequency factor $e^{-E/RT}$. By suitable choice of units of the frequency factor it can be shown that the specific reaction rate, k_r, is given by

$$k_r = Ze^{-E/RT} \qquad . \qquad . \qquad . \quad (5.36)$$

where Z is the total number of molecules colliding per second per unit volume. The above expression holds fairly well for some reactions but for others it does not, the number of effective collisions (i.e. those resulting in reaction) being other than would be expected from equation 5.36. To allow for this discrepancy an extra term was introduced into the equation, the so-called *probability* or *steric factor*, P, which may have any value from unity to zero. This factor is a measure of the probability of the energy that a molecule acquires being distributed in a manner favourable to reaction. Equation 5.36 now becomes

$$k = PZe^{-E/RT} \qquad . \qquad . \qquad . \quad (5.37)$$

In the case of protein denaturation, however, P may turn out to have an astronomical value. For example, in the denaturation of crystalline egg albumin, which has an activation energy of 140,000 calories per mole, P must be of the order of 10^{72}. The reason for this high value is discussed later (p. 171).

The Theory of Absolute Reaction Rates

An alternative approach to the collision treatment is the theory of absolute reaction rates developed by Eyring and his collaborators. This treatment is usually known as the *transition state theory*. Here again activation of the reacting molecules is postulated. It is considered that before molecules react they must pass through a configuration known as the transition state or activated complex which has an energy content greater than either the initial or final states and which are thus separated by an energy barrier. The height of this barrier is the energy of activation and

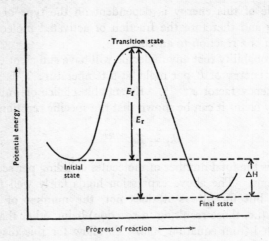

FIG. 5.6

Transition state theory picture of energy of activation.

Note that the value of ΔH shown in this diagram is an approximation: ΔE is the true value but, as explained on page 171, may often be equated with ΔH at biological temperatures.

is illustrated in Fig. 5.6. E_f is the activation energy of the forward reaction, i.e. initial to final state. E_r is the activation energy of the reverse reaction (final to initial) and is greater in value so that the reverse reaction will be much more difficult to accomplish. ΔH is the change in heat content for the reaction. In enzymic reactions the function of the enzyme is to facilitate the reaction by decreasing the energy of activation necessary for the reaction to occur (Fig. 5.7). Enzymic reactions have, as a consequence, lower temperature coefficients than corresponding non-enzymic ones.

The formation of the transition state can be treated in exactly the same manner as ordinary equilibria, but the symbol $\neq$ is used to denote functions associated with it. Thus, the equilibrium

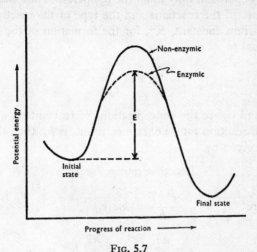

FIG. 5.7

Reduction of energy of activation effected by an enzyme.

constant for the reaction resulting in formation of the transition state, $K^{\neq}$, is related in the following way to the free energy change

$$\Delta G^{\circ \neq} = -RT \ln K^{\neq}$$

$$= \Delta H^{\circ \neq} - T \Delta S^{\circ \neq} \qquad . \qquad (5.38)$$

Consider the reaction :

$$A + B \rightarrow \text{Transition state} \rightarrow \text{Products.}$$

The rate of reaction is equal to the concentration of the activated complex at the top of the energy barrier multiplied by the frequency of crossing the barrier. An approach using partition functions, outside the scope of this book, is made to this problem ; suffice to say that the rate of reaction is equal to $\dfrac{c^{\neq}kT}{h}$, where k is the Boltzmann constant ($1 \cdot 37 \times 10^{-16}$ erg per degree), h is Planck's constant ($6 \cdot 62 \times 10^{-27}$ erg-second), T is the absolute temperature and $c^{\neq}$ the concentration of the transition state. The effective rate of crossing the energy barrier is $\dfrac{kT}{h}$, which is a universal frequency dependent only upon the temperature and independent of the nature of the reactants and the type of the reaction. Now the equilibrium constant, $K^{\ddagger}$, for the formation of the transition state is equal to

$$K^{\ddagger} = \frac{c^{\ddagger}}{c_A c_B} \qquad . \qquad . \qquad . \quad (5.39)$$

where c_A and c_B are the concentrations of reactants A and B. If the specific reaction rate, i.e. rate constant, is k_r, then the rate of reaction is given by

$$\text{Reaction rate} = k_r c_A c_B$$

Therefore

$$\frac{c^{\ddagger} kT}{h} = k_r c_A c_B$$

and

$$k_r = \frac{kT}{h} \cdot \frac{c^{\ddagger}}{c_A c_B} = \frac{kT}{h} K^{\ddagger} \qquad . \qquad . \quad (5.40)$$

Substituting this value of $K^{\ddagger}$ in the expression relating it to the standard free energy change we have

$$\Delta G^{\circ\ddagger} = -RT \ln \frac{k_r h}{kT} = \Delta H^{\circ\ddagger} - T\Delta S^{\circ\ddagger} \qquad . \quad (5.41)$$

Equation 5.41 may be written in exponential form as

$$k_r = \frac{kT}{h} e^{-\Delta G^{\circ\ddagger}/RT} = \frac{kT}{h} e^{\Delta S^{\circ\ddagger}/R} \, e^{-\Delta H^{\circ\ddagger}/RT} \qquad . \quad (5.42)$$

This shows clearly that the specific reaction rate is determined by the *free energy of activation* at a given temperature and emphasizes the fact that the energy or heat of activation is not the only factor to be considered ; what chances exist that the activation energy will be properly used has also to be known.

If it is assumed that $\Delta S^{\circ\ddagger}$ is independent of temperature then, by taking logarithms of equation 5.42 and differentiating, we obtain

$$\frac{d \ln k_r}{dT} = \frac{\Delta H^{\circ\ddagger}}{RT^2} + \frac{1}{T} = \frac{\Delta H^{\circ\ddagger} + RT}{RT^2} \qquad . \quad (5.43)$$

A comparison of equation 5.43 with the Arrhenius equation (5.34) reveals that

$$E = \Delta H^{\circ\ddagger} + RT \qquad . \qquad . \qquad . \qquad (5.44)$$

Thus the difference between the observed change in enthalpy and the activation energy is dependent on temperature. At physiological temperatures E and $\Delta H^{\circ\ddagger}$ are frequently equated, although the difference (RT) of approximately 600 calories is often significant.

If E and $\Delta H^{\circ\ddagger}$ are equated then

$$k_{\mathrm{r}} = \frac{kT}{h}e^{\Delta S^{\circ\ddagger}/R}\,e^{-E/RT} \qquad . \qquad . \qquad (5.45)$$

Comparing this with equation 5.37, we have

$$\frac{kT}{h}e^{\Delta S^{\circ\ddagger}/R} = PZ \qquad . \qquad . \qquad . \qquad (5.46)$$

and since at a given temperature $\dfrac{kT}{h}$ is constant, and the value of Z does not vary greatly for reactions of similar type, it follows that wide variations in the empirically derived probability factor P are related to large changes in entropy involved in the activation reaction.

As we have already seen (p. 112), the entropy change is a measure of the probability of a reaction occurring and the smaller the decrease in entropy involved the faster will the reaction take place. Thus if the formation of the transition state does not demand a decrease in entropy, the reaction may be considered as a very probable one which will proceed at a rapid rate. Herein lies the clue to the very high values of P encountered in protein denaturation. The highly organized protein structure is readily disorganized and this involves very large increases in entropy, both for the activation and for the overall step.

The values of P may be roughly correlated with the entropy changes as follows:

If $\qquad P = 1 \quad \Delta S^{\circ\ddagger} = 0$,

$\qquad\qquad P < 1 \quad \Delta S^{\circ\ddagger}$ is negative and the reaction is 'slow',

$\qquad\qquad P > 1 \quad \Delta S^{\circ\ddagger}$ is large and positive and the reaction is 'fast'.

It will be appreciated, therefore, that while a high value for the activation energy E renders a reaction more difficult to accomplish, a high value of ΔS has the opposite effect ; the relative magnitudes of E ($\simeq \Delta H^{\circ\ddagger}$) and $\Delta S^{\circ\ddagger}$ in equation 5.38 influence the value of

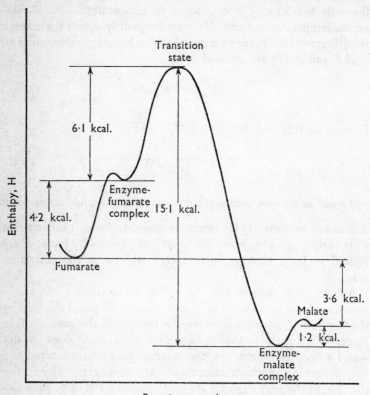

Reaction co-ordinate

FIG. 5.8

Changes in enthalpy for the fumarate hydratase reaction at pH 6·35. (After MASSEY, 1953.)

$\Delta G^{\circ\ddagger}$ which, of course, is the term determining the rate of reaction.

As we shall see in the next chapter, the kinetics of enzyme action can be satisfactorily explained on the basis of the formation of an enzyme-substrate complex. However, this complex is not identical with the transition state and the enzyme and substrate must be activated and pass through a transition state before the enzyme-substrate complex is produced. This compound must then pass through the configuration of another transition state before the products of the reaction are formed. Further, if the products should remain bound to the enzyme and require activation in order to dissociate, then a second transition state would lie between the enzyme-substrate complex and the products; this will certainly be the case for a reversible reaction. Such a situation would correspond to

$$E + S \rightleftharpoons E—S^+ \rightleftharpoons E—S \rightleftharpoons E—X^+ \rightleftharpoons E—P \rightleftharpoons E—P^+ \rightleftharpoons E + P$$

Transition state 1	Enzyme-substrate complex	Transition state 2	Enzyme-product complex	Transition state 3

The activation energy of the enzyme fumarate hydratase, which catalyses the reaction

$$\text{fumarate} + H_2O \rightleftharpoons \text{malate}$$

has been determined by Massey (1953) from measurements of the changes in the initial velocity of reaction in response to temperature. Changes in enthalpy for the forward and reverse reactions were calculated from the activation energies by use of equation 5.44 and these relationships are shown in Fig. 5.8. It will be observed that, with either malate or fumarate as the substrate, the initial step is the formation of an enzyme-substrate complex, presumably via a transition state. This is followed by activation of the complex to a second transition state, corresponding to the highest peak of the diagram, and formation of an enzyme-product complex which is then activated to promote dissociation into enzyme and products.

Example 5.3.—Write down an expression for the effect of change in temperature on the equilibrium constant of a reversible reaction.

Trypsin is reversibly denatured by heating. At 44° C the equilibrium constant for the transformation of native to denatured trypsin is 1·00 and at 50° C it is 7·20. Calculate the changes occurring in heat content and entropy at 44° C.

(Leeds Honours Course Finals, 1951.)

The required expression is $\quad \dfrac{d\ln K}{dT} = \dfrac{\varDelta H}{RT^2}$.

To calculate the change in heat content, this is required in its integrated form

$$\ln\frac{K_2}{K_1} = \frac{\varDelta H}{R}\left(\frac{T_2 - T_1}{T_1 T_2}\right).$$

Substituting the values of K at 317° and 323° absolute (44° and 50° C) and R (1·987 calories per degree per mole)

$$\Delta H = \frac{2 \cdot 303 \times 1 \cdot 987 \times \log 7 \cdot 2 \times 323 \times 317}{(323 - 317)}$$

$$= \frac{2 \cdot 303 \times 1 \cdot 987 \times 0 \cdot 8573 \times 323 \times 317}{6}$$

$$= 67,150 \text{ calories.}$$

Taking the same activity coefficient for native and denatured protein, the relation between equilibrium constant and free energy change becomes $\Delta G = -RT \ln K$ and ΔG is related to the change in heat content and entropy by the expression $\Delta G = \Delta H - T\Delta S$. But at 317° absolute $K = 1$, therefore $\ln K = 0$ and $\Delta G = 0$.

Hence $\Delta H = T\Delta S$.

Therefore $\Delta S = \dfrac{67150}{317} = \underline{211 \cdot 8 \text{ cal deg}^{-1} \text{mole}^{-1}}$

Since 1 cal = 4·185 joules.
$\Delta S = 886 \cdot 5 \text{ J deg}^{-1} \text{ mole}^{-1}$.

REFERENCES AND SUGGESTED READING

BRAY, H. G. & WHITE, K. (1967). *Kinetics and Thermodynamics in Biochemistry*, 2nd ed., Chaps 5 and 6. London : Churchill.

DAWES, E. A. (1964). Enzyme kinetics (optimum pH, temperature and activation energy). In *Comprehensive Biochemistry*, **12**, 89, ed. M. Florkin & E. H. Stotz. Amsterdam : Elsevier.

DIXON, M. & WEBB, E. C. (1964). *Enzymes*, 2nd ed., Chap. IV. London : Longmans Green.

GLASSTONE, S. (1962). *Textbook of Physical Chemistry*, 2nd ed., Chap. XIII. London : Macmillan and Co.

GLASSTONE, S., LAIDLER, K. J. & EYRING, H. (1941). *The Theory of Rate Processes*. New York : McGraw-Hill.

HINSHELWOOD, C. N. (1945). *Kinetics of Chemical Change*. London : Oxford U.P.

JOHNSON, F. H., EYRING, H. & POLISSAR, M. J. (1954). *The Kinetic Basis of Molecular Biology*. New York : Wiley.

KUSTIN, K. (1969). *Fast Reactions. Methods in Enzymology*, **16**. New York and London : Academic Press.

LAIDLER, K. J. (1965). *Chemical Kinetics*, 2nd ed. New York : McGraw-Hill.

MASSEY, V. (1953). Studies on fumarase. *Biochem. J.*, **53**, 72.

PROBLEMS

5.1. A sample of radioactive iron, ^{59}Fe, was assayed in a Geiger-Müller tube at intervals of time and the following data obtained :

Time (days)	Radioactivity (counts per minute)
0	3981
2	3864
6	3648
10	3437
14	3238
20	2965

Determine the half-life period and evaluate the decay constant.

5.2. The kinetics of the aerobic oxidation of enzymatically reduced nicotinamide-adenine dinucleotide ($NADH_2$) have been investigated at pH 7·38 and 30° C. The reaction rate was followed spectrophotometrically by measuring the decrease in extinction at 340 nm over a period of 30 minutes. The catalysed reaction may be represented as

$$NADH_2 + riboflavin \rightarrow NAD + riboflavin\ H_2\ (leuco\text{-}riboflavin)$$

Time (minutes)	Extinction at 340 nm
1	0·347
2	0·339
5	0·327
9	0·302
16·5	0·275
23	0·254
27	0·239
30	0·229

Determine the rate constant and the order of the reaction.

(After SINGER & KEARNEY (1950), *J. biol. Chem.*, **183**, 409.)

5.3. The effect of substrate concentration on the initial rate of the fumarate hydratase reaction, which catalyses

$$fumarate^- + H_2O \rightleftharpoons malate^-$$

has been investigated in both directions at pH 7·29. The experimental figures obtained are recorded below.

Substrate concentration M	Initial rate × 10⁷, mole/sec. Fumarate	Malate
0·40	5·0	2·8
0·20	6·0	3·2
0·10	5·9	3·2
0·05	5·6	4·0
0·025	5·4	2·5
0·0125	6·0	3·0

Determine the order of the reaction in the initial stages. What explanation can you offer for your findings?

(After SCOTT & POWELL (1948), *J. Amer. chem. Soc.*, **70**, 1104.)

5.4. The relationship between the initial rate of reaction and temperature for the fumarate hydratase system (see Problem 5.3) has been investigated in both directions and the following data were obtained at pH 7·29.

Initial rates $\times 10^7$ mole/sec		Temperature
Fumarate	Malate	° C
15·85	9·550	50
14·13	5·623	45
10·00	4·467	40
8·913	3·548	35
6·310	2·818	30
5·012	1·778	25
3·981	1·259	20
3·162	0·8511	15
2·818	0·6310	10

Calculate the apparent activation energies when the substrate is (a) fumarate and (b) malate.

(After SCOTT & POWELL (1948), *J. Amer. chem. Soc.*, 70, 1104.)

5.5.   In an experiment to obtain an estimate of the heat of activation of the breakdown of tryptophan to indole by the tryptophanase enzyme of *Escherichia coli*, indole production was measured at various intervals of time at four different temperatures.   Results obtained are recorded below.

Time (min)	μg Indole produced at temperature (°C)			
	0	14·5	27	37
5			7·0	25·0
10		3·7	15·3	27·0
15			24·5	28·6
20		6·4		30·8
25			25·4	
30	2·1	12·8	26·0	
40		16·0		
50		17·0		
60	4·5	17·0		

Use these data to obtain a value for the heat of activation of the tryptophanase reaction.

(After HAPPOLD & MORRISON, quoted by HAPPOLD (1950), *Advances in Enzymol.*, 10, 66.)

5.6.   Heat denaturation of the proteolytic enzyme trypsin is a reversible process and may be followed by measuring the loss of enzymic activity which occurs when the active native trypsin is denatured to the inactive form.   The reaction may be represented

Trypsin (native) ⇌ Trypsin (denatured)
active   inactive

The effect of temperature on the equilibrium of the reaction was investigated by Anson & Mirsky and some of their results are given below.

Temperature °C	Percentage denaturation
42	32·8
43	39·2
44	50·0
45	57·4
48	80·4
50	87·8

Use this information to determine the heat of reaction of the denaturation process and also obtain a value for the change in entropy which occurs. Comment on the values obtained for ΔH and ΔS.

(After ANSON & MIRSKY (1933-34), *J. gen. Physiol.*, **17**, 393. Glasgow Honours Course Finals, 1954.)

5.7. The energy of activation of the hydrolysis of sodium β-glycerophosphate by bone phosphatase has been determined by Bodansky using both cat and human bone phosphatases.

The phosphatase preparations were incubated at various temperatures with sodium β-glycerophosphate at the optimal pH (9·0 − 9·2) in a suitable buffer solution in the presence of glycine and magnesium ions. The conditions were such that the reaction velocity was directly proportional to the concentration of the enzyme. The reaction velocity was determined by measuring the amount of phosphorus liberated as inorganic phosphate per ml of hydrolysis mixture during the portion of the reaction which is of zero order. Results obtained are given in the following table :

Temperature	Phosphorus liberated as phosphate per ml hydrolysis mixture per minute	Number of experiments performed
Action of cat bone phosphatase		
°C	mg	
12·00	0·000425	2
17·30	0·000625	2
20·90	0·000819	3
27·05	0·00111	1
27·10	0·00117	1
31·70	0·00147	1
32·00	0·00142	1
37·50	0·00195	3
42·40	0·00243	2
Action of human bone phosphatase		
20·00	0·000664	2
25·00	0·000863	2
30·00	0·00103	2
35·00	0·00150	2
40·00	0·00187	2

Use these data to obtain a value for the energy of activation of the hydrolytic reaction.

(After BODANSKY (1939), *J. biol. Chem.*, **129**, 197.)

5.8. The Michaelis constant of the citrate dehydrogenase of cucumber seeds has been determined at two different temperatures. Dann obtained the following values in five determinations at each temperature :

25° C.	142	281	137	130	137	millimolar
35° C.	65	71	70	115	78	millimolar

Calculate the heat of formation of the enzyme-substrate complex from this information.

(After DANN (1931), *Biochem. J.*, 25, 177.)

5.9. The tropine esterase activity of rabbit serum has been investigated by measuring the acid produced when atropine sulphate is hydrolysed by the enzyme. Continuous titration in buffer-free medium with a glass electrode enabled the reaction to be expressed in terms of ml 0.02 N-NaOH required to neutralize the liberated acid. The following data show tropine esterase activity as a function of temperature ; the figures refer to 30 minutes' action by 2.5 per cent. rabbit serum on 0.25 per cent. atropine sulphate, which provides excess substrate, at pH 8.4.

ml. 0.02 N-NaOH per 30 min		*Temperature °C*
Enzyme	*Control*	
0·09	0·05	20
0·14	0·06	28
0·175	0·07	32
0·24	0·08	36
0·255	0·09	38
0·24	0·10	40
0·19	0·125	44·5

Calculate the activation energy of the tropine esterase reaction.

(After GLICK (1940), *J. biol. Chem.*, 134, 617.)

5.10. Kiese determined the effect of temperature on the Michaelis constant of carbonic anhydrase which catalyses the reaction

$$H_2O + CO_2 \rightleftharpoons H_2CO_3.$$

He obtained the following values in 0.04 M-phosphate buffer at pH 7.4

Temperature °C	$10^3 \times K_m$
1·0	1·2
5·0	2·3
8·0	2·9
12·5	5·2

Use this information to determine the heat of formation of the enzyme-substrate complex.

(After KIESE (1941), *Biochem. Z.*, 307, 400.)

5.11. The reaction velocity of the citrate dehydrogenase of cucumber seeds has been measured by the Thunberg tube technique. The citrate concentration used was large enough to ensure complete saturation of the enzyme and measurements were made at 25° and 35° with the same enzyme preparation. The ratio of the rate constants k_{35}/k_{25} in three different experiments gave values of 1.61, 1.70 and 1.64. Determine the heat of activation of the citrate dehydrogenase-citrate complex.

(After DANN (1931), *Biochem. J.*, 25, 177.)

5.12. The action of 0·2 per cent. phenol in killing *B. typhosus* has been investigated by means of viable counts of the organisms at various intervals of time. The following table records the mean viable count of triplicate plates at time intervals following the addition of phenol :

Time (min)	Mean number of micro-organisms surviving
0	20400
2	18000
4	11600
6	8000
8	6400
10	5200
15	2800
20	1500
25	750
30	400
35	250
40	120
45	64

Determine whether the killing process displays well defined kinetics, and, if so its characteristics.

(After LEE & GILBERT (1917-18), *J. phys. Chem.*, 22, 348.)

5.13. The following data were obtained in studies on the rates of spontaneous (non-enzymic) decarboxylation of oxaloacetic and oxalosuccinic acids.

Each manometric flask contained 1·5 ml of 0·3 M-acetate buffer, pH 5·1, made up with keto acid and water to a volume of 2·5 ml. Oxaloacetic acid (19 μmoles) and oxalosuccinic acid (16 μmoles) were tipped into the main compartment after temperature equilibration at 25°. The gas phase was air.

Oxaloacetic acid		Oxalosuccinic acid	
Time (min)	CO_2 evolved (μl)	Time (min)	CO_2 evolved (μl)
15	16	5	47
30	40	10	89
45	60	15	122
60	74	20	150
120	142	25	170
		30	189

Determine the order of the reaction and the velocity constants for the spontaneous decarboxylation of each keto acid.

(After OCHOA (1948), *J. biol. Chem.*, 174, 115.)

5.14. Myosin catalyses the reaction

$$ATP \rightarrow ADP + \text{inorganic phosphate.}$$

The kinetics of the thermal deactivation of myosin have been studied by incubation of identical solutions of myosin at three different temperatures, followed by addition of ATP solution (0·002 M) and determination of the rates of production of inorganic phosphate. These rates, called activities, are recorded below for experiments conducted at pH 7·0 with an enzyme concentration of 0·09 g/l.

G

Temperature °C	Time sec	Activity μmole phosphate/sec
35·2	90	0·873
	320	0·776
	776	0·600
	1400	0·415
	2182	0·272
30·2	70	0·685
	1500	0·600
	2610	0·536
25·0	1000	0·453
	74000	0·175

Determine the heat of activation of the thermal deactivation process.

(After OUELLET, LAIDLER & MORALES (1952), *Arch. Biochem. Biophys.*, **39**, 37.)

5.15. A study of the non-enzymic hydrolysis of adenosine triphosphate (ATP) is of interest for comparison with the enzymic reaction apparently occurring during the functioning of a muscle. The following data on the catalysed hydrolysis of ATP are taken from Friess.
The reaction

$$ATP + H_2O \xrightarrow{\ H^+\ } ADP + H_3PO_4$$

was studied at pH 1·33 in the presence of 0·300 M sodium chloride. At 50·24° C the following values were obtained :

ATP (moles/litre)	Time (sec)
0·0198	0
0·0188	3000
0·0183	5600
0·0166	12300
0·0161	15000
0·0156	17000
0·0150	19200
0·0145	21400

Does the reaction obey first order kinetics with respect to ATP ? What is the rate constant ? Is the reaction monomolecular or bimolecular ? How might you test this experimentally ? With $(ATP)_0 = 0.0198$ M, pH = 1·33, NaCl = 0·300 M the following rate constants were calculated :

Temp. °C	$10^6 \times k_1$ (sec^{-1})
39·94	4·67
43·82	7·22
47·06	10·0
50·24	13·9

Plot these data according to the Arrhenius equation. What is the activation energy ?

Calculate $\Delta G^{\ddagger}$, $\Delta H^{\ddagger}$ and $\Delta S^{\ddagger}$ at the reference temperature of 40·0° C.

(After FRIESS (1953), *J. Amer. chem. Soc.*, **75**, 323. Harvard Medical Sciences 201 ab.)

5.16. Snyder and Snyder give the following observations for the rate of flashing of fireflies as a function of temperature :

Temperature of air °C	Rate of flashings per minute (k)
28·3	15·0
28·8	15·4
26·0	12·6
22·6	10·0
22·3	9·9
23·2	11·1
24·1	11·5
26·5	12·1
19·4	8·1

Calculate the Arrhenius activation energy, E, for the process.

Taking k at 19·4° as the reference standard, use the calculated value of E to compute the value of k at 26·0°, and compare the computed and observed values of k.

(After SNYDER & SNYDER (1920), *Amer. J. Physiol.*, **51**, 536. Harvard Medical Sciences, 201 ab.)

5.17. The thermal denaturation of lobster haemocyanin has been studied by Fleisher using the following technique :

Solutions of haemocyanin were sealed in glass tubes, heated for various time intervals at four different temperatures and then immediately cooled by immersion in ice-water. Under these conditions the denatured protein precipitated as an amorphous white solid which was separated from the remaining soluble protein by centrifugation. The concentration of the soluble protein remaining in solution was determined by means of the specific refractive increment.

The following data were obtained at pH 7·40

Temperature used for denaturation °C	Time (min)	Specific refractive increment
65·6	0	0·0149
	2	0·0136
	4	0·0120
	6	0·0107
	8	0·0096
	10	0·0091
69·6	0	0·0151
	1	0·0116
	2	0·0097
	3	0·0075
	4	0·0061
	5	0·0041
71·5	0	0·0151
	1	0·0080
	2	0·0049
	3	0·0034
	4	0·0023
	5	0·0011
73·6	0	0·0148
	0·5	0·0060
	1·0	0·0027
	1·5	0·0010

Determine the order of reaction for denaturation and the specific reaction rate (rate constant) at each temperature. Evaluate the apparent activation energy for the denaturation process.

(Glasgow Honours Course Finals, 1957. After FLEISHER (1957), *Biochim. Biophys. Acta*, **23**, 28.)

5.18. The reversible formation of oxyhaemoglobin

$$\text{Hb} + \text{O}_2 \rightleftharpoons \text{HbO}_2$$

has been studied by the 'rapid flow' technique.

(*a*) Formulate a possible kinetic equation for this reversible reaction. Give the expression for the equilibrium constant and show its relation to the two velocity constants.

(*b*) In studying the dissociation of oxyhaemoglobin sodium dithionite is used to remove the free oxygen from solution. This effectively eliminates the combination reaction and permits the study of the rate of dissociation. The following data were obtained at 20° C and pH 7·4.

Time (milliseconds)	Oxyhaemoglobin concentration (mM) (as concentration of combined oxygen)
0	0·100
9·2	0·072
13·2	0·062
20·4	0·050
34·4	0·035
78·0	0·021

Write down the equation for the reaction and give a possible kinetic equation.

Does the reaction appear to be first order ? Does this order imply a unimolecular reaction ?

If the reaction is unimolecular will the measured velocity constant depend on the percentage of oxyhaemoglobin initially present in the reaction mixture ? Will the rate of change of oxyhaemoglobin concentration depend on the total haemoglobin concentration ?

The assumption has been tacitly made that the sodium dithionite combines with the free oxygen in the solution and that it does not remove the oxygen directly from the oxyhaemoglobin. How would you test this hypothesis and what results would you expect in the two possible cases ?

(Glasgow Honours Course Finals, 1961.)

5.19. 'The denaturation of proteins may in some circumstances be reversible.' Discuss the implications of this statement.

The half-life of thermal denaturation of haemoglobin in dilute electrolyte at 60° C was 57·7 minutes and at 65° C was 8·80 minutes. Calculate the heat and entropy of activation for this reaction.

$$(R = 1·98 \text{ cal deg}^{-1} \text{ mol}^{-1}; \ kT/h = 0·69 \times 10^{13} \text{ sec}^{-1})$$

(Special Degree of B.Sc., Biological Chemistry, University of Bristol, 1959.)

5.20. The velocities of hydrolysis of carbobenzoxyglycyl-L-phenylalanine by chymotrypsin in aqueous solution and in the presence of excess substrate are given at different temperatures in the table below. Calculate the activation energy of the enzyme-substrate complex.

Temperature (°C)	V_{max} (arbitrary units/μg enzyme)
25	125
32	173
34·5	196
37	219

$$(R = 1·99 \text{ calories/mole/°K} ; \log_e 10 = 2·303.)$$

(Leeds Honours Course Finals, 1959. *Note:* this was only part of the question set ; see also Problem 7.14).

5.21. Derive equations describing concentration versus time curves for each component of the system :

$$(a) \quad A \underset{k_2}{\overset{k_1}{\rightleftharpoons}} B \qquad (b) \quad A \xrightarrow{k_1} B \xrightarrow{k_2} C$$

for the case where the initial concentrations of A, B and C are a_0, 0 and 0 respectively.

5.22. The effects of temperature on the rates of decomposition of hydrogen peroxide in the presence of Fe^{2+} and of catalase respectively are given by the following table :

Fe²⁺		Catalase	
Temp. (°C)	k (l. mole⁻¹ sec⁻¹)	Temp. (°C)	k (l. mole⁻¹ sec⁻¹ × 10⁷)
34·8	110	25·5	3·50
34·8	109	23·5	3·50
25·0	60·0	22·5	3·54
25·0	61·5	7·2	3·11
12·3	29·5	3·0	2·86
12·3	31·5	2·0	2·84

From these data calculate the activation energies of the Fe^{2+}-catalysed and catalase-catalysed decompositions of hydrogen peroxide. Comment on the results of these experiments.

(University of Newcastle upon Tyne, Biochemistry Honours Course, Part I, 1970.)

CHAPTER VI

ENZYME KINETICS

THE study of the kinetics of enzymic reactions is aimed ultimately at obtaining information on the mechanism of the reaction. The kinetics of enzyme action were first explained satisfactorily by the postulate that the enzyme combines reversibly with its substrate to form an intermediate enzyme-substrate complex. Henri (1903) and Michaelis and Menten (1913) were the pioneers of this approach and their concepts were vindicated when evidence for the existence of such complexes was obtained by very delicate spectroscopic techniques in the study of catalase and peroxidase (1936–37). More recently it has proved possible in certain favourable cases actually to isolate covalent enzyme-substrate compounds in amounts sufficient for their further study. The enzyme-substrate complex thus formed may then undergo various reactions.

It is possible to distinguish three stages in any enzymic reaction. First, when an enzyme E is mixed with a substrate S reaction occurs between them to form the enzyme-substrate complex ES. This interaction takes place so rapidly that it is exceedingly difficult to study without special apparatus. Generally, since there is a great excess of substrate relative to enzyme, the disappearance of free substrate is not readily detected at this stage. The concentration of ES[1] increases until it reaches a level at which its rate of formation is exactly balanced by its rate of decomposition to products, i.e. a steady state is attained where $d[ES]/dt = 0$, corresponding to stage two of the reaction (Fig. 6.1).

The pre-steady state exists over a very short time interval (a few milliseconds) and consequently in the very early stages of a reaction [P] may be regarded as approximately zero while [ES] has a small finite value (comparison of the curves for [ES] and [P] in Fig. 6.1 will illustrate this point).

The rate of formation of product P, which has been increasing simultaneously with the increase in concentration of ES, is now constant in this second stage (and equal to $k_{+2}[ES]$) and, provided the concentration of substrate is sufficiently high, may remain

[1] Concentrations are denoted throughout by [ES], [E], [P], etc.

effectively so for a measurable period. This constant rate of formation is the *initial rate of reaction* which is the experimentally determined parameter in the majority of enzyme kinetic work.

As the substrate is used up the concentration of ES decreases and the rate of reaction likewise falls, giving rise to the third stage of the process. Obviously the course of the reaction is only defined completely by investigation of all three stages. However, it is

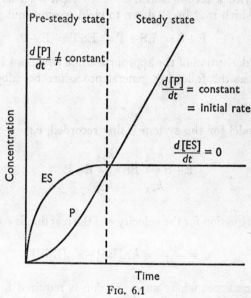

Fig. 6.1

The concentrations of enzyme-substrate complex ES and product P in the approach to and establishment of steady state kinetics. When $d[P]/dt$ becomes constant the 'initial rate' of the reaction is manifest. Note that the pre-steady state phase is of very short duration, usually only a few milliseconds.

generally not profitable to investigate the course of the reaction in the third phase as complications may arise due to factors such as enzyme inactivation (denaturation) and accumulation of inhibitory products, e.g. H^+.

The kinetics of enzyme action have been treated principally in one of two ways. Michaelis and Menten assumed that the enzyme-substrate complex remains in equilibrium with free enzyme and substrate throughout the course of the reaction, whereas Briggs and Haldane (1925) applied the more general steady-state treatment.

Consider as before, an enzyme E reacting with substrate S to give an enzyme-substrate complex ES. The complex may then either :

(*a*) break down to give enzyme and substrate or undergo a reaction to yield enzyme and product(s), P.

$$E + S \rightleftharpoons ES \rightleftharpoons E + P, \text{ or}$$

(*b*) react with a second substrate T to yield a ternary complex EST which may break down to yield enzyme and product(s)

$$E + S \rightleftharpoons ES + T \rightleftharpoons EST \rightleftharpoons E + P$$

For the derivation of the appropriate rate equations it is recommended that the following general procedure be adopted in all cases.

1. A model for the system is first recorded, e.g.

$$E + S \underset{k_{-1}}{\overset{k_{+1}}{\rightleftharpoons}} ES \underset{k_{-2}}{\overset{k_{+2}}{\rightleftharpoons}} E + P \qquad . \qquad . \qquad (6.1)$$

2. The equation for the velocity v of the reaction is set down, i.e.

$$v = \frac{d[P]}{dt} = k_{+2}[ES] - k_{-2}[E][P] . \qquad . \qquad (6.2)$$

In those instances where an expression is required for an *initial* velocity of reaction we have the situation that at time $t = 0$, $[P] = 0$ and therefore

$$v = \frac{d[P]}{dt} = k_{+2}[ES] \qquad . \qquad . \qquad . \qquad (6.3)$$

and the model reduces to

$$E + S \rightleftharpoons ES \rightarrow E + P$$

as a special (but very important) limiting case.

3. *The conservation equation* is next set out, recording the total concentration of enzyme $[E_0]$ as the sum of the free and combined enzyme concentrations, thus

$$[E_0] = [E] + [ES] \qquad . \qquad . \qquad . \qquad (6.4)$$

(In a similar manner the total concentration of substrate, $[S_0]$, is equal to the sum of free and combined substrate, $[S]+[ES]$, although in this instance, since $[S]$ is so very much greater than $[ES]$, the latter term may be neglected and $[S_0]$ equated with $[S]$.)

4. A set of equations is now written which describes the steady state (or other) condition. If we consider the Briggs-Haldane steady-state model then, since the rate of formation of ES is equal to its rate of removal,

$$\frac{d[ES]}{dt} = 0 = k_{+1}[E][S] - (k_{+2}+k_{-1})\,[ES] \qquad (6.5)$$

and

$$\frac{d[E]}{dt} = 0 \qquad . \qquad . \qquad . \qquad . \qquad (6.6)$$

It should be noted that there will be n such equations for a model comprising n enzyme-containing species but only $n-1$ are independent, the nth being deducible from the other $n-1$ equations and the conservation equation.

To deduce the required rate equation we need to obtain $[ES]$ as a function of $[E_0]$. This can be done by using the conservation equation (6.4) provided we can obtain $[E]$ as a function of $[ES]$. This in turn can be obtained from the steady-state relationships. Thus

$$[E] = \frac{(k_{+2}+k_{-1})}{k_{+1}} \cdot \frac{[ES]}{[S]}$$

Hence

$$[ES] = \frac{[E_0]}{1 + \dfrac{(k_{+2}+k_{-1})}{k_{+1}} \cdot \dfrac{1}{[S]}}$$

and

$$v = k_{+2}[ES]$$

i.e.,

$$v = \frac{k_{+2}[E_0]}{1 + \dfrac{(k_{+2}+k_{-1})}{k_{+1}} \cdot \dfrac{1}{[S]}} \qquad . \qquad . \qquad . \qquad (6.7a)$$

This is the desired rate equation.

The maximum velocity of reaction V occurs when the enzyme is saturated with substrate, that is when all of the enzyme is in the form of ES. Thus we may write

$$V = k_{+2}[E_0] \quad \text{and therefore}$$

$$v = \frac{V}{1 + \frac{(k_{+2} + k_{-1})}{k_{+1}} \cdot \frac{1}{[S]}} \quad . \quad . \quad (6.7b)$$

Now when $v = V/2$,

$$[S] = \frac{k_{+2} + k_{-1}}{k_{+1}}$$

and this value of [S] is known as the *Michaelis constant*, denoted by K_m, which may be defined therefore as *the substrate concentration at half maximum velocity of reaction*. K_m is a characteristic constant for a given enzyme, having the dimensions of concentration ; thus

$$K_m = \frac{k_{+2} + k_{-1}}{k_{+1}}$$

Hence we may write $\qquad v = \dfrac{V}{1 + \dfrac{K_m}{[S]}} \quad . \quad . \quad . \quad (6.8)$

k_{-1}/k_{+1} is the equilibrium constant K_s for the dissociation of ES into E and S. K_s is referred to as the *substrate constant*[1] and it is, in fact, the reciprocal of the affinity of an enzyme for its substrate, i.e., the smaller the value of K_s the greater the affinity of enzyme for substrate. Thus

$$K_m = K_s + \frac{k_{+2}}{k_{+1}}$$

and it will be apparent that the Michaelis constant may be equated with the substrate constant only when k_{+2} is very small in comparison with k_{-1}.

The Michaelis-Menten treatment assumes that equilibrium conditions hold throughout the reaction. It may now be seen that the Michaelis-Menten treatment is a limiting case of the more general

---

[1] The Commission of Enzymes of the International Union of Biochemistry has recommended (1961) that all equilibria involving combinations of enzymes with substrates, inhibitors or products should be expressed in terms of dissociation constants rather than as association constants. It will be noted that this practice is at variance with the modern convention that equilibrium constants are always expressed with the species on the right hand side of the equation in the numerator.

Briggs-Haldane steady-state treatment for the special situation where $k_{+2} \ll k_{-1}$. While the Michaelis-Menten equation

$$v = \frac{V}{1 + \dfrac{K_s}{[S]}} \qquad . \qquad . \qquad . \qquad (6.9)$$

holds for a number of enzymes which have been investigated there are many other enzymes for which the more general equation 6.7b applies.

The Michaelis constant is always determined when an enzyme is being characterized. There are several methods available (for full details see Dixon and Webb, 1964), but here we shall confine ourselves to two widely used techniques. It will be observed that equations 6.7 and 6.8 are of the general form

$$v = \frac{C_1}{1 + \dfrac{C_2}{[S]}} \qquad . \qquad . \qquad . \qquad (6.10)$$

where C_1 and C_2 are constants. This is the equation of a rectangular hyperbola and if the initial velocity is plotted as a function of the substrate concentration the maximum velocity may be determined, and hence the substrate concentration at half maximum velocity can be obtained graphically (Fig. 6.2). The greatest uncertainty in this method attends the evaluation of V since the precise asymptotic value to accept is frequently in doubt. For this reason an alternative and more accurate graphical method was introduced by Lineweaver and Burk (1934), who made use of the linear form of the equation for a rectangular hyperbola by plotting the reciprocals of velocity and substrate concentrations and thus obtaining a straight line (Fig. 6.3).

If the reciprocal of equation 6.8 is taken we have

$$\frac{1}{v} = \frac{1}{V} + \frac{K_m}{V[S]} \qquad . \qquad . \qquad . \qquad (6.11)$$

This is the equation of a straight line with slope K_m/V and intercept of $1/V$ on the $1/v$ axis, and from the plotted values K_m can be determined. If the line is produced to the left of the ordinate axis it will give an intercept on the abscissa equal to $-1/K_m$ (Fig. 6.3). This is evident by putting $1/v = 0$ in equation 6.11, which then becomes $1/[S] = -1/K_m$.

It is useful to remember therefore that if any rate equation, no matter how complicated it may appear to be, can be reduced to the form of equation 6.10 then the plot of $1/v$ versus $1/[S]$ will be linear and the intercept on the ordinate will give $1/C_1$, i.e. $1/V_{app}$, while the intercept on the abscissa will give $-1/C_2$, i.e. $1/K_{m\ app}$,

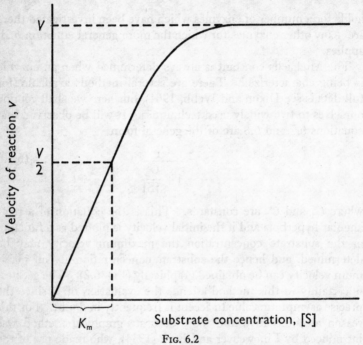

FIG. 6.2

Graphical determination of the Michaelis constant, K_m, of an enzyme. Note that at low substrate concentrations the reaction is of first order with respect to substrate, but changes to zero order when the enzyme is saturated with substrate.

where V_{app} and $K_{m\ app}$ are respectively the apparent maximum velocity and apparent Michaelis constant under the given conditions. Consultation of Table 6.1 (p. 204) for the characteristics of two-substrate reactions and of Table 6.2 (p. 218) for those of different types of inhibition will illustrate this application.

Example 6.1.   The D-serine dehydratase of *Neurospora crassa* has been shown to require pyridoxal phosphate as coenzyme. The enzyme catalyses the reaction

$$CH_2OH.CHNH_2.COOH \rightarrow CH_3CO.COOH + NH_3.$$

The following figures were obtained in an experiment to determine the pyridoxal phosphate saturation curve of the enzyme.

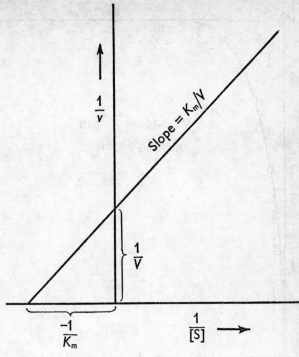

Fig. 6.3

Determination of Michaelis constant by Lineweaver-Burk reciprocal plot.

μMoles pyruvic acid formed in 20 minutes	Pyridoxal phosphate concentration $\times 10^5$ (M)
0·150	0·20
0·200	0·40
0·275	0·85
0·315	1·25
0·340	1·70
0·350	2·00
0·360	8·00

Use these data to determine the apparent Michaelis constant for the serine dehydratase with respect to pyridoxal phosphate.

(After YANOFSKY (1952), *J. biol. Chem.*, **198**, 343.)

From the reaction catalysed it will be seen that pyruvic acid formation is a measure of the velocity of the reaction. The required constant may be obtained graphically by plotting either (1) the velocity of reaction v against coenzyme concentration [S] or (2) $1/v$ against $1/[S]$, the Lineweaver-Burk method.

The graph for method 1 is shown in Fig. 6.4. From this the value of K_m obtained is $3 \cdot 2 \times 10^{-6}$ M.

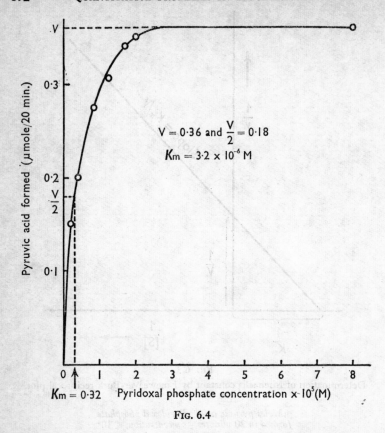

$$V = 0.36 \text{ and } \frac{V}{2} = 0.18$$
$$K_m = 3.2 \times 10^{-6} \text{ M}$$

$K_m = 0.32$ Pyridoxal phosphate concentration $\times 10^5$(M)

FIG. 6.4

Method 2 requires that the reciprocals of the velocity of reaction and co-enzyme concentration be obtained.

Velocity (v) μmoles pyruvic acid formed in 20 min	$\dfrac{1}{v}$	Coenzyme concentration [S] ($\times 10^5$) M	$\dfrac{1}{[S]}$ ($\times 10^{-5}$) M^{-1}
0.150	6.66	0.20	5.0
0.200	5.00	0.40	2.5
0.275	3.64	0.85	1.17
0.315	3.17	1.25	0.80
0.340	2.94	1.70	0.58
0.350	2.86	2.00	0.50
0.360	2.78	8.00	0.125

These values are plotted in the graph in Fig. 6.5 from which measurements of the slope and intercept enable K_m to be evaluated ; in this case a value of 3.56×10^{-6} M is obtained. The intercept on the abscissa yields a value of 3.51×10^{-6} M for K_m. This method gives more reliable results than method 1.

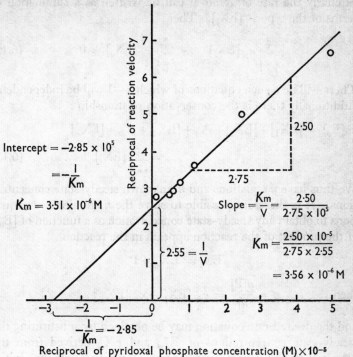

Reciprocal of pyridoxal phosphate concentration (M) $\times 10^{-5}$

FIG. 6.5

Multi-substrate Systems (General Case)

Consider a general model system comprising n enzyme-containing species, each of which can be reversibly converted to any other. For each such species there is a steady-state relationship equating the rate of formation with the rate of removal. The rate of formation is given by a summation of terms of the type $\kappa_{ji}[EX_j]$ where κ_{ji} is the product (a kappa product) of the velocity constant k_{ji} for the reaction in which EX_j is converted to EX_i and any associated substrate concentration, e.g., the conversion of EX_j to EX_i may involve substrate S, thus

$$EX_j + S \underset{\kappa_{ij}}{\overset{\kappa_{ji}}{\rightleftharpoons}} EX_i$$

Similarly the rate of removal can be written as a summation of terms of the type $\kappa_{ij}[EX_i]$. Then

$$\sum_{\substack{j=1 \\ i \neq j}}^{n} \kappa_{ji}[EX_j] \; - \; \sum_{\substack{j=1 \\ i \neq j}}^{n} \kappa_{ij}[EX_i] = 0 \quad . \qquad . \quad (6.12)$$

There will be n such equations of which $n-1$ will be independent. Additionally there is the conservation relationship :

$$[E_0] = [EX_1] + [EX_2] + \ldots + [EX_i] + \ldots + [EX_n]$$

$$= \sum_{i=1}^{n} [EX_i] \qquad . \qquad . \quad (6.13)$$

We thus have n equations and n unknown steady-state concentrations. It is therefore possible to solve the n simultaneous equations to obtain any steady-state concentration as a function of $[E_0]$. If the product of the reaction appears in the reaction

$$EX_l \rightleftharpoons EX_m + P$$

then $\qquad v = \dfrac{d[P]}{dt} = \kappa_{lm}[EX_l] - \kappa_{ml}[EX_m]$

and the desired rate equation may be obtained by substituting the steady-state concentrations of EX_l and EX_m derived from the solution of the above simultaneous equations. If an initial reaction velocity is obtained, i.e., at $t = 0$, when $[P] = 0$, then this equation reduces to

$$v = \kappa_{lm}[EX_l]$$

The solution of the simultaneous equations may be effected algebraically or by a schematic method due to King and Altman (1956). An example of the two approaches is given with reference to the following model of a non-sequential enzyme mechanism, which is characterized by the formation and release of one product before the second substrate is bound.

$$E + A \underset{\kappa_2}{\overset{\kappa_1}{\rightleftharpoons}} EA \overset{\kappa_3}{\longrightarrow} E' + C$$

$$E' + B \underset{\kappa_5}{\overset{\kappa_4}{\rightleftharpoons}} E'B \overset{\kappa_6}{\longrightarrow} E + D$$

This system is the so-called 'ping-pong' model applicable to the pyridoxal phosphate-dependent aminotransferases and certain other two-substrate enzyme systems in which one substrate (A) interacts with a tightly bound enzyme-coenzyme molecule (E) to give a stable, modified enzyme-coenzyme molecule (E') and one product (C). The original enzyme-coenzyme structure is restored in a second half-reaction with a second substrate (B) which produces a second product (D). Such a mechanism may be denoted diagrammatically by the notation of Cleland (1963) in which a horizontal line represents the enzyme and, from left to right along the line, the sequence of formation and conversion of the various enzyme complexes is recorded (beneath the line) and the binding of substrates and release of products is indicated by vertical arrows above the line.

$$
\begin{array}{ccccc}
\text{A} & & \text{C} \quad \text{B} & & \text{D} \\
\downarrow & & \uparrow \quad \downarrow & & \uparrow \\
\hline
\text{E} & \text{(EA} \quad \text{E'C)} & \text{E'} & \text{(E'B} \quad \text{ED)} & \text{E}
\end{array}
$$

(a) ALGEBRAIC METHOD

In the steady-state :

$$v = \frac{d[\text{C}]}{dt} = \frac{d[\text{D}]}{dt} = \kappa_3[\text{EA}] = \kappa_6[\text{E'B}] \qquad . \quad (6.14)$$

Also $$[\text{E}_0] = [\text{E}] + [\text{E'}] + [\text{EA}] + [\text{E'B}] \qquad . \qquad (6.15)$$

There are additionally four steady-state relationships of which three will be used (it is immaterial which three are employed).

$$\frac{d[\text{E}]}{dt} = 0 = \kappa_2[\text{EA}] + \kappa_6[\text{E'B}] - \kappa_1[\text{E}] \qquad . \quad (6.16)$$

$$\frac{d[\text{E'}]}{dt} = 0 = \kappa_5[\text{E'B}] + \kappa_3[\text{EA}] - \kappa_4[\text{E'}] \qquad . \quad (6.17)$$

$$\frac{d[\text{EA}]}{dt} = 0 = \kappa_1[\text{E}] - (\kappa_2 + \kappa_3)[\text{EA}] \qquad . \qquad (6.18)$$

To obtain the necessary rate equation we need [EA] as a function of [E₀]. This can be obtained from equation 6.15 provided we can get [E], [E'] and [E'B] as functions of [EA]. These, in turn, can be obtained from equations 6.16–6.18. Thus, from equation 6.18

$$[E] = \left(\frac{\kappa_2 + \kappa_3}{\kappa_1} \right) [EA]$$

From equation 6.16, or directly from equation 6.14

$$[E'B] = \frac{\kappa_3}{\kappa_6} [EA]$$

and combining with equation 6.17 we have

$$[E'] = \frac{\kappa_3}{\kappa_4} \left(\frac{\kappa_5 + \kappa_6}{\kappa_6} \right) [EA]$$

Substituting these values in the conservation equation 6.15 and rearranging, we have

$$[EA] = \frac{[E_0]}{1 + \left(\dfrac{\kappa_2 + \kappa_3}{\kappa_1} \right) + \dfrac{\kappa_3}{\kappa_6} + \dfrac{\kappa_3}{\kappa_4} \left(\dfrac{\kappa_5 + \kappa_6}{\kappa_6} \right)}$$

Thus the rate equation $v = \kappa_3[EA]$ becomes

$$v = \frac{\kappa_3[E_0]}{1 + \left(\dfrac{\kappa_2 + \kappa_3}{\kappa_1} \right) + \dfrac{\kappa_3}{\kappa_6} + \dfrac{\kappa_3}{\kappa_4} \left(\dfrac{\kappa_5 + \kappa_6}{\kappa_6} \right)}$$

The denominator terms $1 + \dfrac{\kappa_3}{\kappa_6}$ can be collected and expressed as $(\kappa_3 + \kappa_6)/\kappa_6$ and if both numerator and denominator are then divided by this term we obtain

$$v = \frac{\left(\dfrac{\kappa_3 \kappa_6}{\kappa_3 + \kappa_6} \right) [E_0]}{1 + \left(\dfrac{\kappa_2 + \kappa_3}{\kappa_3 + \kappa_6} \right) \dfrac{\kappa_6}{\kappa_1} + \left(\dfrac{\kappa_5 + \kappa_6}{\kappa_3 + \kappa_6} \right) \dfrac{\kappa_3}{\kappa_4}}$$

If we now put $\kappa_1 = k_1[A]$ and $\kappa_4 = k_4[B]$ etc. we have

$$v = \frac{\left(\dfrac{k_3 k_6}{k_3 + k_6}\right)[E_0]}{1 + \left(\dfrac{k_2 + k_3}{k_3 + k_6}\right)\dfrac{k_6}{k_1}\dfrac{1}{[A]} + \left(\dfrac{k_5 + k_6}{k_3 + k_6}\right)\dfrac{k_3}{k_4} \cdot \dfrac{1}{[B]}}$$

which reduces to the form

$$v = \frac{C_1[E_0]}{1 + \dfrac{C_2}{[A]} + \dfrac{C_3}{[B]}} = \frac{V}{1 + \dfrac{C_2}{[A]} + \dfrac{C_3}{[B]}} \qquad . \text{(6.19)}$$

where C_2 and C_3 are the Michaelis constants for substrates A and B respectively.

(b) SCHEMATIC METHOD OF KING AND ALTMAN (1956)

When the number n of enzyme-containing species is greater than three, methods for the solution of n simultaneous linear equations involving successive elimination are tedious. In this situation determinant methods are much superior.[1]

For convenience of treatment equations 6.12 and 6.13 may be divided through by $[E_0]$, and of the n steady-state relationships the equation $\dfrac{d[EX_l]}{dt} = 0$ is omitted for simplicity (EX_l being the species giving rise to the product on conversion to EX_m). In their expanded form the appropriate n equations now become :

$$\frac{d[EX_1/E_0]}{dt} = -\left(\Sigma\kappa_{1j}\right)\left[\frac{EX_1}{E_0}\right] + \kappa_{21}\left[\frac{EX_2}{E_0}\right] + \ldots + \kappa_{l1}\left[\frac{EX_l}{E_0}\right] + \ldots + \kappa_{n1}\left[\frac{EX_n}{E_0}\right] = 0$$

$$\frac{d[EX_2/E_0]}{dt} = \kappa_{12}\left[\frac{EX_1}{E_0}\right] - \left(\Sigma\kappa_{2j}\right)\left[\frac{EX_2}{E_0}\right] + \ldots + \kappa_{l2}\left[\frac{EX_l}{E_0}\right] + \ldots + \kappa_{n2}\left[\frac{EX_n}{E_0}\right] = 0$$

$$\left[\frac{EX_1}{E_0}\right] + \left[\frac{EX_2}{E_0}\right] + \ldots + \left[\frac{EX_l}{E_0}\right] + \ldots + \left[\frac{EX_n}{E_0}\right] = 1 \qquad \text{(6.20)}$$

$$\frac{d[EX_n/E_0]}{dt} = \kappa_{1n}\left[\frac{EX_1}{E_0}\right] + \kappa_{2n}\left[\frac{EX_2}{E_0}\right] + \ldots + \kappa_{ln}\left[\frac{EX_l}{E_0}\right] + \ldots - \left(\Sigma\kappa_{nj}\right)\left[\frac{EX_n}{E_0}\right] = 0$$

[1] See Appendix 7.

When set out in this way all the $(-\Sigma\kappa)$ terms lie on a diagonal. Applying Cramer's Rule the value for $[EX_l/E_0]$ is obtained as follows :

$$\left[\frac{EX_l}{E_0}\right] = \frac{\begin{vmatrix} -\Sigma\kappa_{1j} & \kappa_{21}\ldots\ldots\ldots\ldots\ldots 0\ldots\ldots\kappa_{n1} \\ \kappa_{12} & -\Sigma\kappa_{2j}\ldots\ldots\ldots\ldots\ldots 0\ldots\ldots\kappa_{n2} \\ \vdots & \vdots \qquad\qquad\qquad \vdots \qquad \vdots \\ 1 & 1\ldots\ldots\ldots\ldots\ldots\ \ldots\ldots\ldots 1 \\ \vdots & \vdots \qquad\qquad\qquad \vdots \qquad \vdots \\ \kappa_{1n} & \kappa_{2n}\ldots\ldots\ldots\ldots\ldots 0\ldots\ldots-\Sigma\kappa_{nj} \end{vmatrix}}{\begin{vmatrix} -\Sigma\kappa_{1j} & \kappa_{21}\ldots\ldots\ldots\ldots\kappa_{l1}\ldots\ldots\kappa_{n1} \\ \kappa_{12} & -\Sigma\kappa_{2j}\ldots\ldots\ldots\ldots\kappa_{l2}\ldots\ldots\kappa_{n2} \\ \vdots & \vdots \qquad\qquad \vdots \qquad \vdots \\ 1 & 1\ldots\ldots\ldots\ldots\ldots 1\ldots\ldots 1 \\ \vdots & \vdots \qquad\qquad \vdots \qquad \vdots \\ \kappa_{1n} & \kappa_{2n}\ldots\ldots\ldots\ldots\kappa_{ln}\ldots-\Sigma\kappa_{nj} \end{vmatrix}} \qquad . \ (6.21)$$

Values for every $\left[\dfrac{EX_i}{E_0}\right]$ may be obtained in an analogous manner if so desired.   Substituting from equation 6.21 into the relationship

$$v = \frac{d[P]}{dt} = \kappa_{lm}\left[\frac{EX_l}{E_0}\right]$$

we therefore obtain the desired rate equation giving the initial velocity of reaction. The two determinants may be expanded by the usual algebraic processes. King and Altman have however described a schematic method for deducing the nature of the numerator and denominator terms of equation 6.21. The method depends upon the following properties of the numerator and denominator determinants :

(i)   The denominator of the corresponding expression for each $\left[\dfrac{EX_i}{E_0}\right]$ is the same and is the sum of the n corresponding numerators (this follows from equation 6.20).   In deducing the nature of the terms appearing in the numerator and denominator of equation 6.21 we may therefore consider the form of the numerator terms only.

(ii) the kappa products all contain $n-1$ terms.

(iii) the numerator contains no column index l and hence no κ corresponding to a reaction in which EX_l is the reactant appears in the kappa products.

(iv) all the κ values with the same initial index are in the same column and hence no kappa product contains sequences of the type . . . $\kappa_{ij}\kappa_{im}$ . . ., i.e. the kappa products do not contain κ values originating from the same enzyme-containing species.

(v) every κ_{ij} $(i \neq l, j \neq l)$ occurs in two places in the determinant, once as a diagonal element (i.e. in one of the $-\Sigma\kappa$ terms) and once in a non-diagonal position. Certain conclusions can be drawn regarding kappa products containing sequences of the type . . . $\kappa_{cd} \cdot \kappa_{de} \cdot \kappa_{ec}$ . . . i.e. sequences corresponding to a closed loop or cycle of reactions. The sequences deriving from the expansion of the diagonal elements will have a positive sign if there is an even number of $-\Sigma$ terms lying on the diagonal and a negative sign if there is an odd number of such terms. The sequences deriving from the non-diagonal elements will have a sign which can be determined by interchanging columns to bring the second indices of the kappas into their natural order, i.e. on to the diagonal position. To illustrate the argument, consider the situation where the $-\Sigma$ terms are even in number (and all lie on the diagonal). The diagonal expansion terms will then have a positive sign. If the number of EX species in a closed loop is even then an odd number of column interchanges is required to bring the relevant kappas on to the diagonal, i.e. an even number of $-\Sigma$ terms will have been removed from the diagonal and the terms of the expansion of the new diagonal will now have a negative sign (see Appendix 7). Similarly when there is an odd number of EX species in a closed loop, an even number of column interchanges is required and as a result an odd number of $-\Sigma$ terms has been removed from the diagonal position. The expansion of the new diagonal will consequently lead to negative terms. It therefore follows that terms involving kappa sequences corresponding to closed loops appear twice in the determinant expansion but cancel out.

These deductions can be used to formulate the following simple schematic method for the derivation of the terms of the numerator and denominator of the rate equation.

For a system of reactions possessing n enzyme-containing species, diagrams are drawn showing the total number of combinations of $(n-1)$ reactions leading individually or in sequence to each enzyme-containing species.

If the initial rate of reaction $\dfrac{d[P]}{dt}$ is given by $\kappa_{lm}[EX_l]$, i.e. by $\kappa_{lm}[E_0]\dfrac{[EX_l]}{[E_0]}$ then

$$v = \kappa_{lm}[E_0] \frac{\text{combined sequences leading to } EX_l}{\substack{\text{combined sequences leading to all} \\ \text{the } EX_i \text{ species}}}$$

The sequences disallowed are those involving cycles or in which more than one reaction leads away from any EX_i.

For the aminotransferase reaction (p. 195) the model may be rewritten as

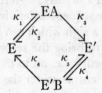

For this model $n-1 = 3$. The following combinations of kappas exist:

for E:

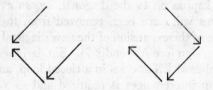

for EA:

for E′ :

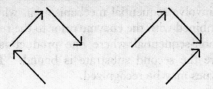

for E′B :

Hence

$$v = \frac{d[C]}{dt} = \kappa_3[E_0] \cdot \frac{[EA]}{[E_0]}$$

$$= \kappa_3[E_0] \frac{\kappa_4\kappa_6\kappa_1}{\kappa_4\kappa_6\kappa_1 + \kappa_1\kappa_3\kappa_4 + \kappa_2\kappa_6\kappa_4 + \kappa_6\kappa_4\kappa_3 + \kappa_6\kappa_1\kappa_3 + \kappa_1\kappa_3\kappa_5}$$

Putting $\kappa_1 = k_1[A]$ etc. and re-arranging we get :

$$v = \frac{\left(\dfrac{k_3 k_6}{k_3 + k_6}\right)[E_0]}{1 + \dfrac{k_6}{k_1}\left(\dfrac{k_2 + k_3}{k_3 + k_6}\right)\dfrac{1}{[A]} + \dfrac{k_3}{k_4}\left(\dfrac{k_5 + k_6}{k_3 + k_6}\right)\dfrac{1}{[B]}} \quad . \quad (6.22)$$

The method is simple in use, particularly as it is usually possible to deduce up to half of the terms from the remainder on considerations of the symmetry of the model. The method has the additional but doubtful advantage of being capable of use in the absence of an understanding of the underlying theoretical considerations.

Derivation of the Parameters of the Rate Equations for Two-substrate Systems

Two-substrate systems embrace many reactions of importance in metabolism as will be appreciated from the fact that one of the substrates may be a coenzyme, as in the case of the nicotinamide nucleotide dehydrogenases. The aminotransferases, phosphotransferases and hydrolytic enzymes such as esterases (where the second substrate is water) are other examples. Two-substrate systems may involve sequential mechanisms in which both substrates are combined with the enzyme prior to the release of either product, or non-sequential where one product is formed and released before the second substrate is bound. The following mechanistic types may be recognized.

(a) SEQUENTIAL MECHANISMS

1. Random-order systems. The enzyme is assumed to have two binding sites, one for each substrate or product. The substrates may be bound independently of each other and in any order. Thus we may write

$$\begin{array}{c} \text{EA} \\ K_a \nearrow \quad \searrow K_b' \\ E \quad {\pm A \; \pm B} \quad {\pm B \; \pm A} \quad \text{EAB} \quad \xrightarrow{\;k\;} \quad E + C + D \\ K_b \searrow \quad \nearrow K_a' \\ \text{EB} \end{array}$$

where K_a, K_a', K_b and K_b' are equilibrium (dissociation) constants for the indicated reactions. In the Cleland notation this becomes

$$\begin{array}{cc} A \quad B & C \quad D \\ \downarrow \quad \downarrow & \uparrow \quad \uparrow \\ \text{EA} & \text{ED} \\ E \qquad \left(\begin{array}{c}\text{EAB}\\ \text{ECD}\end{array}\right) \qquad E \\ \text{EB} & \text{EC} \\ \uparrow \qquad & \downarrow \qquad \\ B \quad A & D \quad C \end{array}$$

2. Ordered systems. In these cases the substrates are bound to the enzyme in a mandatory order, thus

$$E + A \underset{k_{-1}}{\overset{k_{+1}}{\rightleftharpoons}} EA + B \underset{k_{-2}}{\overset{k_{+2}}{\rightleftharpoons}} EAB \xrightarrow{k_{+3}} E + C + D$$

For simplicity of treatment the ordered release of products C and D is not considered and in the Cleland notation :

$$
\begin{array}{cccc}
\text{A} & \text{B} & & \text{CD} \\
\downarrow & \downarrow & & \uparrow \\
\hline
\text{E} \quad \text{EA} & \text{(EAB} & \text{ECD)} & \text{E}
\end{array}
$$

(b) NON-SEQUENTIAL MECHANISMS

' Ping-pong ' systems. In these systems one product is formed and released before the second substrate is bound by the enzyme. As discussed previously (p. 194) the reactions may be formulated as

$$
\text{E} + \text{A} \underset{k_2}{\overset{k_1}{\rightleftharpoons}} \text{EA} \overset{k_3}{\longrightarrow} \text{E}' + \text{C}
$$

$$
\text{E}' + \text{B} \underset{k_5}{\overset{k_4}{\rightleftharpoons}} \text{E}'\text{B} \overset{k_6}{\longrightarrow} \text{E} + \text{D}
$$

When treated according to equilibrium (Michaelis-Menten) theory the random-order model gives a rate equation of the type :

$$
v = \frac{C_1[\text{E}_0]}{1 + \dfrac{C_2}{[\text{A}]} + \dfrac{C_3}{[\text{B}]} + \dfrac{C_4}{[\text{A}][\text{B}]}} \qquad . \qquad . \qquad (6.23)
$$

When treated by the steady-state theory a complex and unusable equation involving squared terms is obtained.

The ordered model treated by steady-state theory gives an equation of the same form as equation 6.23. In the case of the ping-pong system a similar relationship holds except that there is no term containing both [A] and [B], thus

$$
v = \frac{C_1[\text{E}_0]}{1 + \dfrac{C_2}{[\text{A}]} + \dfrac{C_3}{[\text{B}]}}
$$

The significance of the values of C for each model is given in Table 6.1.

TABLE 6.1

The Significance of the Values of the Constants C_n in Two-substrate Enzyme Systems

Model	C_1	C_2	C_3	C_4
Random order	k	K_a'	K_b'	$K_a K_b'$ $(= K_a' K_b)$
Ordered[1]	k_{+3}	$\dfrac{k_{+3}}{k_{+1}}$	$\dfrac{(k_{+3} + k_{-2})}{k_{+2}}$	$\dfrac{k_1(k_{+3} + k_{-2})}{k_{+1}k_{+2}}$
Ping-pong	$\dfrac{k_3 k_6}{k_3 + k_6}$	$\dfrac{k_6(k_2 + k_3)}{k_2(k_3 + k_6)}$	$\dfrac{k_3(k_5 + k_6)}{k_4(k_3 + k_6)}$	0

[1] Note : $C_4/C_3 = k_{-1}/k_{+1} = K_s$ for A, the first bound substrate. For each model the maximum velocity $V = C_1[E_0]$, and C_2 and C_3 may be considered to be Michaelis constants for substrates A and B respectively.

DERIVATION OF THE PARAMETERS

For models where the following rate equation holds :

$$v = \frac{C_1[E_0]}{1 + \dfrac{C_2}{[A]} + \dfrac{C_3}{[B]} + \dfrac{C_4}{[A][B]}}$$

an initial plot of reciprocal velocity versus reciprocal [A] (at constant [B]) is made for a series of experiments in which B is varied from experiment to experiment. This yields a graph of the type shown in Fig. 6.6.

It can be seen from the reciprocal rate equation

$$\frac{1}{v} = \frac{1}{[A]}\left(C_2 + \frac{C_4}{[B]}\right)\frac{1}{C_1[E_0]} + \left(1 + \frac{C_3}{[B]}\right)\frac{1}{C_1[E_0]}$$

that the intercept on the ordinate (I_0) has the significance

$$I_0 = \left(1 + \frac{C_3}{[B]}\right)\frac{1}{C_1[E_0]}$$

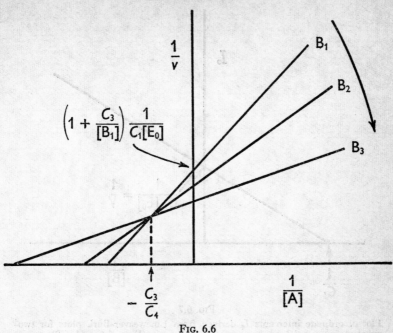

FIG. 6.6

Lineweaver-Burk plots for a two-substrate sequential system in which the concentration of substrate A is varied while that of substrate B is held constant at three different values, where $[B_1] < [B_2] < [B_3]$.

A secondary plot of I_0 versus $1/[B]$ yields a straight line, the intercepts of which on the abscissa and ordinate give $-1/C_3$ and $1/C_1[E_0]$ respectively (Fig. 6.7).

The primary Lineweaver-Burk plots (Fig. 6.6) give a family of straight lines intersecting at a point with abscissa co-ordinate of $-C_3/C_4$. Hence $C_1[E_0]$, C_3 and C_4 can be obtained. C_2 (as well as $C_1[E_0]$ and C_4) may be obtained in an analogous manner from an experiment in which B is the variable substrate and A is kept constant at a series of differing concentrations.

The intersect point of the lines in Fig. 6.6 may be in the upper left-hand quadrant, on the abscissa or in the lower left-hand quadrant depending upon the values of certain constants. For example, in the random-ordered system described by the equation

$$v = \frac{k[E_0]}{1 + \dfrac{K_a'}{[A]} + \dfrac{K_b'}{[B]} + \dfrac{K_a K_b'}{[A][B]}}$$

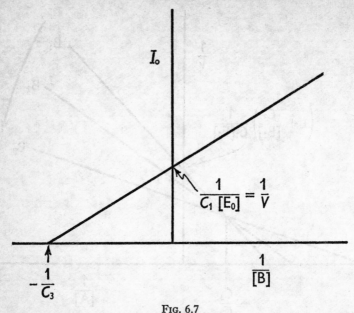

FIG. 6.7

Plot of ordinate intercepts I_0 derived from Lineweaver-Burk plots for two-substrate sequential or non-sequential systems (Figs. 6.6 and 6.8), versus $1/[B]$.

the intersect point has co-ordinates $-\dfrac{1}{K_a}$, $\left(1 - \dfrac{K_a'}{K_a}\right)\dfrac{1}{k[E_0]}$. The intersect will therefore lie in the upper left-hand quadrant if $K_a' < K_a$, on the abscissa if $K_a' = K_a$ (and, it follows $K_b' = K_b$), and in the lower left-hand quadrant if $K_a' > K_a$.

In the case of the ping-pong model it can be seen from the reciprocal rate equation

$$\frac{1}{v} = \frac{1}{[A]} \cdot \frac{C_2}{C_1[E_0]} + \left(1 + \frac{C_3}{[B]}\right)\frac{1}{C_1[E_0]}$$

that a plot of $1/v$ versus $1/[A]$, keeping B constant at a series of differing concentrations, gives a family of parallel straight lines (Fig. 6.8).

The ordinate intercept (I_0) has the significance

$$I_0 = \left(1 + \frac{C_3}{[B]}\right)\frac{1}{C_1[E_0]}$$

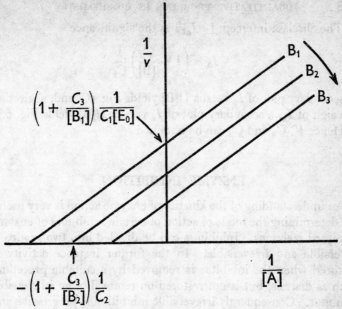

FIG. 6.8

Lineweaver-Burk plots for a two-substrate non-sequential 'ping pong' system in which the concentration of substrate A is varied while that of substrate B is held constant at three different values, where $[B_1] < [B_2] < [B_3]$.

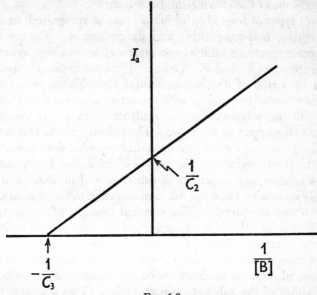

FIG. 6.9

Plot of abscissa intercepts $-I_a$ derived from Lineweaver-Burk plots for two-substrate non-sequential 'ping pong' system (Fig. 6.8) versus $1/[B]$.

The abscissa intercept $(-I_a)$ has the significance

$$-I_a = \left(1 + \frac{C_3}{[B]}\right) \frac{1}{C_2}$$

A secondary plot of I_0 versus $1/[B]$ yields Fig. 6.7 and, neglecting the sign of I_a, a secondary plot of I_a versus $1/[B]$ yields Fig. 6.9. Hence V, C_2 and C_3 can be obtained.

ENZYME INHIBITION

An understanding of the kinetics of enzyme action is very useful for determining the mode of action of certain inhibitors of enzyme catalysed reactions. Inhibitors can be divided into two groups—reversible and irreversible. In the former instance activity is restored when the inhibitor is removed by a suitable procedure such as dialysis, but is not restored on removal of an irreversible inhibitor. Consequently irreversible inhibition is progressive and becomes complete when all the enzyme is combined with the inhibitor, whereas reversible inhibition attains an equilibrium dependent on the value of the inhibitor constant, i.e. the dissociation constant of the enzyme-inhibitor complex.

Three types of reversible inhibition may be recognized, namely competitive, non-competitive and uncompetitive. Competitive and non-competitive inhibition correspond respectively to action on the apparent K_m and V. A competitive inhibitor is a compound which by virtue of its close structural resemblance to the true substrate is able to combine with the active centre and thus compete with the substrate for the available enzyme. It therefore increases the apparent K_m value. The velocity of the reaction is decreased because the enzyme-inhibitor complex does not break down to yield products. The degree of inhibition is dependent on the relative concentrations of substrate and inhibitor and by suitably increasing the substrate concentration the inhibition may be overcome completely. The classical example of competitive inhibition is the inhibition of succinate dehydrogenase by malonic acid, the lower homologue of the true substrate.

Non-competitive inhibition occurs when the reversible inhibition depends only on the inhibitor concentration and is unaffected by variation of the substrate concentration. This indicates that the combination of inhibitor with enzyme occurs at some point essential for activity but not at the active centre where substrate combination occurs, i.e. the affinity of the enzyme for its substrate is unaffected—the action is on V.

Uncompetitive inhibition occurs when the inhibitor cannot combine with the free enzyme but only with the enzyme-substrate complex or some other enzyme species which is unable itself to combine with substrate. It would seem therefore that if complex formation does not alter the configuration of the enzyme then combination of the inhibitor must occur at least partly via the bound substrate. Uncompetitive inhibition thus affects both substrate binding and maximum velocity and, as will be seen, the characteristic feature is that both are decreased by the same factor.

It must be emphasized that the distinction between the various types of inhibition by selected agents applies to individual enzymes only. For example an inhibitor may act competitively with one enzyme and non-competitively with another ; each enzyme must be assessed individually.

The difference between competitive, non-competitive and uncompetitive inhibitors may be determined experimentally and the methods are based on the following derivations.

Competitive Inhibition

The reactions occurring are

$$E + S \rightleftharpoons ES \rightarrow E + P$$
$$E + I \rightleftharpoons EI$$

where I is the inhibitor. It is assumed that EI undergoes no further reaction. The inhibitor concentration may be denoted by [I] and the dissociation constant of the enzyme-inhibitor complex by K_1.

The rate equation for this system may be derived on the assumption that either

(a) Michaelis-Menten kinetics (equilibrium conditions) obtain, i.e. $k_{+2} \ll k_{-1}$, or

(b) Briggs-Haldane kinetics (steady-state conditions) apply.

The following is the derivation where equilibrium conditions hold for the system.

The rate equation is

$$v = k_{+2}[ES]$$

and the conservation equation

$$[E_0] = [E] + [ES] + [EI]$$

Also

$$K_s = \frac{[E][S]}{[ES]}$$

and

$$K_i = \frac{[E][I]}{[EI]}$$

Hence

$$[E_0] = \frac{K_s[ES]}{[S]} + [ES] + \frac{K_s[ES]}{[S]} \cdot \frac{[I]}{K_i}$$

Thus

$$v = \frac{k_{+2}[E_0]}{1 + \dfrac{K_s}{[S]}\left(1 + \dfrac{[I]}{K_i}\right)} \qquad . \quad . \quad (6.24)$$

For steady-state conditions the relationship is of the form

$$v = \frac{k_{+2}[E_0]}{1 + \dfrac{K_m}{[S]}\left(1 + \dfrac{[I]}{K_i}\right)} \qquad . \quad . \quad (6.25)$$

where

$$K_m = \frac{k_{+2}+k_{-1}}{k_{+1}}$$

Both rate equations for competitive inhibition are of the same general form as that for the uninhibited reaction, i.e.

$$v = \frac{C_1}{1 + \dfrac{C_2}{[S]}} \qquad . \qquad . \qquad . \qquad . \quad (6.10)$$

provided that [I] is constant. In both cases the numerator constants are identical ($C_1 = k_{+2}[E_0] = V$). The denominator constants differ by the factor $(1+[I]/K_i)$. These conclusions are reflected by the shape of the Lineweaver-Burk plots. Rearranging equation 6.25 into the form

$$\frac{1}{v} = \frac{K_m\left(1 + \dfrac{[I]}{K_i}\right)}{k_{+2}[E_0]} \cdot \frac{1}{[S]} + \frac{1}{k_{+2}[E_0]} \qquad . \quad (6.26)$$

it can be seen that a linear $1/v$ versus $1/[S]$ plot will be obtained if [I] is kept constant. The effect of the inhibitor is to increase the slope of the line of the reciprocal plot by a factor $(1 + [I]/K_i)$. The intercept on the ordinate gives $V(= k_{+2}[E_0])$ and is independent of [I]. The value of K_i may be evaluated from the intercepts on the abscissa obtained in the absence and presence of inhibitor. These intercepts have the significance $-1/K_m$ and $-1/K_m(1 + [I]/K_i)$ respectively and hence their ratio gives $(1 + [I]/K_i)$ from which K_i may be calculated knowing [I].

An alternative formulation may be obtained from equations 6.9 and 6.26.

$$\frac{v}{v_i} = 1 + \frac{K_m}{K_i}\left(\frac{[I]}{K_m + [S]}\right)$$

where v_i is the velocity in the presence of the inhibitor. In this case if v/v_i is plotted against different values of inhibitor concentration [I], straight lines are obtained with unit intercept and slope dependent on [S].

Non-Competitive Inhibition

When non-competitive inhibition occurs, it is usually assumed that the inhibitor does not interfere with the combination of the enzyme with its substrate but only with the maximum velocity of reaction, V.[1] The following reactions may occur

$$E + S \rightleftharpoons ES \rightarrow E + P$$
$$E + I \rightleftharpoons EI$$
$$ES + I \rightleftharpoons ESI$$
$$EI + S \rightleftharpoons ESI$$

This situation is relatively complicated. It is an example of a model system in which it is possible to go from any one enzyme-containing species to any other by two independent routes. A treatment of the system assuming steady-state conditions yields a complex rate equation containing terms in $[S]^2$ and $[I]^2$ (for an analogous situation see the discussion of the random-order two-substrate system, p. 202).

[1] Note that the inhibitor may interfere with the binding of substrate in the sense that the equilibrium constants for the reactions $E + S \rightleftharpoons ES$ and $EI + S \rightleftharpoons ESI$ may not be identical. They are assumed to be identical in the derivation (assumption (c)) employed.

H

In deriving a rate equation the following limiting or simplifying assumptions are made :

(a) Michaelis-Menten (equilibrium) kinetics hold.

(b) EI and ESI are dead-end products, i.e. they undergo no reaction other than the reversible dissociation shown in the model.

(c) The dissociation constants for ES and ESI (to EI + S) are identical as are the dissociation constants for EI and ESI (to ES + I).

Then it can be shown, in a manner similar to that employed for competitive inhibition, that

$$v = \frac{k_{+2}[E_0]}{\left(1 + \dfrac{K_s}{[S]}\right)\left(1 + \dfrac{(I)}{K_i}\right)} \qquad . \qquad . \quad (6.27)$$

This equation is of the same form as that of the uninhibited reaction (equation 6.10) where the numerator constant $C_1 = k_{+2}[E_0]/(1 + [I]/K_i)$ and the denominator constant $C_2 = K_s$.

Here the effect of a non-competitive inhibitor is to increase *both slope and intercept* by the factor $(1 + (I)/K_i)$ when $1/v$ is plotted against $1/[S]$. This affords a ready means of differentiation of competitive and non-competitive inhibitors since in the former case the intercept does not increase.

For non-competitive inhibition it can be shown that

$$\frac{v}{v_i} = 1 + \frac{[I]}{K_i} \qquad . \qquad . \qquad . \quad (6.28)$$

Thus the plot of v/v_i against [I] gives a straight line with unit intercept and slope of $1/K_i$, independent of [S]. With a competitive inhibitor, as already seen, the slope is dependent on [S] and this affords a second method for distinguishing between the two types of inhibition. Fig. 6.10 illustrates these differences for the plot of $1/v$ against $1/[S]$ and Fig. 6.11 for v/v_i against [I].

Dixon and Webb (1964) also treat the case where the ESI complex does break down to yield products of the reaction but at a rate different from the breakdown of ES, so that the observed velocity is the sum of the two reactions.

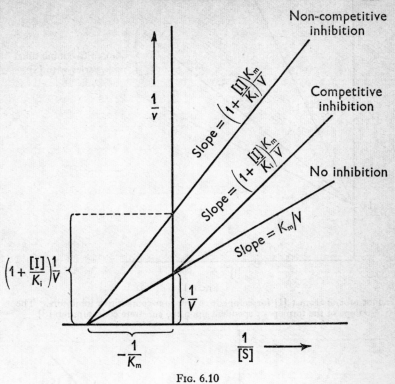

FIG. 6.10

Plot of $1/v$ against $1/[S]$ for competitive and non-competitive inhibition.

A simpler graphical method of obtaining values for K_i in the case of competitive and non-competitive inhibitors has been described by Dixon (1953). With a competitive inhibitor, as previously seen, the effect of varying independently both [S] and [I] must be determined in order to obtain K_i. However, if $1/v_i$ is plotted against [I], keeping [S] constant, a straight line will be obtained, and if this is done at two different substrate concentrations, $[S_1]$ and $[S_2]$, the resultant lines will intersect at a point to the left of the ordinate axis, as shown in Fig. 6.12. This point lies at a value of $-K_i$ which can thus be read directly from the graph. This may be proved from equation 6.26 (replacing $k_{+2}[E_0]$ by V) as follows:

$$\frac{1}{v_i} = \frac{K_m}{V}\left(1 + \frac{[I]}{K_i}\right)\frac{1}{[S]} + \frac{1}{V}$$

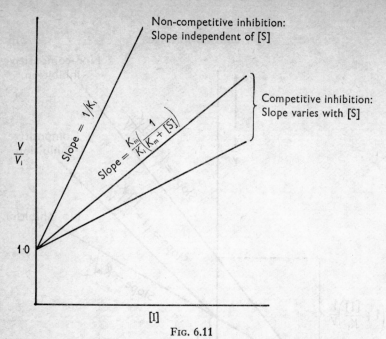

FIG. 6.11

Plot of v/v_i against [I] for competitive and non-competitive inhibitors. The slope of the former is dependent upon the substrate concentration [S].

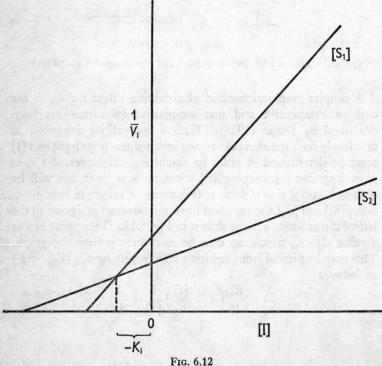

FIG. 6.12

Determination of K_i for a competitive inhibitor (DIXON, 1953).

The equation represents each line and at the point of intersection $1/v_i$ and $[I]$ will be the same for both lines as also will be V. Consequently

$$\left(1 + \frac{[I]}{K_i}\right) \frac{K_m}{[S_1]} = \left(1 + \frac{[I]}{K_i}\right) \frac{K_m}{[S_2]}.$$

This can be true only if either $[S_1] = [S_2]$ or if $[I] = -K_i$, and since the former is not true, $[I] = -K_i$. When K_i has been evaluated, the same graph may be employed to determine K_m since

each line cuts the abscissa at a value equal to $-K_i\left(\dfrac{[S]}{K_m} + 1\right)$.

An alternative method is given by Dixon for use if K_m has already been determined by the Lineweaver-Burk plot in the absence of an inhibitor. In this case it is necessary only to determine the inhibitory effects at one substrate concentration. The lines intersect at a point giving K_i at a height of $1/V$ and, as this value will already have been obtained in the original plot, a horizontal line drawn at a height of $1/V$ will intersect the inhibitor line at a point equal to $-K_i$.

With non-competitive inhibitors in the plot of $1/v_i$ against $[I]$ at two different substrate concentrations $[S_1]$ and $[S_2]$, the lines do cross but they meet at a point on the abscissa which gives an intercept of $-K_i$. This is proved by putting $1/v_i = 0$ in equation 6.28, and the type of graph obtained is shown in Fig. 6.13.

Uncompetitive Inhibition

An uncompetitive inhibitor combines with species of the enzyme which are themselves unable to combine with substrate, e.g. ES or some subsequent compound. The simplest model system is

$$E + S \rightleftharpoons ES \rightarrow E + P$$
$$ES + I \rightleftharpoons ESI$$

where ESI is a dead-end product. Thus by combining with such enzyme species, the inhibitor prevents the regeneration of the enzyme in a form capable of reacting with the substrate. The binding of the inhibitor is unaffected by increase of substrate concentration, which is also without effect on the release of the form of the enzyme capable of combining with substrate. Consequently, as with non-competitive inhibition, increasing the substrate concentration does not relieve the inhibition.

The rate equation is

$$v = k_{+2}[ES]$$

and the conservation equation

$$[E_0] = [E] + [ES] + [ESI]$$

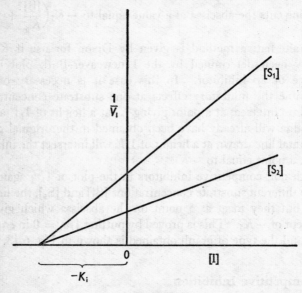

FIG. 6.13

Determination of K_I for a non-competitive inhibitor (DIXON, 1953).

By applying the general procedure it can be shown that

$$v = \frac{\dfrac{k_{+2}[E_0]}{\left(1 + \dfrac{[I]}{K_I}\right)}}{1 + \dfrac{K_m}{\left(1 + \dfrac{[I]}{K_I}\right)} \cdot \dfrac{1}{[S]}} \qquad . \qquad . \quad (6.29)$$

Comparison with the general form of equation 6.10 reveals that

$$C_1 = \frac{k_{+2}[E_0]}{\left(1+\dfrac{[I]}{K_i}\right)} = \frac{V}{\left(1+\dfrac{[I]}{K_i}\right)}$$

and

$$C_2 = \frac{K_m}{\left(1+\dfrac{[I]}{K_i}\right)}$$

Thus both V and K_m for an uncompetitive inhibitor are *decreased by the same factor*, namely $1/(1+[I]/K_i)$. In the plot of $1/v$ versus $1/[S]$, as $[I]$ is increased the slope of the line remains constant but the intercepts on both ordinate and abscissa increase by the constant factor $(1+[I]/K_i)$. This is illustrated in Fig. 6.14.

Uncompetitive inhibition is rarely encountered with single substrate reactions but is quite common in more complex cases involving multi-substrates.

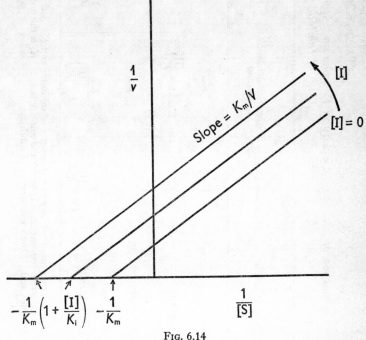

FIG. 6.14
Lineweaver-Burk plot for uncompetitive inhibition.

TABLE 6.2

Comparison of Maximum Velocity of Reaction and Apparent Michaelis Constant for Various Types of Inhibition

Type of inhibition	Maximum velocity	Apparent Michaelis constant	Lineweaver–Burk plot		
			Intercept on ordinate	Intercept on abscissa	Slope
None	V	K_m	$\dfrac{1}{V}$	$-\dfrac{1}{K_m}$	$\dfrac{K_m}{V}$
Competitive	V	$K_m\left(1+\dfrac{[I]}{K_i}\right)$	$\dfrac{1}{V}$	$-\dfrac{1}{K_m}\left(1+\dfrac{[I]}{K_i}\right)$	$\dfrac{K_m}{V}\left(1+\dfrac{[I]}{K_i}\right)$
Non-competitive	$\dfrac{V}{\left(1+\dfrac{[I]}{K_i}\right)}$	K_m	$\dfrac{1}{V}\left(1+\dfrac{[I]}{K_i}\right)$	$-\dfrac{1}{K_m}$	$\dfrac{K_m}{V}\left(1+\dfrac{[I]}{K_i}\right)$
Uncompetitive	$\dfrac{V}{\left(1+\dfrac{[I]}{K_i}\right)}$	$\dfrac{K_m}{\left(1+\dfrac{[I]}{K_i}\right)}$	$\dfrac{1}{V}\left(1+\dfrac{[I]}{K_i}\right)$	$-\dfrac{1}{K_m}\left(1+\dfrac{[I]}{K_i}\right)$	$\dfrac{K_m}{V}$

V is the maximum velocity in the absence of inhibitor and K_m is the Michaelis constant for the uninhibited reaction, i.e. the substrate concentration at which the velocity is $V/2$. K_i is the dissociation constant of the appropriate enzyme-inhibitor complex and $[I]$ the concentration of inhibitor.

Thus, by graphical means, the type of inhibition produced by given substances may be decided and it is normally possible to determine the dissociation constant K_i of the enzyme-inhibitor complex. Table 6.2 compares the parameters for the various types of inhibition.

Example 6.2.—It has been shown that the action of α-amylase on starch and on dextrins of high molecular weight is approximately 100 times as rapid as its action on the smaller dextrins formed subsequently. Schwimmer (1950) has investigated the kinetics of malt α-amylase action, including the effect of digestion products on the velocity of dextrinization of soluble starch by the enzyme, in an effort to elucidate this problem. His findings are given in the following table.

The hydrolysis of starch proceeds in the order

$$\text{Starch} \rightarrow \text{α-dextrin} \rightarrow \text{limit dextrin} \rightarrow \text{maltose.}$$

From the data provided it is possible to obtain, first, the value of K_m, the dissociation constant for the enzyme-substrate complex. This is obtained from the values of the relative hydrolysis velocities at different substrate concentrations in the absence of inhibitor by plotting $\frac{1}{[S]}$ against $\frac{1}{v}$. Thus

$[S]$	$\frac{1}{[S]}$	v	$\frac{1}{v}$
11·52	0·087	100	0·010
5·76	0·174	82	0·012
2·88	0·348	56	0·018
1·44	0·695	39	0·026

Inhibitor	Inhibitor concentration mg/ml	Substrate concentration, mg/ml			
		11·52	5·76	2·88	1·44
		Relative hydrolysis velocities			
None . .	0·00	100	82	56	39
Maltose . .	6·35	90	77	50	36
	12·70	82	66	46	31
	25·40	67	57	38	27
Limit dextrin .	6·68	102	80	44	31
	13·35	106	81	38	27
	26·70	100	84	29	19
α-Dextrin .	1·67	105	78	49	32
	3·34	100	76	45	28
	6·68	100	70	37	23

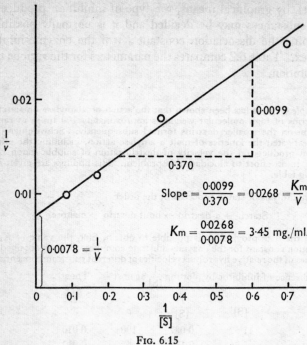

FIG. 6.15

These reciprocals are plotted in Fig. 6.15, from which a value of $K_m = 3.45$ mg per ml is obtained. As the molecular weight of the starch is not given (it is still a matter of some dispute), the value for K_m cannot be given in terms of the customary moles per litre.

To determine the type of inhibition produced by maltose, limit dextrin and α-dextrin, and the K_I values involved, it is necessary to plot either $1/v$ against $1/[S]$ or, better in this instance, v/v_I against $[I]$.

As the velocities in the presence and absence of varying inhibitor concentrations are given, it is possible to calculate the v/v_I ratios for each substrate concentration. Thus, for maltose, the following figures are obtained:

Maltose concentration mg/ml	Substrate concentration, mg/ml											
	11·52			5·76			2·88			1·44		
[I]	v	v_I	$\frac{v}{v_I}$	v	v_I	$\frac{v}{v_I}$	v	v_I	$\frac{v}{v_I}$	v	v_I	$\frac{v}{v_I}$
0·00	100	—	—	82	—	—	56	—	—	39	—	—
6·35	—	90	1·11	—	77	1·065	—	50	1·12	—	36	1·08
12·70	—	82	1·20	—	66	1·24	—	46	1·22	—	31	1·26
25·40	—	67	1·50	—	57	1·44	—	38	1·48	—	27	1·44

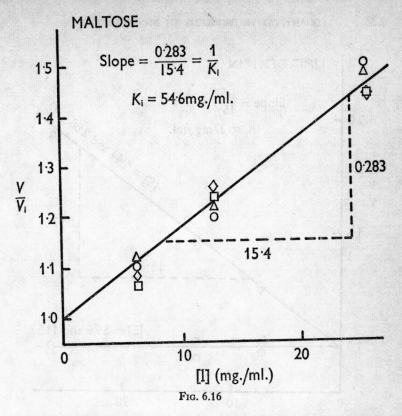

$$\text{Slope} = \frac{0\cdot283}{15\cdot4} = \frac{1}{K_i}$$

$$K_i = 54\cdot6\text{mg./ml.}$$

FIG. 6.16

Fig. 6.16 shows the plot of these v/v_I values against [I], from which it will be observed that the sets of points for each inhibitor concentration are such that one straight line suffices for all, and there is an intercept on the ordinate of 1·0. This means that the slope of the line is independent of the substrate concentration [S], and therefore the inhibition is non-competitive. Consequently the slope of the line is equal to $1/K_I$ and this enables K_I to be evaluated. A figure of 54·6 mg per ml is obtained.

Exactly the same procedure is carried out for the limit dextrin and α-dextrin inhibitions, but, below, the full working out of the v/v_I values is omitted and the figures for them are simply recorded in the table.

Inhibitor concentration [I] mg/ml		v/v_I values for substrate concentration mg/ml			
		11·52	5·76	2·88	1·44
Limit dextrin	6·68	0·98	1·03	1·27	1·26
	13·35	0·945	1·02	1·47	1·44
	26·70	1·00	0·98	1·93	2·05
α-Dextrin	1·67	0·95	1·05	1·11	1·22
	3·34	1·00	1·08	1·24	1·39
	6·68	1·00	1·17	1·51	1·70

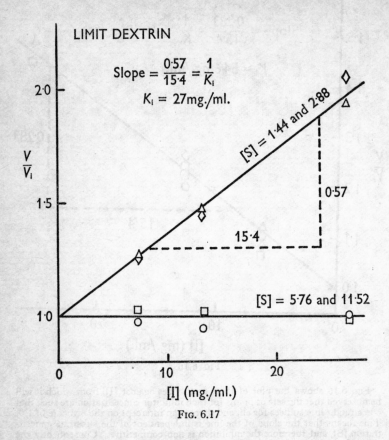

FIG. 6.17

Figs. 6.17 and 6.18 show the plots of v/v_i against [I]. From Fig. 6.17 it will be noticed that at the highest substrate concentrations the limit dextrin does not inhibit over the concentration range employed, i.e. a straight line parallel to the [I] axis is obtained. At the two lowest substrate concentrations, however, inhibition does occur, but is the same in both cases ; one straight line suffices for all the points. This indicates that the inhibition is non-competitive since the slope is obviously independent of substrate concentration in the range where inhibition occurs. Here again, K_i can be evaluated from the slope and a value of 27 mg per ml obtained. The behaviour of limit dextrin in inhibiting non-competitively at lower substrate concentrations and not at all at higher concentrations is unusual and no explanation has been offered to account for it. In Fig. 6.18 the results with α-dextrin are shown and a difference is at once noticeable. At the highest substrate concentration there is no inhibition, but with the other three concentrations inhibition does occur and the slopes of the lines increase with decreasing substrate concentration, i.e. increased inhibition. Here, therefore, the inhibition is competitive in nature. The values of K_i may be obtained for the different substrate concentrations by applying equation 6.18. The slope of the line where competitive inhibition

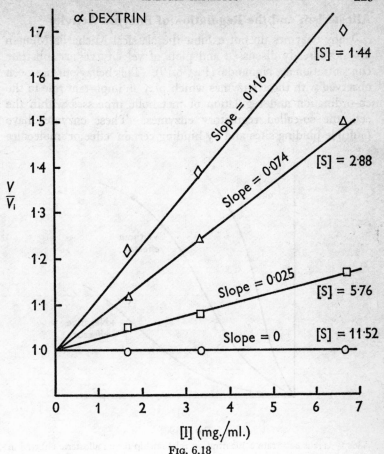

α DEXTRIN

FIG. 6.18

occurs is equal to $\dfrac{K_m}{K_I}\left(\dfrac{1}{K_m + [S]}\right)$. The value of K_m was obtained previously from the plot of $1/v$ against $1/[S]$ and K_I therefore can be evaluated.

$$K_m = 3 \cdot 45 \text{ mg/ml}$$

Substrate concentration, [S] mg/ml	Slope of line	$\dfrac{K_m}{slope}$	$K_m + [S]$	$\dfrac{1}{K_m + [S]}$	K_I mg/ml
1·44	0·116	29·75	4·89	0·205	6·1
2·88	0·074	46·6	6·33	0·158	7·4
5·76	0·025	138	9·21	0·108	14·9

These results show that the enzyme has less affinity for the inhibitor as the substrate concentration increases, i.e. K_I increases as the substrate concentration increases.

Allosterism and the Regulation of Enzyme Activity

Many enzymes do not exhibit the classical Michaelis-Menten kinetics already discussed and plots of velocity versus substrate concentration are sigmoidal (Fig. 6.19). This behaviour has been observed with those enzymes which play an important role in the co-ordination and regulation of metabolic processes within the cell, the so-called regulatory enzymes. These enzymes have multiple binding sites and by binding certain 'effector' molecules

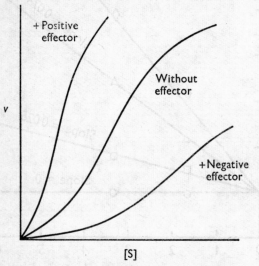

FIG. 6.19

Velocity versus substrate concentration relationship for an allosteric enzyme in the presence and absence of added positive and negative effectors.

to specific sites other than the active site the activity of the enzyme may be either increased or decreased, i.e. the effectors serve either as reversible activators or inhibitors. Frequently there is no structural similarity between effectors and substrates.

It is believed that the combination of an effector with an enzyme produces conformational changes in the protein which alter the affinity of the enzyme for its substrate. Often combination with an effector changes the observed kinetics from sigmoidal to hyperbolic. To distinguish between effectors which bind directly to the active site and those that bind to other sites, they are referred to respectively as *isosteric* and *allosteric effectors* and the binding sites other than the active site are termed *allosteric sites*.

It is useful to classify allosteric effects as being either :

1. *Homotropic*, where interactions, which may be either co-operative or antagonistic, occur between identical ligands, e.g. between substrate or co-enzyme or inhibitor ; or

2. *Heterotropic*, where interactions, which may be either co-operative or antagonistic, occur between different ligands.

In the case of homotropic interaction between substrate molecules the sigmoidal curve reflects a situation where two molecules of substrate react with the enzyme and the binding of the first molecule in some way facilitates the binding of the second or alters the rate of reaction of the second bound molecule, i.e. there is a co-operative effect in binding more than one molecule of substrate to the enzyme. This behaviour implies that a threshold concentration of substrate exists, below which changes in concentration have relatively little effect on enzyme activity, yet when exceeded slight changes in concentration have profound effects. Thus, over quite a narrow range of substrate concentration the activity of the enzyme responds in a very sensitive fashion, a desirable feature in a regulatory enzyme.

Present evidence indicates that many of the enzymes subject to allosteric control are composed of sub-units, a finding which supports a model proposed by Monod, Wyman and Changeux (1965). They suggested that the allosteric protein is an oligomer, comprising a symmetrical aggregate of protein structural units or protomers. Each protomer possesses one site capable of forming a stereospecific complex with a particular effector, and when binding of this effector occurs inter-protomer binding is modified and the quaternary structure of the oligomer is in some way affected. It is suggested that the oligomer exists in at least two catalytically active states and that the equilibrium between them is influenced by the binding of the effector. These authors have developed a full mathematical treatment of their model to which the reader is referred.

A sigmoidal binding curve was first observed for the combination of oxygen with haemoglobin

$$\text{Hb}_n + n\text{O}_2 \rightleftharpoons [\text{HbO}_2]_n$$

The kinetics are described by the Hill equation, which is also applicable to enzymic reactions of the general type

$$E + nS \rightleftharpoons ES_n \rightarrow E + nP$$

where it is assumed that ES_n is the product-forming complex. The Hill equation may be written in the form

$$v = \frac{V}{1 + \dfrac{K'}{[S]^n}} \qquad \cdot \qquad \cdot \qquad \cdot \qquad (6.30a)$$

where n is the number of substrate binding sites, provided that there is infinite interaction between these binding sites. Clearly when $n = 1$, and there is no interaction between the sites, the curve is hyperbolic. There are also important cases of known enzymes for which $n < 1$ and where, consequently, there are antagonistic effects.

Rearrangement of equation 6.30a gives

$$\frac{v}{V} = \frac{[S]^n}{[S]^n + K'} \qquad \cdot \qquad \cdot \qquad \cdot \qquad (6.30b)$$

and

$$[S]^n \frac{(V - v)}{v} = K'$$

Taking logarithms

$$n \log [S] + \log \frac{(V - v)}{v} = \log K'$$

and

$$\log \frac{v}{(V - v)} = n \log [S] - \log K' . \qquad . \qquad (6.31)$$

This is the equation of a straight line if $\log v/(V-v)$ is plotted versus $\log [S]$, the slope being n and the intercept on the ordinate $-\log K'$.

When $v = V/2$, $v/(V-v) = 1$ and $\log v/(V-v) = 0$. Thus the value of $\log [S]$ corresponding to $\log v/(V-v) = 0$ gives the logarithm of the substrate concentration at half maximum velocity. Obviously for sigmoidal kinetics, i.e. when $n \neq 1$, K' is not equivalent to the substrate concentration at half maximum velocity of reaction. When $\log v/(V-v) = 0$ then

$$\log K' = n \log [S_{\frac{1}{2}}]$$

where $[S_{\frac{1}{2}}]$ denotes the substrate concentration at $V/2$, and therefore

$$K' = [S_{\frac{1}{2}}]^n$$

The value of n, which is referred to as the *interaction coefficient*,
can be determined from the slope of the linear Hill plot (Fig. 6.20).
While clearly n must be an integer, experimentally determined
values are often fractional and the next highest integer is then
usually interpreted as the number of binding sites.

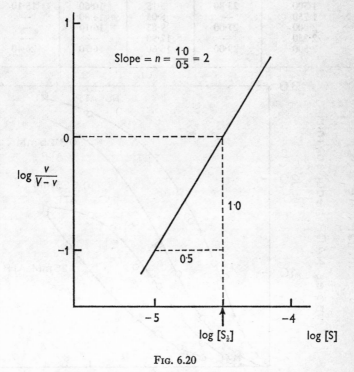

FIG. 6.20

Linear plot of the Hill equation and determination of the interaction coefficient, n.

Example 6.3.—The kinetics of the glucose 6-phosphate dehydrogenase of
the nitrogen-fixing bacterium *Azotobacter beijerinckii* have been investigated and
the action of ATP on the activity of the enzyme studied. The following data
were obtained in a series of assays (each using 0·75 μg of protein) for the effect
of glucose 6-phosphate concentration, in the absence and presence of three
concentrations of ATP, on the initial velocity of reaction :

Glucose 6-phosphate + NADP$^+$ ⇌ Gluconate 6-phosphate + NADPH + H$^+$

| Glucose 6-phosphate concentration mM | Initial velocity of NADP⁺ reduction (nmoles/min) | | | |
| | ATP concentration (mM) | | | |
	0	1·25	0·625	0·3125
0·375	5·15	0·35	0·64	1·48
0·500	9·65	0·43	0·97	1·93
0·625	10·95	0·58	—	—
0·750	14·15	0·97	2·75	5·15
0·875	15·45	1·29	—	—
1·000	18·65	2·25	5·47	9·97
1·250	22·70	3·54	8·69	13·00
1·500	23·80	5·15	10·60	15·10
1·750	—	8·05	14·80	—
2·000	27·00	9·83	16·10	—
2·250	—	12·90	—	—
2·500	29·00	16·60	16·70	26·40

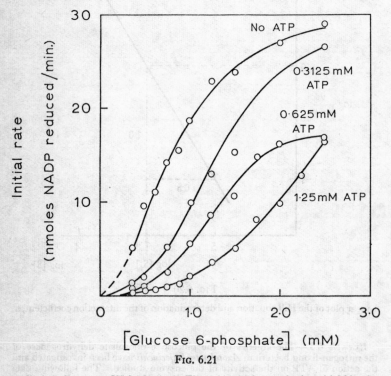

Fig. 6.21

A plot of v versus [S] is shown in Fig. 6.21 which reveals sigmoidal kinetics in the absence of ATP, suggesting that the enzyme is an allosteric protein. A Hill plot, in which log $[v/(V-v)]$ is plotted against log [glucose 6-phosphate], is now carried out. The value of the maximum velocity of reaction V is taken from the v versus [S] plot; it is, of course, subject to the errors already discussed. Assume, then, that $V = 29 \cdot 5$ nmoles/min for the system employed. (Note that, for comparison with other enzymes, it is customary to record maximum velocities of reaction as μmoles/min/mg protein. Thus in the present case, since $0 \cdot 75$ μg of protein was used in each assay

$$29 \cdot 5 \text{ nmoles/min}/0 \cdot 75 \text{ }\mu\text{g protein}$$

$$= \frac{29 \cdot 5 \times 10^3}{0 \cdot 75 \times 10^3} \text{ }\mu\text{moles/min/mg protein}$$

$$= 39 \cdot 3 \text{ }\mu\text{moles/min/mg protein.}$$

For the present purpose it is not necessary to convert all the values recorded into these units, since they were all obtained with the same enzyme preparation used at $0 \cdot 75$ μg per assay).

The operation for the reaction in the absence of ATP is detailed fully below, and for the three ATP concentrations is presented in summary form.

Glucose 6-phosphate concentration mM		Reaction in absence of ATP $V = 29 \cdot 50$ nmoles/min			
[S]	log [S]	v	$V-v$	$\dfrac{v}{V-v}$	$\log \dfrac{v}{V-v}$
0·375	$\bar{1}$·574	5·15	24·35	0·212	$\bar{1}$·325
0·500	$\bar{1}$·699	9·65	19·85	0·486	$\bar{1}$·687
0·625	$\bar{1}$·796	10·95	18·55	0·590	$\bar{1}$·771
0·750	$\bar{1}$·875	14·15	15·35	0·921	$\bar{1}$·964
0·875	$\bar{1}$·942	15·45	14·05	1·100	0·041
1·000	0·000	18·65	10·85	1·720	0·236
1·250	0·097	22·70	6·80	3·340	0·524
1·500	0·176	23·80	5·70	4·175	0·621
1·750	0·243	—	—	—	—
2·000	0·301	27·00	2·50	10·80	1·033
2·250	0·352	—	—	—	—
2·500	0·398	29·00	0·50	58·00	1·763

A summary of values of $\log v/(V-v)$ for all the experiments is tabulated below.

log [S]	$\log \dfrac{v}{V-v}$ for ATP concentrations (mM) of			
	0	1·25	0·625	0·3125
$\bar{1}$·574	$\bar{1}$·325	$\bar{2}$·084	$\bar{2}$·348	$\bar{2}$·719
$\bar{1}$·699	$\bar{1}$·687	$\bar{2}$·170	$\bar{2}$·529	$\bar{2}$·845
$\bar{1}$·796	$\bar{1}$·771	$\bar{2}$·322	—	—
$\bar{1}$·875	$\bar{1}$·964	$\bar{2}$·529	$\bar{2}$·980	$\bar{1}$·325
$\bar{1}$·942	0·041	$\bar{2}$·655	—	—
0·000	0·236	$\bar{2}$·917	$\bar{1}$·357	$\bar{1}$·708
0·097	0·524	$\bar{1}$·135	$\bar{1}$·620	$\bar{1}$·897
0·176	0·621	$\bar{1}$·325	$\bar{1}$·749	0·021
0·243	—	$\bar{1}$·574	0·002	—
0·301	1·033	$\bar{1}$·699	0·079	—
0·352	—	$\bar{1}$·890	—	—
0·398	1·763	0·109	0·116	0·930

The Hill plot is shown in Fig. 6.22 from which the four lines give an average slope of $+2\cdot55$. The interaction coefficient is thus 2·55 and it may be concluded that the enzyme has at least three binding sites for glucose 6-phosphate. (Unpublished data of P. J. SENIOR.)

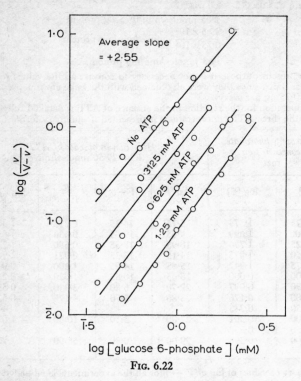

FIG. 6.22

Isoenzymes

It has been discovered that certain enzymes exist in more than one structural form within the same species, or even within the same cell. These multiple forms of an enzyme have been termed *isoenzymes* (or, less desirably, isozymes).

Lactate dehydrogenase was one of the first enzymes of this category to be investigated and it was found to exist in five different forms which could be separated by gel electrophoresis. Each form had the same molecular weight of about 134,000 and comprised four polypeptide chains, each of molecular weight 33,500. Two types of polypeptide chain were discovered and designated M and H respectively, since the lactate dehydrogenase of skeletal muscle consisted of four of the former sub-units whereas the enzyme from heart muscle comprised four of the latter type. The other three isoenzymes are composed of different combinations of the M and H chains. Thus the five different isoenzymes may be represented as :

$$M_4 \qquad M_3H \qquad M_2H_2 \qquad MH_3 \qquad H_4$$

The separated, individual M and H chains are devoid of enzymic activity and differ significantly in their amino acid content and sequence and in their immunological response. By mixing the M and H chains together in appropriate proportions it is possible to obtain all the isoenzymes of lactate dehydrogenase *in vitro*.

The five isoenzymes all catalyse the same reaction but have quite different K_m values for their substrates and exhibit different turnover numbers. Clearly, then, the activity of lactate dehydrogenase can be controlled by virtue of its molecular structure, and present evidence indicates that the two different polypeptide chains, M and H, are coded by two separate genes and that the relative proportions of the two types of chain synthesized, and hence the particular isoenzyme produced in a given cell, are under genetic control.

Many other enzymes have now been shown to exist in the form of isoenzymes and in each case they have been found to consist of various combinations of different polypeptide chains.

Molecular and Catalytic Centre Activity

The activity of an enzyme was formerly expressed in terms of its *turnover number*, which was defined as the number of molecules of substrate transformed per minute when the enzyme is working at its maximum rate, i.e. optimal substrate concentration, pH and temperature. However, in view of the fact that 'turnover number' was used in several different senses by various authors, the Commission on Enzymes has recommended (1961) that the term should be discontinued and replaced by two more explicit terms which cover the cases where confusion previously occurred.

MOLECULAR ACTIVITY.—This is defined as the number of molecules of substrate (or equivalents of the group concerned) transformed per minute by one molecule of enzyme at optimal substrate concentration.

CATALYTIC CENTRE ACTIVITY.—When the enzyme has a

distinguishing prosthetic group or catalytic centre whose concentration can be measured, the catalytic power can be expressed as the number of molecules of substrate transformed per minute per catalytic centre.

Thus the molecular activity and the catalytic centre activity will be identical if the enzyme molecule possesses one active centre ; if the enzyme molecule has n centres then

$$\text{molecular activity} = n \times \text{catalytic centre activity.}$$

SPECIFIC ACTIVITY.—The determination of molecular and catalytic centre activities demands a knowledge of the molecular weight of the enzyme and in many instances this is not known. For this reason it is convenient to have some other basis for recording enzyme activity, especially where enzyme isolation and purification is being attempted and information concerning the purity of the enzyme preparation is required. The accepted method in these circumstances is to measure the *specific activity* of the enzyme preparation, which is defined as enzyme units per milligram of protein. One *unit* (U) of any enzyme is defined as that amount which will catalyse the transformation of one micromole of substrate per minute, or, where more than one bond of each substrate molecule is attacked, one micro-equivalent of the group concerned per minute, under defined conditions, e.g. temperature stated (25° where practicable) and optimum pH and substrate concentration. Where two identical molecules react together, the unit will be the amount of enzyme which catalyses the transformation of two micromoles per minute.

REFERENCES AND SUGGESTED READING

ALBERTY, R. A. (1966). *Adv. Enzymol.*, **17**, 1.

BOYER, P. D., LARDY, H. & MYRBÄCK, K., Editors (1959). *The Enzymes*, vol. 1, 2nd ed., Chaps 1, 2, 3 and 4. New York : Academic Press.

BRAY, H. G. & WHITE, K. (1967). *Kinetics and Thermodynamics in Biochemistry*, 2nd ed. Chap 7. London : Churchill.

CLELAND, W. W. (1963). The kinetics of enzyme-catalysed reactions with two or more substrates or products. *Biochim. biophys. Acta*, **67**, 104, 173, 188.

CLELAND, W. W. (1970). Steady state kinetics. *The Enzymes*, vol. 2, 3rd ed. New York : Academic Press.

DIXON, M. (1953). *Biochem J.*, **55**, 170.

DIXON, M. & WEBB, E. C. (1964). *Enzymes*, 2nd ed., Chap IV. London : Longmans, Green.

KING, E. L. & ALTMAN, C. (1956). *J. phys. Chem.*, **60**, 1375.

LINEWEAVER, H. & BURK, D. (1934). *J. Am. chem. Soc.*, **56**, 658.

MAHLER, H. R. & CORDES, E. H. (1966). *Biological Chemistry*, Chap 6. New York : Harper & Row.

MICHAELIS, L. & MENTEN, M. L. (1913). Dukinetik der Invertinwirkung. *Biochem. Z.*, **49**, 333.

MONOD, J., WYMAN, J. & CHANGEUX, J.-P. (1965). On the nature of allosteric transitions: a plausible model. *J. molec. Biol.*, **12**, 88.

REINER, J. M. (1959). *Behaviour of Enzyme Systems*. Minneapolis : Burgess.

REINER, J. M. (1964). Quantitative aspects of enzymes and enzyme systems. In *Comprehensive Biochemistry*, **12**, 126, ed. M. Florkin & E. H. Stotz. Amsterdam : Elsevier.

SCHWIMMER, S. (1950). Kinetics of malt α-amylase action. *J. biol. Chem.*, **186**, 181.

PROBLEMS

6.1. The following rates of evolution of carbon dioxide were obtained in the decarboxylation of an amino acid at various concentrations.

$\mu l.$ $CO_2/5$ min	Amino acid concentration millimolar
9·65	0·6
15·2	1·0
22·3	1·5
25·6	2·0
34·5	3·0
40·0	4·0

Calculate the Michaelis constant of the enzyme.

(Leeds General Studies Course, 1954.)

6.2. The deamination of both L-serine and L-threonine by *Escherichia coli* is believed to be catalysed by one and the same enzyme. The saturation curves for enzyme with substrate were obtained, using purified enzyme preparations, in the following manner :

Enzyme, substrate, co-factor additions (glutathione and adenosine 5-phosphate) and phosphate buffer pH 7·8, to give a final volume of 1 ml, were incubated at 37° C for 10 minutes and then the keto acid production determined in each case. Results obtained were as follows :

Substrate concentration mg/ml	Keto acid formed per 10 minutes μmoles	
	L-serine substrate	L-threonine substrate
0·05	0·14	0·29
0·15	0·37	0·45
0·25	—	1·00
0·50	0·58	1·25
1·00	0·71	1·55
5·00	0·89	1·60

From these data determine the Michaelis constant for the enzyme with respect to each substrate. What would you expect to happen if the enzyme were added to an equimolar mixture of L-serine and L-threonine ?

(After WOOD & GUNSALUS (1949), *J. biol. Chem.*, **181**, 171.)

6.3. The effect of pH on the Michaelis constant of the 6-phosphogluconate dehydrogenase of rat liver has been investigated by Glock and McLean. Using glycylglycine buffer of pH 7·6 and 9·0, the activity of the dehydrogenase was measured spectrophotometrically by following the rate of reduction of nicotina-mide-adenine dinucleotide phosphate at 340 nm.

Substrate concentration	Increase in extinction in initial 5 minutes	
$\times 10^4$ M	pH 7·6	pH 9·0
0·174	0·074	0·034
0·267	0·085	0·047
0·526	0·098	0·075
1·666	0·114	0·128
4·00	—	0·167

Determine at which pH value the enzyme has its greatest affinity for the substrate.

(After GLOCK & McLEAN (1953), *Biochem. J.*, **55**, 400.)

6.4.   The following data were obtained in an experiment to determine the coenzyme saturation curve for a partially purified sample of tryptophanase, which catalyses the reaction

Tryptophan → Indole + Pyruvic acid + Ammonia.

The coenzyme used was barium pyridoxal phosphate.   In each determination 2 mg L-tryptophan were incubated with 0·03 ml apoenzyme in 2 ml buffer.

Indole formed in 10 minutes (μg)	Barium pyridoxal phosphate present μg/ml
1·7	0·3
2·7	0·5
4·3	1·0
6·5	2·0
7·3	3·0
7·8	4·0
7·9	5·0
8·1	10·0

Calculate the dissociation constant of the pyridoxal phosphate-tryptophanase complex.

(After WOOD, GUNSALUS & UMBREIT (1947), *J. biol. Chem.*, **170**, 313 ;   Glasgow Honours Course Finals 1953.)

6.5.   The substrate saturation curve for tryptophanase (see Problem 6.4) has been determined and the following data obtained :

L-tryptophan concentration (mg per 10 ml)	Indole produced per 15 min (μg per 10 ml)
0·25	23·9
0·50	28·6
1·0	37·7
1·5	41·0
2·0	43·2
3·0	43·2
4·0	41·9
5·0	44·4

From these figures obtain a value for the Michaelis constant for tryptophanase using the Lineweaver and Burk method of evaluation.

(After DAWES & HAPPOLD (1949), *Biochem. J.*, **44**, 349.)

6.6. Pig heart sarcosomes catalyse the oxidative phosphorylation of α-oxoglutaric acid according to the equation

$$\alpha\text{-oxoglutarate} + \tfrac{1}{2}O_2 + x\text{ADP} + x\text{H}_3\text{PO}_4 \rightarrow \text{succinate} + CO_2 + x\text{ATP} + x\text{H}_2\text{O}.$$

This reaction can be coupled to the phosphorylation of glucose to hexosemonophosphate (HMP) by the addition of glucose and yeast hexokinase, e.g.

$$x\text{ATP} + x\text{glucose} \rightarrow x\text{HMP} + x\text{ADP}.$$

ADP is thus regenerated and plays a catalytic role in the overall reaction, while the amount of phosphorylation occurring is measured by determining the increase in HMP concentration. The following information has been obtained regarding the effect of concentration of sarcosomes on the rate of oxidation of α-oxoglutarate and on phosphorylation of glucose to HMP.

	Sarcosomal protein (mg/ml)			
	0·12	0·31	0·62	1·25
Oxygen uptake (μg atoms)				
Without ADP .	0·26	0·64	1·68	4·15
With excess ADP	0·70	1·76	3·56	6·26
ΔHMP (μmoles)				
Without ADP .	0	0·19	1·24	4·47
With excess ADP	1·82	4·36	8·46	15·8

Use the Michaelis-Menten equation to determine the average Michaelis constants of the endogenous ADP of the sarcosomes for oxidation and for phosphorylation. The sarcosome content of energy-rich phosphate (ADP) in μmoles is given by mg sarcosomal protein × 0·018.

(After SLATER & HOLTON (1953), *Biochem. J.*, **55**, 530.)

6.7. What factors affect the rate of enzymic reaction? In an enzyme catalysed reaction the following data were obtained :

Velocity in moles/hour	Substrate concentration molar
0·1230	5·000
0·1212	2·500
0·1192	1·667
0·1177	1·250
0·1160	1·000
0·1143	0·833
0·1127	0·715
0·1111	0·625
0·1096	0·555
0·1081	0·500

Calculate the Michaelis constant of the enzyme and the maximum attainable reaction velocity.

(Leeds Honours Course Finals, 1954.)

6.8. The effect of temperature on the Michaelis constant of an enzyme has been investigated using the luciferin-luciferase system of the luminescent ostracod crustacean *Cypridina*.

The initial velocities of the reaction were followed by measuring photo-electrically the amount of light produced per unit time. Experimental conditions restricted the measurements to two temperatures viz. 15° and 22° C, and four series of experiments were performed, therefore, at each temperature to compensate for this limitation. Results from one such experiment are given below.

Ml of luciferin solution per 20 ml reaction mixture	Initial velocity in millivolts/minute	
	15°	22°
0·04	15	18
0·06	19	22
0·08	23	33
0·10	26	37
0·20	44	62
0·40	56	91
0·90	76	123

The workers estimated the concentration of the luciferin solution to be 8×10^{-5} M. Assuming this molarity, determine the effect of change of temperature on the Michaelis constant and comment on your results.

(After KAUZMANN, CHASE & BRIGHAM (1949), *Arch. Biochem.*, **24**, 281.)

6.9. Carbobenzoxy-L-glutamyl-L-tyrosine is hydrolysed by swine kidney pepsinase (cathepsin) in accordance with the equation :

Carbobenzoxy-L-glutamyl-L-tyrosine →

carbobenzoxy-L-glutamic acid + L-tyrosine.

The reaction may be followed manometrically by the addition of tyrosine decarboxylase when CO_2 is evolved and tyramine formed,

i.e.   L-tyrosine → tyramine + CO_2.

It was discovered that the carbobenzoxy-L-glutamic acid formed in the hydrolysis had an inhibitory effect on the progress of the reaction and hence caused a decrease in reaction rate. The nature of this inhibition was investigated by Frantz and Stephenson, who added carbobenzoxy-L-glutamic acid, at two different concentration levels, to the reaction mixture. Some of their data are recorded in the following table. From these figures demonstrate graphically that carbobenzoxy-L-glutamic acid is a competitive inhibitor of the cathepsin reaction and obtain values for the Michaelis constant of cathepsin and also K_I, the dissociation constant of the enzyme-inhibitor complex.

Concentration of carbobenzoxy-L-glutamyl-L-tyrosine μmoles per ml	Concentration of carbobenzoxy-L-glutamic acid μmoles per ml	Rate of CO_2 production μmoles per ml per min
4·7	0·0	0·0434
4·7	7·58	0·0285
4·7	30·3	0.0133
6·5	0·0	0·0526
6·5	7·58	0·0357
6·5	30·3	0·0147
10·8	0·0	0·0713
10·8	7·58	0·0512
10·8	30·3	0·0266
30·3	0·0	0·1111
30·3	7·58	0·0909
30·3	30·3	0·0581

(After FRANTZ & STEPHENSON (1947), *J. biol. Chem.*, 169, 359.)

6.10. The action of various compounds on crystalline carboxypeptidase activity has been studied by following their effects on the hydrolysis of L-carbobenzoxy-glycyl-D-phenylalanine. The course of the reaction was followed colorimetrically by use of ninhydrin and compounds investigated included phenylacetate, hydrocinnamate, phenylbutyrate and benzoate. Data obtained are recorded in the following table, from which determine the types of inhibition involved.

Inhibitor	Initial substrate concentration M	Reaction rate arbitrary units
None	0·0713	166
	0·0581	142·6
	0·0384	111
	0·0285	111
	0·0125	66
0·002 M Hydrocinnamate	0·100	16
	0·050	11·1
	0·040	7·6
	0·025	5·55
0·002 M Phenylacetate	0·047	40
	0·033	33·3
	0·0166	18·1
0·002 M Phenylbutyrate	0·055	90·9
	0·040	57·1
	0·025	50·0
	0·0125	28·5
0·05 M Benzoate	0·100	40·8
	0·050	38·4
	0·025	33·3
	0·0175	30·3

(After ELKINS-KAUFMAN & NEURATH (1949), *J. biol. Chem.*, 178, 645.)

6.11. The following data were obtained in an experiment to determine the molecular activity ('turnover number') of catalase :

The enzyme preparation, which was 85 per cent. pure, had a dry weight of 2·95 mg per litre. Manometrically it was ascertained that 0·1 ml of this preparation liberated 340 μl oxygen, measured at N.T.P., from excess hydrogen peroxide in 10 minutes. The molecular weight of catalase may be taken as 225,000.

Use these data to calculate the molecular activity of catalase. (The transition from α to β activity of the catalase preparation may be ignored and a uniform rate of oxygen evolution assumed.)

(Glasgow Honours Course Finals, 1952.)

6.12. The β-glucuronidase optimally active at pH 3·4 is inhibited by mucic acid. The following data were obtained in an experiment where the effect of variation of substrate concentration on reaction velocity was studied in the absence and presence of 0·0001 M-mucate. The substrate was phenolphthalein glucuronide and the reaction velocity is expressed in μg phenolphthalein liberated in 60 minutes at 37°. Determine the type of inhibitor and the enzyme inhibitor dissociation constant.

Substrate concentration millimolar	Reaction velocity	
	No inhibitor	0·0001 M-mucate
1	32·0	3·0
2	43·6	6·3
3	50·4	9·0
4	53·0	11·8
5	56·5	14·5
10	62·1	24·4

(Glasgow Honours Course Finals, 1954.)

6.13. The following data were obtained in an experiment to determine the affinity between tropine esterase and atropine sulphate as substrate. The reaction was followed manometrically in bicarbonate buffer by measuring the CO_2 evolution resulting from acid liberated by the hydrolysis of the atropine sulphate. Rabbit serum (2·5 per cent.) was the source of the enzyme and the reaction was carried out at 30° and pH 7·4.

Concentration of atropine sulphate		CO_2 liberated in 100 min
Per cent.	10^3 M	mm³
0·250	7·4	28·5
0·100	3·0	28·0
0·050	1·5	28·0
0·025	0·74	26·6
0·0125	0·37	25·0
0·010	0·30	23·5
0·005	0·15	20·3

Determine the Michaelis constant of tropine esterase.

(After GLICK (1940), *J. biol. Chem.*, 134, 617.)

6.14. Derive an expression relating the initial velocity of an enzyme reaction to the concentration of substrate present.

The enzymic decomposition of a substrate A is inhibited by compounds B and C. The following table gives initial velocities of decomposition of A (in micromoles/ml/min) in the presence of B and C, separately, at concentrations stated (in millimoles/ml). The amount of enzyme present was the same in every case.

(a) Conc. of A : 10

Conc. of B :	0	0·5	2·0	3·0	4·0	8·0
Velocity :	10	8·0	5·3	4·2	3·6	2·2
Conc. of C :	0	1·0	2·0	3·0	4·0	5·0
Velocity :	10	4·5	2·9	2·1	1·68	1·39

(b) Conc. of A : 30

Conc. of B :	0	1·5	2·5	3·5	5·0	7·0
Velocity :	20	11·8	9·0	7·8	6·2	4·8
Conc. of C :	0	1·0	2·0	3·0	4·0	8·0
Velocity :	20	15·4	11·8	10·0	8·4	5·3

What conclusions can you draw about the action of the inhibitors ? Give your reasons.

(Leeds Honours Course Finals, 1955.)

6.15. The influence of ATP concentration on the rate of dephosphorylation of ATP by myosin, which catalyses the reaction

$$ATP \rightarrow ADP + \text{inorganic phosphate}$$

has been studied at 25° and pH 7·0. The following data were obtained :

Velocity of reaction	ATP concentration
μmoles inorganic phosphate produced/litre/sec	μM
0·067	7·5
0·095	12·5
0·119	20·0
0·149	32·5
0·185	62·5
0·191	155·0
0·195	320·0

Determine the Michaelis constant of myosin.

(After OUELLET, LAIDLER & MORALES (1952), *Arch. Biochem. Biophys.*, 39, 37.)

6.16. The citrate-oxaloacetate lyase (citratase or citridesmolase) enzyme of *Aerobacter aerogenes* carries out the reaction :

$$Citrate \rightarrow Oxaloacetate + Acetate$$

and requires magnesium as a cofactor.

The progress of the reaction may be followed by determining the keto acid produced. The following data were obtained in experiments to investigate the effect of calcium ions on the reaction, and at the times of analysis keto acid was being produced linearly with time. The reaction mixture consisted of phosphate buffer pH 7·0, enzyme and citrate; magnesium and calcium ions were added to various concentrations.

1. Concentration of Mg^{2+} in reaction mixture, 8·35 mM. Incubation time, 100 sec.

ml. $CaCl_2$ (0·025 M) added to 3 ml reaction mixture :	0	0·1	0·2	0·3	0·4
μmoles/litre of keto acid produced:	1175	965	725	615	540

2. Concentration of Mg^{2+} in reaction mixture, 2·88 mM. Incubation time, 120 sec.

ml $CaCl_2$ (0·025 M) added to 3 ml reaction mixture :	0	0·1	0·2	0·3	0·4
μmoles/litre of keto acid produced :	1020	676	480	370	316

3. Concentration of Mg^{2+} in reaction mixture, 0·96 mM. Incubation time, 180 sec.

ml $CaCl_2$ (0·025 M) added to 3 ml reaction mixture :	0	0·1	0·2	0·3	0·4
μmoles/litre of keto acid produced :	865	420	286	226	166

4. Concentration of Mg^{2+} in reaction mixture, 0·32 mM. Incubation time, 210 sec.

ml $CaCl_2$ (0·025 M) added to 3 ml reaction mixture :	0	0·1	0·2	0·3
μmoles/litre of keto acid produced :	590	220	130	105

Use these data to determine

(a) the Michaelis constant of citrate-oxaloacetate lyase for Mg^{2+},

(b) the type of inhibition produced by Ca^{2+},

(c) the dissociation constant of the enzyme-inhibitor complex.

(After DAGLEY & DAWES (1955), *Biochim. biophys. Acta*, 17, 177.)

6.17. In a study of the action of phosphatase on glucose 1-phosphate inorganic phosphate was liberated at the following rates at the ester concentrations stated :

Inorganic phosphate liberated μg	Concentration of glucose 1-phosphate μmole/ml
6·8	0·5
10·1	1·0
12·2	1·5
14·1	2·0
15·1	3·0
16·4	4·0
17·6	5·0
18·2	10·0

Determine the Michaelis constant (K_m) of the enzyme.

(Glasgow Double Science Course, 1959.)

6.18. In the study of the enzyme-catalysed decarboxylation of an amino acid the following rates of evolution of CO_2 were observed :

Amino acid concentration (millimolar)	CO_2 evolved (μl/10 minutes)
0·10	8·2
0·20	14·7
0·33	21·7
0·67	34·5
1·00	43·4

Estimate

(a) the probable reaction rate at an amino acid concentration of 2·00 mM,

(b) the maximum attainable reaction rate.

(Glasgow Double Science Course, 1960.)

6.19. What methods can be used to estimate the number of substrate-binding sites per molecule of enzyme?

In a study of the effect of $NADH_2$ concentration on the binding of $NADH_2$ by an alcohol dehydrogenase the following results were obtained with a total enzyme concentration of $3·00 \times 10^{-5}$ M :

Concentration of bound $NADH_2$	Concentration of free $NADH_2$
10^5 M	10^5 M
9·00	5·76
8·40	4·12
7·65	2·83
5·52	1·35
2·91	0·49

Estimate

(a) the number of $NADH_2$ binding sites per molecule of enzyme,

(b) the dissociation constant for the enzyme-$NADH_2$ complex.

(Glasgow Double Science Course, 1960.)

6.20. Show how the velocity of enzyme reactions is affected by change of temperature.

In an enzyme reaction the velocity, v (moles litre^{-1} min^{-1}), was measured as a function of substrate concentration, s (moles litre^{-1}), with the following results :

$s \times 10^3$	0·25	0·5	1	2·5	5	10	20	40
$v \times 10^3$	0·39	0·74	1·32	2·49	3·24	3·34	2·69	1·77

How do you account for these results ? Show how you would proceed to analyse them, deriving any equations you would use. (Do not attempt a detailed numerical solution.)

(Special Degree of B.Sc., Biological Chemistry, University of Bristol, 1957.)

6.21. Explain how the behaviour of enzyme inhibitors can be elucidated by kinetic measurements.

The velocity of an enzyme hydrolysis was measured in arbitrary units, at 20° C and constant enzyme concentration, in the absence of inhibitor (v_0) and in the presence of 0·005 M inhibitor (v_I), over a range of substrate concentrations (C_S, moles litre^{-1}). Calculate the dissociation constants of enzyme-substrate and enzyme-inhibitor complexes.

C_s	0·00125	0·00100	0·00075	0·00050
v_0	151	138	118	93
v_i	83·9	72·4	58·8	41·9

(Special Degree of B.Sc., Biological Chemistry, University of Bristol, 1959.)

6.22. Values for V_m and K_s (dissociation constant) were determined for the hydrolysis of acetylcholine bromide by the bovine esterase over a range of temperatures :

$T°$ C	20	25	30	35
$V_m \times 10^6$ (moles litre^{-1} sec^{-1})	1·84	1·93	2·04	2·17
$K_s \times 10^4$ (moles litre^{-1})	4·03	3·75	3·35	3·05

(a) Assuming that the enzyme concentration was 10^{-11} M, calculate the heat, free energy and entropy of activation of the enzyme-substrate complex.

(b) Without attempting a numerical solution, show how you would calculate the heat, free energy and entropy of formation of the enzyme-substrate complex. Discuss the significance of these quantities.

($kT/h = 6·10 \times 10^{12}$ sec^{-1} at 20° C, $R = 1·98$ cal degree^{-1} mole^{-1}, $e = 2·718$, base of natural logarithms.)

(Special Degree of B.Sc., Biological Chemistry, University of Bristol, 1958.)

6.23. Show that the steady-state velocity for the model

$$E + S \underset{k_{-1}}{\overset{k_{+1}}{\rightleftharpoons}} ES' \overset{k_{+2}}{\rightarrow} ES'' \overset{k_{+3}}{\rightarrow} E + P$$

is given by a relationship of the form

$$v = \frac{V_{max.}}{1 + K/S}$$

where $\quad V_{max.} = E_0 k_{+2} k_{+3}/(k_{+2} + k_{+3})$

$\qquad K = k_{+3}(k_{-1} + k_{+2})/(k_{+2} + k_{+3})k_{+1}$

$\qquad S$ = initial substrate concentration

$\qquad E_0$ = initial enzyme concentration

In a study of the hydrolysis of benzoyl-L-arginine ethyl ester under the influence of ficin, a proteolytic enzyme, the initial rates of hydrolysis were measured for various pH values and substrate concentrations. The values obtained, expressed as micromoles of ester hydrolysed per minute, were as follows :

pH	Substrate concentration		
	0·01 M	0·005 M	0·0025 M
3·5	0·084	0·049	0·027
4·0	0·210	0·123	0·068
4·5	0·392	0·235	0·129
5·0	0·555	0·327	0·180
5·5	0·633	0·373	0·205
6·5	0·675	0·398	0·218

Calculate values for the Michaelis constant and maximum velocity for each pH and from these values deduce what information you can about the mechanism of the enzyme's action.

(Adapted from Glasgow Honours Course Finals, 1960.)

6.24. The decarboxylation of glyoxylate by mitochondria is inhibited by malonate. In a kinetic study the following results were obtained:

Glyoxylate concentration mM	Rate of evolution of CO_2 (arbitrary units)		
	In absence of malonate	In presence of 1·26 mM malonate	In presence of 1·95 mM malonate
1·0	2·50	2·17	1·82
0·75	2·44	1·82	1·39
0·60	2·08	1·41	1·28
0·50	1·89	1·30	1·00
0·40	1·67	1·09	—
0·33	1·39	—	—
0·25	1·02	—	0·56

Is the inhibition by malonate competitive?

(Glasgow Double Science Course, 1963.)

6.25. Derive an expression relating the initial velocity of an enzyme-controlled reaction to the concentration of substrate in the presence of a non-competitive inhibitor.

Glutamate dehydrogenase catalyses the following reaction:

$$\begin{array}{c}
\text{COOH} \\
| \\
\text{(CH}_2)_2 \\
| \\
\text{HC.NH}_2 \\
| \\
\text{COOH}
\end{array}
+ \text{NAD}^+ \rightleftharpoons
\begin{array}{c}
\text{COOH} \\
| \\
\text{(CH}_2)_2 \\
| \\
\text{C=NH} \\
| \\
\text{COOH}
\end{array}
+ \text{NADH} + \text{H}^+$$

The course of the reaction can be followed spectrophotometrically by measuring the increase in the extinction of the system at 340 nm. The following data were obtained in an experiment to determine the effect of sodium salicylate on the above reaction:

Glutamate concentration (mM)	ΔE_{340}/minute	
	in absence of inhibitor	in presence of 40 mM salicylate
1·5	0·21	0·08
2·0	0·25	0·10
3·0	0·28	0·12
4·0	0·33	0·13
8·0	0·44	0·16
16·0	0·40	0·18

From these figures demonstrate graphically that the inhibition is non-competitive. Evaluate K_m for the enzyme and K_i, the dissociation constant of the enzyme-inhibitor complex.

(University of Wales, Cardiff, Honours Course Finals, 1964.) (Data from GOULD, HUGGINS & SMITH (1963), *Biochem. J.* **88**, 346.)

I

6.26. A substance AH_2 undergoes two successive ionizations

$$AH_2 \rightleftharpoons AH^- \rightleftharpoons A^{2-}$$

Derive the Michaelis function for AH^-.

In a study of the action of α-chymotrypsin on acetyl-L-tryptophan ethyl ester the maximum velocity of hydrolysis (V_{max}) obtained by saturating the enzyme with substrate was found to vary with pH as follows :

pH	Velocity (arbitrary units)	pH	Velocity (arbitrary units)
5·4	10	6·8	192
5·6	19	7·2	260
5·8	32	7·4	280
6·4	99	7·8	300
6·5	112	8·0	328
6·6	141	8·4	331
		9·0	327

What conclusion can be drawn from this information ?

(Glasgow Honours Course Finals, 1964.)

6.27. Draw on one graph concentration versus time curves for the components S, ES and P of the enzyme system :

$$E + S \underset{k_{-1}}{\overset{k_{+1}}{\rightleftharpoons}} ES \xrightarrow{k_{+2}} E + P$$

for the period from instantaneous mixing of substrate (at high concentration) with enzyme (at low concentration) to completion of the reaction. Discuss the factors determining the shape of each curve.

Outline the principles of the methods available for the determination of k_{+1}, k_{-1}, and k_{+2}.

(Hull Honours Course Finals, Part I, 1965.)

6.28. Derive expressions relating enzyme and substrate concentrations and reaction velocity in the cases of (a) competitive and (b) non-competitive inhibition of enzyme activity.

In the normal pathway for methionine biosynthesis in *Escherichia coli*, methionine is formed by enzymic methylation of homocysteine by 5-methyltetrahydropteroyl-triglutamate. In an experiment in which serine was the donor of the C_1 unit, the effect of the addition of tetrahydropteroylmonoglutamate on the cofactor activity of the triglutamate was investigated. Methionine formed was estimated by microbiological assay. The following results were obtained :

Triglutamate added (μmoles/litre)	Methionine formed (n-moles)	
	Monoglutamate absent	Monoglutamate present (30 μmoles/litre)
10	500	143
12·5	526	172
16·7	555	218
25	589	286
50	714	455

Determine the nature of the inhibition by the monoglutamate and K_1 for the inhibition. Write structural formulae for the two reduced folic acid derivatives.

(Hull Honours Course Finals, Part I, 1966.)

6.29. Derive the expression

$$v = V_{max} - K_m(v/S)$$

which describes the relationship between the substrate concentration (S) and the velocity (v) of an enzyme-controlled reaction.

The following data were obtained in an experiment to investigate the hydrolysis of starch to maltose by the enzyme amylase :

Starch concentration (%)	Velocity of production of maltose (mg/min)
1·078	0·445
0·647	0·435
0·431	0·400
0·216	0·345
0·129	0·305
0·086	0·260
0·05	0·180
0·04	0·165
0·03	0·140

Using the expression given above, deduce K_m and V_{max} for the reaction.

(University of Wales, Cardiff, Final, 1965.)

6.30. Derive the equation

$$\frac{v}{v_1} = 1 + \frac{K_m}{K_1}\left[\frac{I}{K_m + S}\right]$$

which compares the velocity (v) of an enzymic reaction in the absence of any inhibitor to the velocity (v_1) of the same reaction in the presence of a competitive inhibitor. How does this equation differ from the one obtained for a noncompetitive inhibitor?

The following data were obtained in an experiment to determine the effect of urea on the hydrolysis of propionamide by an amidase enzyme extracted from *Pseudomonas aeruginosa* :

Propionamide concentration (M)	Initial velocity of hydrolysis of propionamide (arbitrary units)		
	In the absence of urea	In the presence of 1 mM urea	In the presence of 2 mM urea
0·25	1·00	0·59	0·31
0·02	0·50	0·27	0·15
0·007	0·30	0·17	0·10

Using the equations discussed in the first part of your answer, deduce the type of inhibition of this reaction by urea and evaluate K_1, the dissociation constant for the enzyme-urea complex.

(University of Wales, Cardiff, Honours Course Finals, 1965.)

6.31.   Show that if an enzyme is capable of acting on either of two substrates A and B singly the rate v_t at which it will act on a mixture of the two is given by

$$v_t = \frac{V_a(a/K_a) + V_b(b/K_b)}{1 + (a/K_a) + (b/K_b)}$$

Where a and b are the concentrations of A and B, K_a and K_b are their respective Michaelis constants and V_a and V_b are the maximum velocities with excess of A and B singly.   Show further that if v_a is the reaction velocity when the enzyme acts on A singly at concentration a and if v_b is the reaction velocity when the enzyme acts on B singly at concentration b then

$$v_t = \frac{v_a[1 + (a/K_a)] + v_b[1 + (b/K_b)]}{1 + (a/K_a) + (b/K_b)}$$

A preparation of arylsulphatase from ox liver was found to hydrolyse both nitrophenyl sulphate ($K_m = 4.4$ mM) and nitrocatechol sulphate ($K_m = 10.0$ mM).   At a constant enzyme concentration the following reaction rates were obtained with the substrates shown.

Substrate and concentration	Inorganic sulphate liberated μg per 30 min	
	Expt. 1	Expt. 2
Nitrocatechol sulphate (8·3 mM)	121·5	126
Nitrophenyl sulphate (2·0 mM)	63	70·5
Nitrochatechol sulphate (8·3 mM) and nitrophenyl sulphate (2·0 mM) together	105	120

How far are these observations consistent with the view that a single enzyme is hydrolysing both substrates ?

(Glasgow Honours Course Finals, 1965.)

6.32.   An attempt to characterize the alkaline phosphatase of human small intestine resulted in the partial purification of two fractions X and Y both of which showed phosphatase activity.   The relationship between substrate concentration and rate of reaction for the two fractions was as follows :

Substrate concentration (mM)	μmoles inorganic phosphate liberated per minute	
	X	Y
1·00	0·18	0·053
1·25	0·20	0·062
1·40	0·22	0·067
1·70	0·24	0·074
2·00	0·27	0·082
2·50	0·30	0·093
3·30	0·34	0·105
5·00	0·42	0·122
10·00	0·48	0·147
20·00	0·53	0·164

Are these observations consistent with the assumption that the phosphatase activities of both fractions are due to the same enzyme ?

(Glasgow Biochemistry [Higher Ordinary] Course, 1966.)

6.33. An enzyme E acts on its substrate S only when activated by combination with an activator M. The process can be described by the following set of equations :

$$E + S \rightleftharpoons ES \qquad (1)$$
$$E + M \rightleftharpoons EM \qquad (2)$$
$$EM + S \rightleftharpoons EMS \qquad (3)$$
$$ES + M \rightleftharpoons EMS \qquad (4)$$
$$EMS \longrightarrow E + M + products \qquad (5)$$

K_s is the dissociation constant for reactions (1) and (3).
K_a is the dissociation constant for reactions (2) and (4).
k is the rate constant for reaction (5).

If reaction (5) is so slow that it does not significantly affect the equilibria of the other four reactions, show that the velocity (v) of the overall reaction

$$S \longrightarrow products$$

is given by

$$v = \frac{k.e}{\left(1 + \dfrac{K_s}{s}\right)\left(1 + \dfrac{K_a}{a}\right)}$$

where

e = total concentration of enzyme,
s = concentration of substrate,
a = concentration of activator.

It may be assumed that both s and a are much larger than e.

In an experiment on an enzyme which was thought to act by a mechanism of the form described above, the following results were obtained at constant substrate concentration :

Activator concentration (mM)	Reaction rate (arbitrary units)
0·100	74
0·125	88
0·170	107
0·250	141
0·330	160
0·500	200
1·000	260

Determine K_a for this reaction.

(Glasgow Honours Course Finals, 1966.)

6.34. The following data describe an enzymic reaction proceeding at constant pH and temperature.

Substrate concentration Moles/litre ($\times 10^3$)	Initial velocity Moles/sec ($\times 10^5$)
2	13
4	20
8	29
12	33
16	36
20	38

The molecular weight of the enzyme is 50,000 ; the concentration of the enzyme is 2 mg/ml.

From the data calculate the Michaelis constant (K_m), the maximum velocity (V_{max}) of reaction that can be achieved and the velocity constant k_{+2} of the reaction in which the enzyme-substrate complex dissociates to give the final product. Derive any equations that you use.

(Leeds Honours Course Finals, 1966.)

6.35. The hydrolysis of p-nitrophenylacetate by α-chymotrypsin may be formulated :

$$E + S \underset{}{\overset{k}{\rightleftharpoons}} ES \rightarrow E + P$$

and a value for the velocity constant, k, evaluated in the usual manner. The reaction may also be formulated:

$$E + S \overset{k_2}{\rightleftharpoons} ES \overset{k_3}{\rightarrow} ES' \rightarrow E + P_2$$
$$\searrow P_1$$

where ES' is acetyl-α-chymotrypsin, P_1 is p-nitrophenol and P_2 is acetate.
Show that

$$\frac{1}{k} = \frac{1}{k_2} + \frac{1}{k_3}$$

(Hull Special Biochemistry II Course, 1966.)

6.36. Derive a rate equation for the following model assuming steady-state conditions :

$$E + S \rightleftharpoons ES \rightleftharpoons E + P$$

In a study of the hydrolysis of acetyl-L-tryptophan ethyl ester by chymotrypsin, initial velocities, expressed as μmoles substrate hydrolysed per minute, were determined in the presence and absence of acetyl-L-tryptophan (0·005 M). From the following data determine the nature of the inhibition and evaluate K_m and K_i :

	Reaction velocities			
Substrate concentration (mM)	1·25	1·00	0·75	0·50
Inhibitor present	83·9	72·4	58·8	41·9
Inhibitor absent	151	138	118	93

(Hull Pass Course Finals, 1967.)

6.37. Show that the steady-state rate equation for the simplest model of an ordered two-substrate enzyme reaction is of the form :

$$v = \frac{C_1[E_0]}{1 + \dfrac{C_2}{[A]} + \dfrac{C_3}{[B]} + \dfrac{C_4}{[A][B]}}$$

where C_1, C_2, C_3 and C_4 are constants and the other symbols have their usual meaning. What are the Michaelis constants for substrates A and B ? Show that C_4/C_3 gives K_s for A, the first-bound substrate. How can the ordered situation be distinguished from random ordered systems ?

(Hull Honours Course Finals, Part II, 1969.)

6.38. (a) For the reaction scheme :

$$E + S \underset{k_{-1}}{\overset{k_1}{\rightleftharpoons}} ES \overset{k_2}{\rightarrow} ES' \overset{k_3}{\rightarrow} E + P$$

show that the steady-state velocity is given by an equation of the form :

$$v = \frac{V_{max}}{1 + K/[S]}$$

where $[S]$ = initial substrate concentration
$[E_0]$ = initial enzyme concentration
$$V_{max} = \frac{k_2 k_3 [E_0]}{k_2 + k_3}$$

(b) The following figures were obtained for the action of a phosphodiesterase on its substrate.

Substrate concentration ($M \times 10^8$)	Initial rate ($\mu mole\ min^{-1}$)
2·5	0·0212
1·66	0·0198
1·0	0·0176
0·66	0·0148
0·5	0·0137
0·4	0·0117

Determine the values of V_{max} and K_m.

(Hull Honours Course Finals, Part I, 1969).

6.39. Show that the steady-state rate equation for a simple ordered two-substrate enzyme reaction is of the form :

$$v = \frac{C_1 [E_0]}{1 + \dfrac{C_2}{[A]} + \dfrac{C_3}{[B]} + \dfrac{C_4}{[A][B]}}$$

where C_1, C_2, C_3 and C_4 are constants and the other symbols have their normal meaning. How can this ordered situation be distinguished from random order systems ?

(Hull Joint Biological Chemistry Course Finals, Part II, 1969.)

6.40. Iodoacetate reacts with sulphydryl compounds according to the following equation :

$$R.S^- + CH_2I.COO^- \rightarrow R.S.CH_2.COO^- + I^-$$

This reaction can be followed very accurately with the silver-silver iodide electrode which measures the iodide ion concentration.

In an experiment iodoacetate was reacted with a nicotinamide nucleotide dehydrogenase at 25°, the reaction being monitored with the silver-silver iodide electrode. The reaction mixture initially contained borate buffer pH 9·0, iodoacetate (5×10^{-4} M) and enzyme (0·252 mg/ml).

The results were as follows :

Time after mixing (min)	[I⁻] (μM)
0	0
2	1·20
4	2·30
8	3·63
10	4·08
15	4·87
22	5·65
30	6·05
40	6·28
50	6·39
60	6·47
70	6·57
80	6·65
90	6·72

The enzyme had a molecular weight of 81,000 and contained 17-SH groups per molecule. Interpret the above data and suggest experiments to support your conclusions.

(Hull Honours Course Finals, Part II, 1970.)

6.41. (a) Derive a rate equation for a competitively inhibited enzyme reaction assuming that steady-state conditions obtain.

(b) The rate equations for the ordered Bi-Bi mechanism :

$$\begin{array}{ccccc} A & B & & P & Q \\ \downarrow & \downarrow & & \uparrow & \uparrow \\ \hline E & EA & EAB & EQ & E \end{array}$$

and for the Theorell-Chance mechanism :

$$\begin{array}{cccc} A & B & P & Q \\ \downarrow & \searrow \nearrow & & \uparrow \\ \hline E & EA & & EQ & E \end{array}$$

are, respectively, of the forms :

$$\frac{v}{E_0} = \frac{c_1 AB - c_2 PQ}{c_3 + c_4 A + c_5 B + c_6 P + c_7 Q + c_8 AB + c_9 PQ + c_{10} AP + c_{11} BQ + c_{12} ABP + c_{13} BPQ}$$

and

$$\frac{v}{E_0} = \frac{c_1 AB - c_2 PQ}{c_3 + c_4 A + c_5 B + c_6 P + c_7 Q + c_8 AB + c_9 PQ + c_{10} AP + c_{11} BQ}$$

where $c_1 - c_{13}$ are complex constants.

Show that product P behaves as a competitive inhibitor with respect to substrate B for a 'Theorell-Chance' enzyme and that this finding can form a basis for the differentiation of the above two mechanisms.

(Hull Honours Course Finals, Part II, 1970.)

6.42. Having obtained a reasonably pure solution of an enzyme, what kinetic experiments would you perform to characterize it ?

Enzyme X breaks down substrate A to produce B. Compounds C and D inhibit this reaction. The following table records an experiment wherein a constant amount of X was allowed to break down A in the presence of C or D. Concentrations are in micromoles/ml, initial reaction velocities are expressed as micromoles A broken down/ml/min. From this information deduce as far as you can the type of inhibition exhibited by C and D in the reaction. State reasons clearly.

mM C	Velocity	
	(with 8 mM A)	(with 24 mM A)
0	5·0	10·0
1·0	3·4	—
2·5	2·3	4·5
4·0	—	3·6
5·0	1·6	3·1
7·0	1·2	2·4

mM D		
0	5·0	10·0
1·0	2·3	—
2·5	1·3	5·4
4·0	—	4·2
5·0	0·7	3·5
7·0	0·6	2·7

(St. Andrews, Queen's College, Honours Finals, 1967.)

6.43. Several esterases (and peptidases) are commonly believed to act by a mechanism of the following form :

$$\text{ester} + \text{enzyme} \underset{k_{-1}}{\overset{k_{+1}}{\rightleftharpoons}} \text{ester-enzyme complex}$$

$$\text{ester-enzyme complex} \underset{k_{-2}}{\overset{k_{+2}}{\rightleftharpoons}} \text{acyl enzyme} + \text{alcohol}$$

$$\text{acyl enzyme} + \text{water} \underset{k_{-3}}{\overset{k_{+3}}{\rightleftharpoons}} \text{acid} + \text{enzyme}$$

Show that the initial velocity (v) of such a reaction is related to the substrate concentration (s) by an equation of the form

$$v = \frac{V_{max}}{K_m/s + 1}$$

where $\qquad V_{max} = k_{+2}e/(1 + k_{+2}/k_{+3})$

and $\qquad K_m = (k_{-1} + k_{+2})/k_{+1}(1 + k_{+2}/k_{+3})$

(Glasgow Honours Course Finals, 1969.)

6.44. Analyse the kinetics of the following enzyme and state what predictions can be made about its nature. How would you test to see if your conclusions are valid ? (Candidates are under no obligation to use all the data provided.)

Substrate concentration (*arbitrary units*)	*Velocity of reaction*			
	No additions	*In the presence of X*	*In the presence of Y*	*In the presence of Z*
0	0	0	0	0
0·5	0·002	0·07	0·017	—
0·7	—	0·18	0·094	0·001
0·8	0·024	—	0·23	—
0·9	0·05	—	0·38	—
1·0	0·091	0·39	0·55	0·01
1·1	0·15	—	0·67	—
1·2	0·23	0·52	0·77	0·04
1·3	0·33	—	0·85	—
1·4	0·43	—	0·90	—
1·5	0·55	0·68	0·93	0·20
1·6	0·63	—	0·95	—
1·8	0·77	0·79	—	0·52
2·0	0·88	—	0·99	0·72
2·4	0·95	—	—	—
2·5	—	—	—	0·94
3·0	0·99	—	—	0·985
10·0	1·0	1·0	1·0	1·0

(Glasgow Honours Course Finals, 1969.)

6.45. Describe briefly how you would proceed to determine the parameters K_m and V_{max} for the action of an enzyme on its substrate.

A certain enzyme was studied in a crude form and also after purification and crystallization, the data given below being obtained. How would you account for these data ?

100 ml of crude enzyme containing 10 mg protein per ml yields 50 mg of pure enzyme.

Substrate concentration (mM)	1	2	4	8	16
Reaction velocity with pure enzyme (μmoles of substrate liberated per minute per mg crystalline protein)	2·7	3·6	4·0	4·4	4·8
Reaction velocity with crude enzyme (μmoles of substrate liberated per minute per 0·1 ml crude enzyme preparation)	0·76	1·14	1·37	1·67	1·76

(University of Manchester, Biological Chemistry Honours Course Part I, 1967.)

6.46. Describe how you would determine the type of inhibition and the inhibitor constant for an inhibitor acting on a given enzyme-substrate system. You need *not* derive any equations you require in your answer.

Acetylcholinesterase exhibits towards acetylcholine a K_m of 1 mM. The enzyme is inhibited by glycine according to the data given below. What is K_I for glycine? How would you expect K_I to be affected by variation in pH?

Concentration of acetylcholine = 1 mM
pH = 7·4, t = 25°

Concentration of glycine (moles/litre)	0	0·05	0·10	0·25	0·5	1·0	1·5
Rate of reaction (μmole acetylcholine hydrolysed/min)	2·5	2·2	1·8	1·45	0·95	0·65	0·4

(University of Manchester, Biological Chemistry Ordinary Degree, 1967.)

6.47. The velocity v of an enzyme reaction depends on the concentration I of a competitive inhibitor according to the equation

$$v = \frac{SV_{max}}{S + K_m(1 + I/K_I)}$$

where S is the concentration of substrate, K_m is the enzyme-substrate Michaelis constant, K_I is the inhibitor constant and V_{max} is the maximum velocity of the reaction. Show how to obtain K_I from a study of the dependence of v on I at two substrate concentrations. In a determination of the inhibitor constant for acetate ions acting on a hydrolytic enzyme, the data given below were obtained for acetic acid and for sodium acetate respectively, What qualitative and quantitative conclusions can be drawn from these data?

Substrate concentration = 3 mM

Concentration of acetic acid or sodium acetate (M)	0	0·03	0·06	0·09	0·12	0·15	0·18	0·21	0·24
Reaction velocity with acetic acid as inhibitor	1·64	1·33	1·14	0·99	0·85	0·64	0·50	0·425	0·30
Reaction velocity with sodium acetate as inhibitor	1·64	1·33	1·15	1·00	0·89	0·81	0·735	0·675	0·625

Substrate concentration = 5 mM

Concentration of acetic acid or sodium acetate (M)	0	0·03	0·06	0·09	0·12	0·15	0·16	0·21	0·24
Reaction velocity with acetic acid as inhibitor	2·1	1·79	1·59	1·40	1·23	1·08	0·83	0·61	0·42
Reaction velocity with sodium acetate as inhibitor	2·1	1·79	1·60	1·45	1·30	1·18	1·09	1·01	0·94

Reaction is carried out at 25° in 0·1 M pH 7·4 phosphate buffer, the velocity being determined as the number of micro-moles of substrate hydrolysed per min per mg protein.

(University of Manchester, Biological Chemistry Honours Course Finals, 1967.)

6.48. Derive relations connecting the concentrations of substrate and competitive inhibitor with (a) the velocity of the enzymic reaction and (b) the concentration of the enzyme-inhibitor complex.

A certain competitive inhibitor when combined with an enzyme forms a strongly coloured complex of molar extinction coefficient 400 at 435 nm, the enzyme itself being transparent at this wavelength. The enzyme-substrate complex is also colourless, K_m is 100 mM while K_i for this inhibitor is 300 mM. Plot a curve showing how increasing concentrations of inhibitor (in the range zero to 1 M) will determine the absolute value of the extinction of solutions containing enzyme at 1 mM (i) without substrate, (ii) in the presence of 200 mM-substrate measured at the instant of mixing. How could you use this colour reaction with inhibitor to determine the concentration of enzyme within a tissue?

(University of Manchester, Biological Chemistry Honours Finals, 1968.)

6.49. Derive a mathematical procedure for determining the overall Michaelis constant for the interaction between ligand and enzyme, for the case where two molecules of ligand bind co-operatively to the enzyme. You should assume that, for this system, the Michaelis constant is a product of the relevant dissociation constants and that the dissociation constant for the binding of the first molecule of ligand is substantially greater than that for the second molecule.

The following data were found for an enzyme-ligand system (1) before and (2) after treatment with p-chloromercuribenzoate. Interpret the data.

Concentration of ligand (mM)	1	2	4	8	16	32	64	128	256	512
Moles of ligand bound per 78 mg protein in condition (i)	0·02	0·08	0·28	0·78	1·44	1·82	1·95	1·99	1·99	2·00
condition (ii)	0·06	0·10	0·18	0·34	0·58	0·90	1·22	1·52	1·72	1·86

In condition (i), the molecular weight of the protein species is found to be 78,000, while in condition (ii) it is 39,000.

(University of Manchester, Biological Chemistry Honours Course Finals, 1969.)

6.50. You are given the relation

$$v_i = \frac{V_{max}S}{K_m\left(1 + \dfrac{I}{K_i}\right) + S}$$

where v_i is the initial velocity of an enzymic reaction in the presence of inhibitor of concentration I and S is the concentration of substrate.

Derive an expression for the ratio $\dfrac{v}{v_i}$ where v is the initial velocity in the absence of inhibitor.

Find K_i, the dissociation constant of the enzyme-inhibitor complex. $K_m = 10$ μmoles/ml and S is constant at 30 μmoles/ml.

Inhibitor concn. (μmoles/ml)	Rate of formation of product (μmoles/ml/min)
0	0·1000
6	0·0835
12	0·0715
18	0·0625
30	0·0500

(University of Manchester, Biological Chemistry, Second B.Sc. examination, 1969.)

CHAPTER VII

OPTICAL AND PHOTOMETRIC ANALYSIS

BIOCHEMISTS make extensive use of photometric analysis and, indeed, it represents one of the most valuable analytical techniques available to them. It is now accepted as a general principle that optical methods should be used whenever possible, always provided they furnish the desired degree of accuracy, in preference to other types of analysis on account of the rapidity, simplicity and sensitivity of measurement which modern apparatus permits. Sensitivity is of particular importance since it enables very small quantities of metabolites and other substances of vital significance in biological material to be determined. Further, optical analysis frequently permits kinetic studies of enzyme reactions to be carried out and can be extremely valuable in the study of very rapid reactions. Oxidation-reduction reactions of whole unicellular organisms, such as bacteria and yeasts, can be studied spectrophotometrically and one of the great virtues of the technique is that the observations made are non-destructive to the system or samples under investigation.

The different optical apparatus available is related to the particular type of analysis to be effected. Colorimeters and spectrophotometers measure the amount of light absorbed by coloured or colourless solutions, turbidimeters and nephelometers measure the light scattered by suspensions, and fluorimeters and spectrophotofluorimeters determine the fluorescence produced by absorbed light. Flame photometers analyse the alkali and alkaline earth metal constituents of biological material by means of their emission spectra and are widely used in clinical and agricultural analysis. Infrared spectrometers permit the determination of infrared spectra and yield information concerning the particular groups present in a molecule. Ultraviolet, visible and infrared absorption spectra afford information concerning proteins, nucleic acids, coenzymes, cytochromes and other pigments, and fluorescent spectra are used for the analysis of the vitamins thiamine and riboflavin and the nucleotides of adenine and nicotinamide. Methods of analysis based on the absorption or scattering of light are therefore extremely versatile.

Theoretical Background

The energy of a molecule may be divided into three main categories, namely electronic, vibrational and rotational. The energy is quantized in discrete levels, as represented diagrammatically for the case of a simple diatomic molecule in Fig. 7.1.

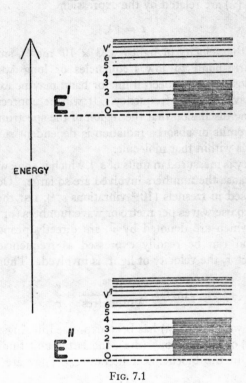

FIG. 7.1

Energy levels of a diatomic molecule. E' and E'' are upper and lower electronic energy levels respectively. V' and V'' are vibrational quantum numbers. Rotational quantum numbers have been omitted for clarity.

The energy difference between electronic levels is very much greater than that between the vibrational levels, and rotational levels are the closest of all. When some change within the molecule entails transition from one energy level (E_1) to another

(E_2), radiation is either emitted or absorbed in accordance with the equation :

$$E_1 - E_2 = h\nu \qquad . \qquad . \qquad . \qquad (7.1)$$

where h is Planck's constant (6.62×10^{-34} Js) and ν the frequency of the radiation emitted or absorbed. If E_1 is greater than E_2, the radiation is emitted, but if the transition should be from E_2 to E_1 then the radiation is absorbed. Since frequency and wavelength (λ) are related by the expression

$$c = \nu\lambda,$$

where c is the velocity of light (ca. 3×10^8 ms^{-1}), small energy changes correspond to low frequencies or long wavelengths. Thus the position of a spectral line or band may be expressed in terms of wavelength or frequency. It will be appreciated from the foregoing discussion that the region of the spectrum in which a molecule emits or absorbs radiation is dependent solely on the energy levels within that molecule.

Frequency is measured in units of s^{-1}, which is somewhat inconvenient because the numbers involved are so large. Occasionally it is expressed in fresnels (10^{12} vibrations s^{-1}), but the common practice is to use waves per metre or wave numbers (m^{-1}). Wave numbers, which are denoted by $\tilde{\nu}$, are directly proportional to energies and can be readily expressed as frequencies since a constant factor, the velocity of light, is involved. Thus

$$\tilde{\nu}\,\text{m}^{-1} = \frac{10^{10}}{\lambda \text{ in metres}} = \frac{\nu}{c}$$

In the past wavelength (λ) has been expressed in Ångstroms (Å), microns (μ) or millimicrons (mμ), the latter unit finding widest application in the biochemical literature. They are related as follows :

$$1\text{Å} = 10^{-8} \text{ cm} \quad 1 \text{ m}\mu = 10^{-7} \text{ cm} \quad 1 \ \mu = 10^{-4} \text{ cm}$$

On the new International System of Units (SI) the unit of wavelength adopted is the metre, m, and the appropriate prefix is used with it (see Appendix 2). In biochemistry the unit formerly used, the millimicron (mμ, i.e. 10^{-7} cm), becomes 10^{-9} m and is correctly termed the *nanometre*, abbreviated to nm. However, the Ångstrom is a very convenient unit for expressing bond lengths and at the present time it seems that X-ray crystallographers are not keen to discontinue its use in favour of the nanometre.

The main regions of the electromagnetic spectrum are shown in Fig. 7.2. The division into regions is rather arbitrary and is mainly decided by the type of experimental technique required. In this discussion attention will be confined to the near ultraviolet and visible regions, which are most commonly employed by biochemists as a routine, and where transitions between electronic, vibrational and rotational levels occur. This is not intended to imply that the other regions are without interest. Crystal analysis by X-ray techniques is of value in connexion with problems of molecular structure and the spectacular results obtained with

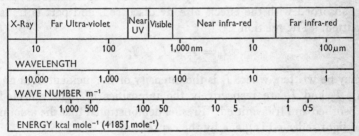

FIG. 7.2

The main regions of the electromagnetic spectrum.

proteins by Kendrew, Perutz and their collaborators are of tremendous importance to biochemists. The far ultraviolet is the region of characteristic absorption by simple molecules, and the near infrared is the region of vibrational and rotational (but not electronic) transitions, where it is possible to assign absorption maxima to particular groups in the molecule and thus elucidate problems of molecular structure.

When light is absorbed, the molecule responsible for the absorption is excited to a higher energy state, in which it usually remains for approximately 10^{-8} second. The excited molecule can lose its energy in several ways, namely

1. by dissociation of the molecule,
2. as heat,
3. as light of different wavelengths (fluorescence, Raman effect),
4. as light of different wavelengths after a time lag (phosphorescence).

A graph showing the amount of light absorption as a function of the wavelength is known as an *absorption spectrum*, and, because it depends on chemical structure, a different spectrum is obtained with each different structure. A means of detecting different chemical structures is thereby afforded and absorption spectra have been likened to chemical 'fingerprints'. It is perhaps worth emphasizing that, although similarity in chemical structure can be ascertained by absorption spectra, absolute identity with a given compound cannot be established on these grounds alone. Nevertheless, it is an exceptionally valuable investigational method and determinations of absorption spectra play an important role in modern biochemistry.

When light falls on any body or solution, part is reflected, part is absorbed with the effects already discussed, and part is transmitted. The relation

$$I_0 = I_r + I_a + I_t$$

may be written, where I_0 is the intensity of the incident light and I_r, I_a and I_t are respectively the intensities of light reflected, absorbed and transmitted. Fresnel demonstrated that the amount of light reflected is given by the expression

$$I_r = \left(\frac{n-1}{n+1}\right)^2 I_0 = kI_0 \qquad . \qquad (7.2)$$

where n is the refractive index of the medium. Aqueous solutions are generally used in biochemistry, in which case k becomes negligible and I_r may be neglected.

The Laws of Lambert and Beer

Colorimetry and spectrophotometry are based on the laws of Lambert and Beer which define the relationship between the intensities of radiation incident on and transmitted by a layer of an absorbing medium.

Lambert's law states that the proportion of radiation absorbed by a homogeneous medium is independent of the intensity of the incident radiation. An alternative expression of the law is the statement that each successive layer of thickness dx of the medium absorbs an equal fraction dI/I of the radiation of intensity I incident upon it, i.e. $dI/I = -b\,dx$, where b is a constant characteristic of the absorbing medium, known as the Napierian *absorption* or *extinction coefficient*.[1]

Integration of this expression gives

$$I = I_0 e^{-bx} \qquad . \qquad . \qquad . \qquad . \quad (7.3)$$

where I_0 and I are the intensities of the incident and transmitted light respectively and x is the thickness of the absorbing medium. Now $bx = B$, where B is the *Napierian absorbance* or *extinction*, and therefore

$$\frac{I}{I_0} = e^{-B} \text{ or } \ln \frac{I_0}{I} = B \qquad . \qquad . \quad (7.3a)$$

Equation 7.3a may be converted to logarithms to the base 10 when

$$\log \frac{I_0}{I} = A \text{ (or } E) = ax \qquad . \qquad . \quad (7.4)$$

$$\text{or } I = I_0 \, 10^{-ax}$$

where A or E represents the *decadic absorbance* or *extinction* which, as with the Napierian term, is related to the *decadic absorption* or *extinction coefficient a* by the expression A (or E) $= ax$. It will be apparent therefore, that the Napierian and decadic absorption coefficients are related by the factor 2·303, i.e. $b = 2·303 \, a$.

The decadic absorbance, A, is the term now recommended for general use by the SI to supersede *extinction* (E) and *optical density* $(O.D.)$, the alternative terms previously employed. However, while absorbance is already commonly used in the United States it has not been widely adopted by biochemists in Britain and *The Biochemical Journal* continues to use extinction. In biochemical work it is customary to refer simply to ' extinction ' or ' absorbance ', but it should be clearly borne in mind that it is the decadic and not Napierian extinction or absorbance which is being so designated.

The ratio of the intensities of transmitted to incident light is termed the *transmittance*, denoted by T, i.e.

$$\frac{I}{I_0} = T \qquad . \qquad . \qquad . \qquad . \quad (7.5)$$

[1] A table listing the symbols and units in spectrophotometry is given on page 278.

whence
$$\log \frac{1}{T} = E \qquad . \qquad . \qquad . \quad (7.6)$$

The percentage light transmission ($T\%$) is given by

$$T\% = 100 \frac{I}{I_0} = 100 \, T$$

Thus
$$\log T\% = \log 100 + \log T = 2 - E \quad . \quad (7.7)$$

Beer studied the influence of the concentration of a substance in solution upon the monochromatic light transmission and found the same relationship between transmission and concentration that Lambert had established between transmission and thickness of layer. This can be expressed by the equation

$$E = \log \frac{I}{I_0} = acl \qquad . \qquad . \qquad . \quad (7.8)$$

where a is the decadic extinction (absorption) coefficient of the absorbing solute, c the concentration and l the length of the light path in the solution. When the concentration is expressed in molarity, a becomes the *molar extinction* or *absorption coefficient*, ϵ, of the solute for the particular wavelength in question. In biochemical work the depth of solution, equal to the light path of the spectrophotometer cuvette, is commonly 1 cm. By definition the molar extinction coefficient ϵ is the extinction of a molar solution in a 1 cm light path and has the dimensions litre mol^{-1} cm^{-1}. This is the form recommended by the International Union of Pure and Applied Chemistry ; the reader will perceive that this unit is equivalent to 10^3 cm^2 mol^{-1}, which unit is also used in the biochemical literature. Molar extinction coefficients quoted in the two units therefore differ by a factor of 10^3.

Comparison of equations 7.4 and 7.8 reveals that $x = cl$, i.e. the light path x of a solid absorber has been separated into two terms for a solution, namely light path and concentration. The light path of the solid may, in a sense, be regarded as including a concentration term, since it is a uniform solid, whereas the solution contains an absorbing solute dissolved and dispersed in a non-absorbing solvent, and light is absorbed by the solute molecules only. (This is, of course, an ideal case; some solvents do absorb light at certain wavelengths.)

In the case of some natural products the molecular weight may not be known with certainty and it is customary to use a 1 per cent. (w/v) solution with a 1 cm light path and to record the extinction as a *specific extinction coefficient*, denoted by $E_{1cm}^{1\%}$.

Example 7.1.—An aqueous solution of uridine-5′ -triphosphate (UTP), at a concentration of 57·8 mg per litre of the trisodium dihydrate (molecular weight 586), gave an extinction (optical density) of 1·014 at pH 7·0 in a spectrophotometer cuvette of 1 cm light path. Calculate the molar extinction coefficient. What would be the percentage light transmission of a 10 μM solution under the same conditions?

$$E = \log \frac{I_0}{I} = 1\cdot014 = acl$$

$$c = \frac{57\cdot8 \times 10^{-3}}{586} = 9\cdot86 \times 10^{-5}\text{M and } l = 1 \text{ cm.}$$

Therefore $$a = \frac{1\cdot014}{9\cdot86 \times 10^{-5} \times 1} = \underline{1\cdot03 \times 10^4.}$$

For a 10 μM solution

$$E = 1\cdot03 \times 10^4 \times 10^{-5} \times 1$$
$$= 0\cdot103.$$

Now the percentage light transmission $T\%$ is related to the extinction by the expression

$$\log T\% = 2 - E$$
$$= 2 - 0\cdot103 = 1\cdot897$$
$$\text{Whence } \underline{T\% = 78\cdot9\%.}$$

DEVIATIONS FROM BEER'S LAW.—Beer's Law is always considered valid, but where deviations from it are encountered in experimental data some other reason is sought such as chemical change, e.g. hydration, dissociation and complex formation. For instance, certain salts form complexes the colours of which are different from those of the simple compounds and, as the concentration of the complex form decreases with dilution, Beer's Law does not hold. The concentrations of acid-base pairs will alter if the pH should change, and since the two species have different absorption characteristics pH variation can lead to deviation. Furthermore, suspensions do not obey the Law as light is scattered by the particles in suspension. For this reason care must be taken to exclude dust particles from solutions being subjected to spectrophotometric analysis.

Apparent deviations from Beer's Law may be instrumental in origin, as when too wide a slit width is used in spectrophotometry and the light is, in consequence, not monochromatic.

Colorimeters for use in the visible region of the spectrum employ filters to select light of the appropriate wavelength, but in practice the light transmitted by the filter is not monochromatic and embraces a broad waveband.

Colorimetry

For the visible region of the spectrum relatively simple instruments called colorimeters may be used to measure the light absorbed by a coloured solution. In its original form (the Duboscq colorimeter) such an instrument was visual, the operator matching by eye the unknown and standard solutions, but these instruments have now been largely superseded by photoelectric colorimeters. However, the visual instrument does illustrate clearly the application of Beer's Law and therefore will be considered first.

Consider two solutions of a coloured compound having concentrations c_1 and c_2. These are placed in a Duboscq colorimeter which permits the thickness of the layers of liquid to be changed and measured and which allows the amounts of transmitted light to be compared. When the system is optically balanced, i.e. when the two layers have the same colour intensity, then the two beams of transmitted light have the same intensity and $I_1 = I_2$ (see Fig. 7.3). Furthermore, since I_0 is the same for each,

$$ac_1l_1 = ac_2l_2,$$

and under the conditions where Beer's Law holds

$$c_1l_1 = c_2l_2.$$

Consequently a colorimeter permits the determination of the concentration of a coloured substance by comparison with a solution of known concentration provided Beer's Law holds and the system is optically balanced. Let c_X be the unknown concentration, then,

$$c_X = \frac{cl}{l_X} \qquad . \qquad . \qquad . \qquad . \qquad (7.9)$$

l and l_X are determined experimentally by the colorimeter, c is known and hence c_X may be obtained.

Example 7.2.—A solution treated with Nessler's reagent and containing 20 μg ammonia per 10 ml was placed in a Duboscq colorimeter and set at a depth of 15 mm. A second solution, similarly treated and containing an unknown amount of ammonia, was matched against the former and the mean of ten readings was 12·5 mm.

Let c be the concentration of the unknown, then

$$12·5 \times c = 15 \times 20$$

and

$$c = 24 \ \mu g \ NH_3 \text{ per 10 ml.}$$

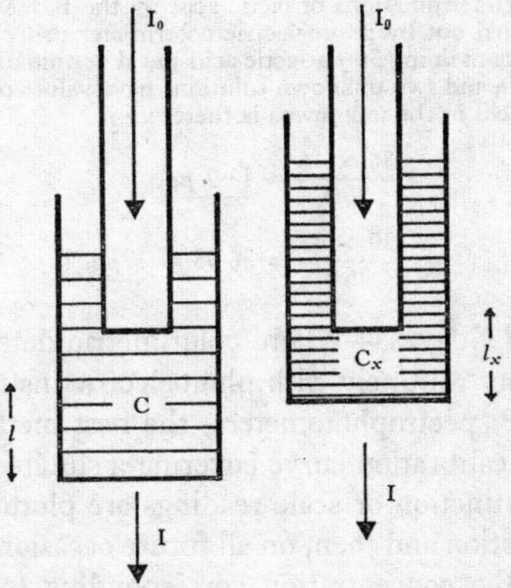

FIG. 7.3

Principle of the Duboscq colorimeter, which permits adjustment of length of light path through the liquids so that the intensities of transmitted light are equal.

The light source for a colorimeter is ordinary white light, but this does not alter the validity of the method provided no attempt is made to match colours of different hue but of apparently equal intensity. The only remedy for this latter situation is the use of monochromatic light, although by the use of suitable filters it is possible to select light of a fairly narrow waveband.

Photoelectric colorimeters are now used almost exclusively for routine laboratory purposes. They usually consist of a light source from which light passes through a suitable filter, an adjustable diaphragm and a tube containing the solution under investigation, to impinge finally on a photocell. The current generated is fed to a milliammeter usually calibrated to give an extinction reading directly. In use the blank solution, e.g. solvent, is placed in position and the diaphragm adjusted to give a zero scale reading; standard and unknown solutions are then, in turn, placed in the light path and the scale readings observed. By simple proportion the concentration of the unknown solution may be calculated. It will be appreciated that colorimeter tubes or cells for use with these instruments must be accurately matched to ensure that the length of the light path is identical in all cases.

Example 7.3.—Determinations of lactic acid by the Barker and Summerson method were carried out by photoelectric colorimeter using a green filter. A standard solution containing 30 μg lactic acid per determination gave a reading of 36·5 on the scale and two unknown solutions gave values of 16 and 45. The amount of lactic acid in the unknowns is therefore :

(a) $$\frac{30 \times 16}{36 \cdot 5} = \underline{13 \cdot 2}\ \mu\text{g.}$$

(b) $$\frac{30 \times 45}{36 \cdot 5} = 36 \cdot 98 \approx \underline{37\,\mu\text{g.}}$$

CALIBRATION CURVES.—Where colorimetric determinations are being carried out routinely with photoelectric instruments, either colorimeters or spectrophotometers, the best method of analysis is to prepare a calibration curve covering a suitable range of concentration. Extinction or scale readings are plotted against substrate concentration and then, on all future occasions, by reference to this graph the concentration corresponding to any observed colorimeter reading may be ascertained without the use of standard solutions (Fig. 7.4(a)). With such a graph the range over which Beer's Law holds (i.e. the linear portion) is readily appreciated and it also permits analyses to be extended outside this range provided the deviation from linearity is not too great. Errors which might otherwise arise through using the proportionality calculation (Example 7.3) in a region where Beer's Law does not apply are thereby eliminated. This latter consideration emphasizes the general principle that when standard solutions are used for colorimetric comparison they should be as near the extinction of the unknown solution as possible.

It has been noted that Beer's Law does not apply to suspensions. However, for turbidimetric work, it is possible to prepare for suspensions a calibration which often displays slight curvature throughout. The density of bacterial suspensions is usually measured in this manner. A blue filter is used with colorimeters for bacterial suspensions because short wavelengths are more strongly scattered than long wavelengths (Rayleigh's $1/\lambda^4$ law, see equation 1.51, page 46). It follows from this consideration that

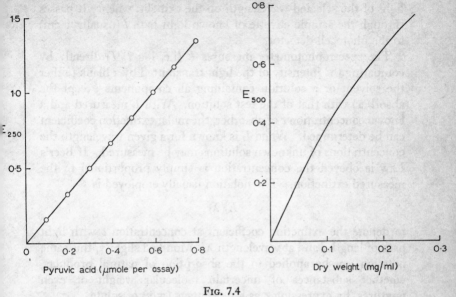

FIG. 7.4

Calibration curves for

(a) determination of pyruvic acid as its semi-carbazide at E_{250}, and

(b) determination of bacterial suspension density of *Azotobacter beijerinckii* at E_{500}.

if solutions are not free from suspended matter the error will increase as extinction measurements are taken further into the ultraviolet region. Spectrophotometers are increasingly used for this type of work, however, and Fig. 7.4(b) illustrates a calibration curve for a bacterial suspension.

Spectrophotometry

A spectrophotometer is an instrument which uses monochromatic light and enables the extinction, E, to be measured for various solutions at different wavelengths. It consists of a suitable light source, a monochromator (either prism or diffraction grating), a cuvette to hold the sample and a detecting device. Light from the source, which is a hydrogen-discharge lamp for the ultraviolet and a tungsten filament lamp for the visible region of the spectrum, is focused on the adjustable entrance slit of the monochromator. The monochromator disperses the light and focuses light of the selected wavelength on the exit slit, whence it passes through the sample cuvette of known light path l (usually 1 cm) to the photocell detector.

The spectrophotometer measures E (i.e. $\log I_0/I$) directly by comparing the intensity of the light transmitted by a blank (either the solvent or a solution containing all components except the absorber) with that of the test solution. With E measured and a known concentration c of absorber, the molar extinction coefficient can be determined. When E is known for a given wavelength the concentrations of unknown solutions may be measured. If Beer's Law is obeyed the concentration is simply proportional to the measured extinction. The notation usually employed is

$$E_l^c(\lambda)$$

to denote the extinction coefficient at concentration c with light path of length l and at wavelength λ. Modified slightly, this same notation can be applied to the absorption of natural products, whether substances of uncertain molecular weight or even mixtures, by expressing c as a percentage (w/v) of solute.

Example 7.4.—If a 0·5 cm layer of 0·2 per cent. (w/v) solution transmits one-tenth of the incident light at 300 nm, then $E = 1\cdot0$, since $E = \log I_0/I = \log 10$.

The measurement may be specified as

$$E_{0\cdot5\text{ cm}}^{0\cdot2\%\text{ (w/v)}}(300\text{ nm}) = 1\cdot0.$$

For convenience of comparison 1 per cent. (w/v) solution and 1 cm light path are usually taken as standard and the previous measurement is converted to this basis as follows :

$$E_{1\text{ cm}}^{1\%\text{ (w/v)}}(300\text{ nm}) = \frac{1\cdot0}{0\cdot5 \times 0\cdot2} = 10.$$

The slit width of a spectrophotometer determines the extent to which the light selected is monochromatic. Most instruments record the slit width in millimetres, which is directly related to the spectral slit width in nanometres for any given wavelength. The accuracy of a spectrophotometric measurement depends on the relationship between the spectral slit width and the *spectral band width*. The latter parameter is defined as the waveband width at half maximum absorption for a particular absorption peak. This will be clear by reference to Fig. 7.5 where the absorbing compound shows a symmetrical absorption curve with E_{max} of 1·0 at 350 nm. The spectral band width thus spans the region between

$E = 0.5$ on the ascending and descending curves, i.e. between 333 and 367 nm, equal to a span of 34 nm.

The effect of the instrumental slit width on the accuracy of measurement will be apparent from Fig. 7.5 where slit widths of 1 and 10 nm are indicated. The spectrophotometer averages the

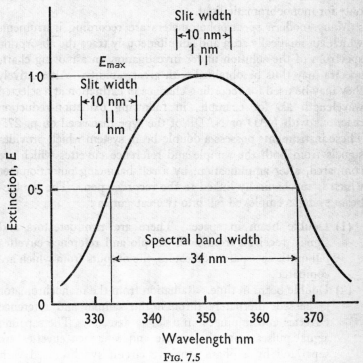

FIG. 7.5

The spectral band width of a compound having a symmetrical absorption curve with $E_{max} = 1.0$ at 350 nm. The effect of the slit width of a spectrophotometer on the accuracy of measurement is illustrated by comparison of slit widths of 1 nm and 10 nm centred on 333 nm on the steeply rising limb of the curve, where $E = 0.5$, and on 350 nm, the adsorption maximum, where $E = 1.0$.

extinction values of the light beam admitted by the slit, consequently as the absorption spectrum is scanned, the error will be greatest where the spectrum is steepest. For example, in Fig. 7.5, if a slit width of 1 nm were centred on a wavelength of 333 nm the instrument would be averaging extinction values of 0.48 at 332.5 nm and 0.52 at 333.5 nm, whereas with a slit width of 10 nm centred on the same wavelength the extinctions averaged would be 0·21 at 228 nm and 0·75 at 338 nm. Clearly the less linear the absorption curve in the region examined, the greater will be the error of using a wide slit width ; maximum accuracy is obtained with measurements at an absorption peak and with a narrow slit width. Further, it must be remembered that Beer's Law holds only for monochromatic light.

Many modern spectrophotometers are recording instruments which automatically scan and simultaneously trace the absorption spectrum of the solution under investigation on a moving chart; spectra may thus be obtained with great rapidity. Alternatively they may be used for recording changes in extinction at a selected wavelength as, for example, in following oxidation-reduction reactions with NAD or NADP of the type discussed on p. 275. These instruments possess a double-beam system which provides signals from both the sample and reference cuvettes which are compared, after amplification, by a self-balancing potentiometer which is mechanically linked to the recorder pen. The double-beam systems employed fall into two categories :

(1) Double beam in space. There are two detectors and signals received from both sample and reference cuvettes fall on these separate detectors, the outputs from which are compared.

(2) Double beam in time. Radiation from the monochromator is passed alternately through the sample and reference cuvettes and impinges on a single detector. The separate signal pulses from the sample and reference cuvettes are separated by a phase sensitive circuit synchronized with the beam-splitting device, rectified and compared in a self-balancing potentiometric recorder.

Analysis of Mixtures

Mixtures of several absorbing substances can be analysed in certain cases. Where the absorption bands of the individual substances of the mixture do not overlap, the analysis is simple and straightforward. Wavelengths are selected where each component in turn displays an absorption adequate for measurement and the other components display negligible absorption. Thus one measurement of extinction at each chosen wavelength suffices to determine each individual substance in the mixture. An important application of this method is the determination of tyrosine and tryptophan in unhydrolysed proteins. Both amino acids absorb strongly at 280 nm, the only amino acids to do so. Since tryptophan is easily destroyed by hydrolytic procedures, the spectroscopic technique affords the most reliable method of analysis for this amino acid. As most proteins contain tyrosine residues measurement of the extinction at 280 nm is a rapid and convenient means of determining the protein content of a solution.

Where the absorption bands overlap and there are n components of the mixture, the absorption must be measured at n wavelengths, chosen so that at each the specific extinction coefficient of one component is much greater than the others. The solution of n simultaneous equations enables the concentrations to be determined. To illustrate the method consider a mixture containing three non-interacting components, X, Y and Z, of concentrations C_X, C_Y and C_Z. At three chosen wavelengths λ_1, λ_2 and λ_3 the specific extinction coefficients are respectively x_1, x_2, x_3, y_1, y_2, y_3 and z_1, z_2, z_3. Assuming Beer's Law to be obeyed, the extinction coefficient of the mixture at each wavelength is the sum of the values for the individual components. Thus

$$K_1 = C_X x_1 + C_Y y_1 + C_Z z_1$$
$$K_2 = C_X x_2 + C_Y y_2 + C_Z z_2$$
$$K_3 = C_X x_3 + C_Y y_3 + C_Z z_3$$

and the three simultaneous equations are solved for C_X, C_Y and C_Z.

This type of analysis can present problems, with a conventional spectrophotometer, however, because it may prove difficult to determine accurately the extinction at each of the analytical wavelengths used. The recent introduction of a double-beam spectrophotometer which possesses a double wavelength mode of operation, permitting the reference and sample beams at different wavelengths (λ_1 and λ_2 respectively) to be combined in space prior to passing through the sample cuvette, overcomes these difficulties. The transmitted beams are time-shared on the detector so that, effectively, the difference in extinction at two different wavelengths is recorded. To analyse a two-component mixture the instrument may be used in two ways. In the first method a baseline is selected where the wavelength of each beam is held constant and the extinction difference $\Delta E = E_{\lambda 2} - E_{\lambda 1}$ is measured. With the sample in the light beam, two wavelengths are selected such that the major component in the mixture, which interferes with the analysis, shows no extinction difference, even though its concentration varies. After these two wavelengths are found, various concentrations of the component to be determined are added to the sample cell, in turn, and ΔE measured to construct the calibration curve.

The second method of operation for a two-component mixture is to record the *derivative curve* of the absorption spectrum of each sample, a technique particularly valuable when the component of interest has absorption peaks on the limb of a strong absorption band of the interfering component. In this method the extinction difference, ΔE, at a constant wavelength interval, $\Delta \lambda (= \lambda_2 - \lambda_1)$, is monitored continuously as the absorption spectrum is scanned. Usually $\Delta \lambda$ is selected to be 1 to 5 nm, depending on the contour of the absorption curve. The recorder traces a curve, the derivative spectrum, relating $\Delta E / \Delta \lambda$ to wavelength. The derivative curve often has the effect of flattening steep absorption curves caused by the interfering compound and thus facilitates the analysis.

In certain cases the absorption spectra of two compounds at equal concentrations may intersect and the point of intersection is known as an *isosbestic point*. This is illustrated in Fig. 7.6, which shows the spectra of alkaline solutions of the 2,4-dinitrophenyl-hydrazones derived from equimolar solutions of pyruvic and α-oxoglutaric acids. The isosbestic point is at 431 nm and it follows, therefore, that at this particular wavelength the solutions of these different hydrazones have the same extinction, i.e. a calibration curve relating extinction to molar concentration is applicable equally to either compound. This observation can be made the basis of a method for determining the absolute concentrations of pyruvic and α-oxoglutaric acids in mixtures of two. In this procedure the extinction is measured at the isosbestic point and at another wavelength, where the spectra show divergent extinctions, e.g. at 390 nm. The first measurement gives the total molar concentration of the mixture while the ratio of E_{390}/E_{431} depends on the relative molar proportions of the two keto acids. As the proportion of α-oxoglutaric acid in the mixture is increased the E_{390}/E_{431} ratio increases in a linear manner from the

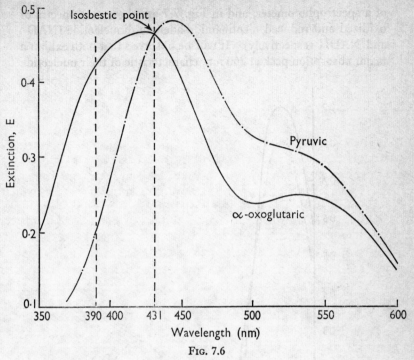

FIG. 7.6

Absorption spectra of the 2,4-dinitrophenylhydrazones of pyruvic and α-oxoglutaric acids in alkaline solution, showing an isosbestic point at 431 nm. The hydrazones were prepared from 0·4 μmole of each keto acid.

value characteristic of pure pyruvic acid to that for pure α-oxoglutaric acid. Thus the relative proportions of the compounds can be ascertained and, since the total molar concentration is known (E_{431} measurement), the absolute amounts of each acid present can be calculated.

A point to notice in connexion with techniques of this type is the importance of using constant slit widths for all extinction measurements made at one given wavelength. Ideally, these should be as narrow as possible. This precaution is particularly vital at wavelengths where the extinction is changing rapidly, for a change in spectral slit width in such regions will affect the extinction measurement appreciably (see p. 269).

ABSORPTION SPECTRA.—The measurement of absorption spectra, to which reference has already been made, is carried out by means of a spectrophotometer, and in Fig. 7.7 are shown the spectra of oxidized and reduced nicotinamide-adenine dinucleotide (NAD$^+$ and NADH respectively). It will be observed that both exhibit a major absorption peak at 260 nm, characteristic of their nucleotide

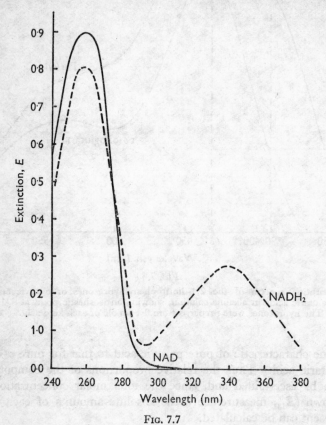

FIG. 7.7

Absorption spectra of oxidized (NAD) and reduced (NADH$_2$) nicotinamide-adenine dinucleotide.

structure, but the spectrum of the reduced form shows a smaller absorption band with a maximum at 340 nm whereas the oxidized form does not. This difference forms the basis of a widely used method for following oxidation-reduction reactions involving NAD (or nicotinamide-adenine dinucleotide phosphate, NADP, which displays similar spectra in oxidized and reduced forms). By carrying out the reaction in a spectrophotometer cuvette with wavelength selected for 340 nm, the reduction of NAD may be observed by the increase in extinction that occurs as the reaction proceeds. Conversely, oxidation of NADH results in decrease of extinction at 340 nm. Thus, in the case of the L-malate : NAD oxidoreductase (malate dehydrogenase)

$$\text{Malate}^= + \text{NAD}^+ \rightleftharpoons \text{Oxaloacetate}^= + \text{NADH} + \text{H}^+$$

the extinction at 340 nm increases as the reaction proceeds from left to right and decreases as the reverse reaction takes place. Examples of such experiments will be found in some of the problems of Chapter IV.

Solutions of native ribonucleic acid (RNA) and of deoxyribonucleic acid (DNA) display the property of *hyperchromicity*, i.e. an increase in extinction, when they are heated. For example, if ribosomal RNA in 0·01 M-phosphate buffer is heated, the extinction at 260 nm increases by about 27 per cent. between 25° and 85°; on cooling the extinction reverts to its original value. This behaviour is attributed to the reversible rupture of hydrogen bonds, and the degree of hyperchromicity observed can be used to determine the extent of the helical structure present in the molecule. In this way it has been calculated that 40–60 per cent. of transfer RNA has helical secondary structure. The study of hyperchromic effects has permitted the determination of the structure of the DNA from bacteriophage ϕX-174 as a single-stranded molecule. The non-reversible breakage of hydrogen bonds which occurs in the early stages of hydrolysis of nucleic acids also leads to a hyperchromic effect.

DIFFERENCE SPECTRA.—To accentuate the finer details of a change in the absorption spectrum produced by a reaction such as oxidation or reduction, or by a change in environment, it is often desirable to obtain a difference spectrum. This is achieved by using as the optical blank a solution which contains the same concentration of absorbing material as does the test cuvette but which is not subjected to the reaction under investigation, or to the change of environment. The absorption relative to the blank is then measured over the appropriate range of the spectrum. In practice the technique usually employs as the blank two optical cells in line containing solutions of the reactants. Solutions of the reactants are mixed in another cell, of light path equivalent

K

to that of the sum of the light paths of the two blank cells, and the absorption compared. The method has been widely used in the study of the oxidation and reduction of cytochromes.

Nephelometry

Nephelometry enables the turbidity of a suspension to be measured by means of the Tyndall effect, i.e. the light scattered by the particles in suspension. The nephelometric method has been used by T. W. Richards to determine the true end-point of the silver nitrate-sodium chloride titration in high precision work for determining the atomic weight of silver. It is also widely used for determining the density of bacterial suspensions. One type of nephelometer passes a beam of light vertically through an iris diaphragm to the hemispherical base of the test tube, which when containing a clear liquid functions as a condenser lens and produces a parallel beam passing centrally up the tube. The focusing action of the tube base is decreased by a turbid solution which also scatters light from the vertical column of liquid. This scattered light is received by a number of photocells which are so arranged round the base of the test tube that they do not receive light directly from the lamp. The current generated is fed to a milliammeter which indicates a measure of the turbidity of the solution. The method is advantageous for following bacterial growth because there is no interference with the contents of the growth tube.

Fluorimetry

Substances which fluoresce in ultraviolet light may be assayed by the techniques of fluorimetry, which are usually very much more sensitive than the corresponding absorption spectrophotometric methods. Recent advances in instrumentation have rendered fluorimetry a very powerful technique and it is, for example, being increasingly applied in the kinetic studies of nicotinamide adenine nucleotide enzymes. The Michaelis constants of NAD- and NADP-linked dehydrogenases for their coenzymes are often very low and useful kinetic data can only be obtained at very low concentrations of substrate and coenzyme. At these low concentrations the spectrophotometric technique is insufficiently sensitive whereas fluorimetric techniques enable reduction of the coenzymes to be monitored since the reduced forms are fluorescent.

Two important members of the B group of vitamins can be assayed easily and rapidly by fluorimetry. Riboflavin fluoresces and thiamin can be oxidized by alkaline potassium ferricynanide to form thiochrome which fluoresces in ultraviolet light. Other compounds of biochemical interest which fluoresce include aromatic amino acids, chlorophyll and various proteins such as fibrinogen, pepsin, serum albumin and trypsin.

Fluorescent molecules attached to enzymes either covalently or by adsorption can, under appropriate conditions, serve as useful indicators for the detection of conformational changes induced by ligands, as in the phenomenon of allosterism (see p. 224). Provided the emission of the fluorophore is sensitive to its environment the method can be successfully applied not only to proteins but also to the conformational changes occurring with lipoprotein membranes. For example, Chance and his collaborators found that submitochondrial preparations elicit a fluorescence with 8-anilino-naphthalene-1-sulphonate which is increased when the particles are brought into the energized state ; this property can thus be utilized for the study of the energized states of mitochondria.

Fluorimeters are so designed that ultraviolet light enters the sample solution and a portion of the fluorescent radiation produced at an angle normal to the irradiating beam enters a phototube which is protected by a filter from the exciting ultraviolet light. Because the amount of fluorescent radiation is relatively small it is necessary to have a photomultiplier which amplifies the emission (often by a factor of the order of 10^6) and, with ultraviolet sources such as the xenon arc or high pressure mercury arc, this makes possible the determination of fluorescent substances down to a concentration of 10^{-8} M. Most photomultipliers display their maximum photosensitivity in the region of 400 nm with a marked decline towards longer wavelengths. Consequently fluorimetric work at wavelengths greater than 500 nm presents difficulties.

A recent development in instrumentation is the spectrophotofluorimeter, which extends the range of the spectrum for analysis to cover the ultraviolet and infrared regions. Two monochromators are introduced, one for the exciting source and the other for the fluorescent emission, thereby eliminating the use of filters. The instrument can also be used in conjunction with an automatic recorder.

Flame Photometry

Flame photometry depends on the emission spectra obtained when alkali or alkaline earth metals are subjected to heat excitation in a non-luminous gas flame.   In one type of flame photometer

TABLE 7.1

Terms used in Spectrophotometry

Term	Symbol	Significance	SI Unit
Napierian absorbance (extinction)	B	$B = \ln I_0/I$	1
Napierian absorptivity (Napierian absorption or extinction coefficient)	b	$b = B/l$	m^{-1}
Molar Napierian absorptivity (molar Napierian absorption or extinction coefficient)	κ	$\kappa = B/lc$	m^2mol^{-1}
Decadic absorbance (extinction)	$A(E)$	$A = \log I_0/I$	1
Decadic absorptivity (decadic absorption or extinction coefficient)	a	$a = A/l$	m^{-1}
Molar decadic absorptivity (molar decadic absorption or extinction coefficient)	ε	$\varepsilon = A/lc$	m^2mol^{-1}
Transmittance	T	$T = I/I_0$	1

light emitted by the flame is focused through suitable optical filters on to a barrier layer photocell and the current generated is taken through a calibrated potentiometer to operate a taut suspension galvanometer. The instrument is calibrated with standard metal solutions, and full-scale deflections of the galvanometer may be obtained with as little as 5 parts per million (p.p.m.) sodium and 10 p.p.m. potassium. It is also possible to obtain flame photometer attachments for most commercially produced spectrophotometers. In flame photometry there is the possibility of considerable interference by ions other than those being determined.

Optical Activity

OPTICAL ROTATION.—Optical activity is related to molecular asymmetry. Molecules which do not have a plane of symmetry exist in two stereochemical forms, which are mirror images of each other and which exhibit optical activity. One form, the *dextro*, rotates the plane of polarized light clockwise and the other, the *laevo* form, rotates it counterclockwise. When both forms are simultaneously present in equal concentration, termed a *racemic* mixture, the two opposite rotations nullify each other and no optical activity is observed.

The rotation of plane polarized light is measured experimentally with a polarimeter and the following relationship holds :

$$[\alpha] = [\alpha]_{\lambda}^{t} \, lC \qquad . \qquad . \qquad . \quad (7.10)$$

where $[\alpha]$ is the observed rotation, l the length in decimeters of the light path through the test solution, C the concentration of solute in grams per millilitre and $[\alpha]_{\lambda}^{t}$ a proportionality constant, termed the *specific rotation*, which depends on the wavelength of the light, λ, and the temperature, t. The specific rotation is a characteristic property of a substrate. The *molar rotation* $[M]_{\lambda}^{t}$ is the product of the specific rotation and the molecular weight, i.e.

$$[M]_{\lambda}^{t} = M[\alpha]_{\lambda}^{t} \qquad . \qquad . \qquad . \quad (7.11)$$

The monochromatic light source usually employed in polarimetric work is the D-lines of sodium (589·0 and 589·6 nm) and the specific rotation is then recorded as $[\alpha]_{D}^{t}$.

Example 7.5.—An optical rotation of $+6\cdot32°$ was observed with a solution of D-glucose in a 10 cm cell at 20° with sodium D-light. What is the molar concentration of D-glucose ? $[\alpha]_{D}^{20}$ for D-glucose is $+52\cdot7°$.

As 10 cm = 1 decimetre, l is unity in the relationship

$$[\alpha] = [\alpha]_{D}^{20} \, lC$$

$$\text{Hence } C = \frac{[\alpha]}{[\alpha]_{D}^{20} \, l} = \frac{6\cdot32}{52\cdot7 \times 1}$$

where C is the concentration in g per ml. As the molecular weight of glucose is 180, the molar concentration will be

$$\frac{6\cdot32 \times 1000}{52\cdot7 \times 180} = \underline{0\cdot66 \text{ M.}}$$

In the past, studies of optical rotation have proved extremely valuable in the investigation of the structure of sugars. More recently the measurement of the variation in optical rotation as a function of the wavelength of monochromatic light, known as *optical rotatory dispersion*, has been successfully applied to the study of steroids and helical structures.

OPTICAL ROTATORY DISPERSION.—Since optical rotation is a consequence of molecular asymmetry, it offers a method for the study of helical structures. Helices are inherently asymmetric although in the case of a polypeptide the observed rotation depends not only upon its helical content but also on the constituent amino acids, which are themselves optically active. The problem posed, therefore, is to assign the appropriate contribution of the helical structures to the overall observed rotation.

The plot of the specific rotation as a function of wavelength sometimes displays maxima and minima and this anomalous behaviour is referred to as a Cotton effect. The Cotton effect displayed by the enzyme phosphoglucomutase is illustrated in Fig. 7.8. With other compounds $[\alpha]$ either increases positively or negatively with decreasing wavelength and no Cotton effect is observed.

The dependence of optical rotation on the wavelength of light for the simplest systems can be expressed over a considerable range of wavelengths by a single term Drude equation of the form

$$[\alpha]_\lambda = \frac{K}{\lambda^2 - \lambda_0^2} \qquad . \qquad . \qquad . \qquad (7.12)$$

where K and λ_0 are constants characteristic of the compound. λ_0 is related to the wavelength at which the absorption of light occurs and K to the index of refraction. The values of these constants can be obtained by appropriate plots of the experimental data, e.g. if $[\alpha]_\lambda \lambda^2$ is plotted versus $[\alpha]_\lambda$, λ_0 can be obtained from the slope and K from the intercept. Systems which can be described by the single term Drude equation are said to give plain dispersion curves. However many systems, including polypeptides, do not give linear plots according to the single term Drude equation and a two-term equation, due to Moffitt and Yang, is necessary to describe their behaviour, e.g.

$$[m']_\lambda = a_0 \frac{\lambda_0^2}{\lambda^2 - \lambda_0^2} + b_0 \frac{\lambda_0^4}{(\lambda^2 - \lambda_0^2)^2} \qquad . \qquad . \qquad (7.13)$$

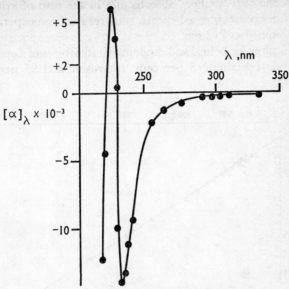

FIG. 7.8

Effect of wavelength on the specific rotation of phosphoglucomutase. A Cotton effect is exhibited and the specific rotation has a minimum of $-14,700°$ at 232 nm.

(After JIRGENSONS (1962), *Biochemistry* 1, 917.)

Here $[m']_\lambda$ is the *reduced mean residue rotation* and is defined by the equation

$$[m']_\lambda = \frac{3}{n^2 + 2} \frac{MRW}{100} [\alpha]_\lambda \qquad . \qquad . \quad (7.14)$$

where n is the index of refraction and MRW the mean residue weight, which for most polypeptides and proteins is about 115. The parameters a_0 and b_0 are determined experimentally from plots of

$$[m']_\lambda \frac{\lambda^2 - \lambda_0^2}{\lambda_0^2} \quad \text{against} \quad \frac{\lambda_0^2}{\lambda^2 - \lambda_0^2}$$

If the Moffitt equation is obeyed such plots should be linear, with a slope of b_0 and intercept on the ordinate of a_0. The constant a_0 represents both the intrinsic residue rotations and the interactions within the helix and varies with the environment, while b_0 is attributed to interactions within the helix. λ_0 is the wavelength at which the chromophore absorbs and in the case of synthetic polypeptides possessing α-helical structures has been experimentally determined as 212 nm.

Moffitt plots for helical and random coil structures of a synthetic polypeptide containing 5 per cent. L-tyrosine and 95 per cent.

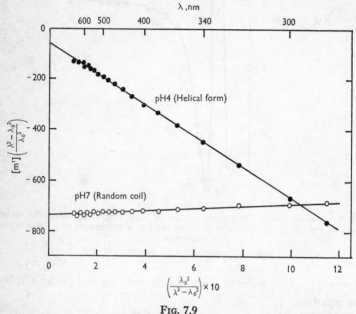

FIG. 7.9

A Moffitt plot (equation 7.13) with $\lambda_1 = 212$ nm for the rotatory dispersions of a synthetic polypeptide, a copolymer of 5 per cent. L-tyrosine with L-glutamic acid. At pH 4·0 in 0·1 M-phosphate buffer the polypeptide exists in the helical form whereas at pH 7·0 in the same solvent it exists as a random coil.

(After URNES & DOTY, 1961.)

L-glutamic acid are shown in Fig. 7.9. It will be observed that both slope and intercept are profoundly affected by structure. Since a transition from the helical to the random coil structure of synthetic polypeptides can often be effected by changing the solvent from a non-polar to a polar type, the effect of solvent on the constant b_0 has been investigated. The right-handed α-helix structure in non-polar solvents is associated with b_0 values of about -630; this value changes to zero for the random coil structure in polar solvents. The values of a_0 correlate less satisfactorily with the helical content. Polypeptides which possess both helical and random coil contents display intermediate values of b_0. The use of b_0 values for the estimation of the degree of helicity in native proteins has not been attended by the same success as with synthetic polypeptides, but nonetheless useful results have been obtained. The problems are difficult and complex, however, and the reader is referred to the treatments by Djerassi (1960), Urnes & Doty (1961) and Jirgensons (1969) for full details.

SUGGESTED READING

BEAVEN, G. H., JOHNSON, E. A., WILLIS, H. A. & MILLER, R. G. J. (1961). *Molecular Spectroscopy.* London: Heywood & Co.

DJERASSI, C. (1960). *Optical Rotatory Dispersion.* New York: McGraw Hill.

GILLAM, A. E. & STERN, E. S. (1970). *An Introduction to Electronic Absorption Spectroscopy in Organic Chemistry.* 3rd ed. Revised by E. S. Stern & C. J. Timmons. London: Arnold.

HEILMEYER, L. (1943). *Spectrophotometry in Medicine,* Trs. A. Jordan & T. L. Tippell. London: Adam Hilger.

JIRGENSONS, B. (1969). *Optical Rotatory Dispersion of Proteins and Other Macromolecules.* Berlin: Springer-Verlag.

LOTHIAN, G. F. (1958). *Absorption Spectrophotometry.* 2nd ed. London: Hilger & Watts.

MORTON, R. A. (1942). *The Application of Absorption Spectra to Vitamins, Hormones and Co-enzymes.* London: Adam Hilger.

OSTER, G. & POLLISTER, A. W. (1955). *Physical Techniques in Biological Research,* vol. 1. *Optical Techniques.* New York: Academic Press.

URNES, P. & DOTY, P. (1961). Optical rotation and the conformation of polypeptides and proteins. *Adv. Protein Chem.* **16**, 401.

PROBLEMS

7.1. What percentage of the incident light is reflected when a light beam is directed perpendicularly to the surface of benzene, ethyl acetate and acetone ? The refractive indices of these compounds are respectively $1 \cdot 5011$, $1 \cdot 3722$ and $1 \cdot 3589$.

7.2. Calculate the extinctions corresponding to the following values of the percentage of transmitted light ($100I/I_0$) :

(*a*) 95 ;   (*b*) 88 ;   (*c*) 71 ;   (*d*) 50 ;   (*e*) $17 \cdot 5$;   (*f*) $1 \cdot 0$.

7.3. An aqueous solution of sodium fumarate ($0 \cdot 454$ mM) gave an extinction of $0 \cdot 65$ at 250 nm in a spectrophotometer cuvette of 1 cm light path. Determine the molar extinction coefficient. What would the percentage light transmission be for a $250 \, \mu$M solution of sodium fumarate under the same conditions?

7.4. Protein concentration has been estimated from the intensity of biuret colours. What experiments would you carry out to ascertain the conditions of maximum accuracy in a given case ? At a certain wavelength a density reading of $0 \cdot 16$ is obtained from a $0 \cdot 18$ per cent. solution in a cell with depth of 1 cm. Calculate an extinction coefficient.

(Leeds Honours Course Finals, 1954.)

7.5. Pyruvic acid may be determined colorimetrically by conversion to its 2,4-dinitrophenylhydrazone followed by the addition of alkali. To calibrate a photoelectric colorimeter the reaction was carried out on solutions containing various amounts of pyruvic acid. The following results were obtained:

Pyruvic acid μg/3 ml	20	40	60	80	100	125	150	175
Scale reading	0·130	0·257	0·390	0·515	0·628	0·750	0·855	0·940

A series of unknown solutions of pyruvic acid was analysed by this method together with a standard solution containing 50 μg pyruvic acid per 3 ml. Scale readings recorded were (a) 0·280, (b) 0·555, (c) 0·690, (d) 0·773 and (e) 0·910 for the unknowns and 0·325 for the standard. What percentage of error is introduced in the analysis by using the proportional method of calculation based on the 50 μg standard instead of the previously established calibration curve?

7.6. A solution of a compound (0·001 M) was placed in a spectrophotometer cuvette of light path 1·05 cm and the percentage light transmission recorded was 18·4 at 470 nm. Determine the molar extinction coefficient. If the molecular weight of the compound was 215, what is the specific extinction coefficient?

7.7. The molar extinction coefficient of reduced nicotinamide-adenine dinucleotide phosphate (NADPH$_2$) at 340 nm is $6·22 \times 10^3$. 3 ml of solution containing 0·2 μmole NADPH$_2$ were placed in a cuvette of 1·05 cm light path. Calculate the percentage light transmission of this sample at 340 nm.

7.8. 3 ml of a solution of partially reduced nicotinamide-adenine dinucleotide were placed in a 1 cm spectrophotometer cuvette and the extinction determined at 340 and 260 nm. The values obtained were 0·207 and 0·900 respectively. Calculate the molar concentrations of oxidized and reduced forms of the dinucleotide.

The molar extinction coefficient of NADH$_2$ at 340 nm is $6·22 \times 10^3$ and the molar extinction coefficients of NAD and NADH$_2$ may be assumed to have the same value of $18·0 \times 10^3$ at 260 nm.

7.9. A mixture of *ortho*, *meta* and *para* cresols dissolved in cyclohexane may be analysed spectrophotometrically in straightforward manner because each exhibits an absorption band in a region where absorption due to the other cresols is negligible. The absorption maxima occur at 752, 776 and 815 cm^{-1} for *ortho*, *meta* and *para* cresols respectively. To test the validity of Beer's Law for solutions of cresols each was made up in cyclohexane at a series of concentrations and the extinctions measured. Data obtained are recorded below.

ortho		*meta*		*para*	
Concentration g/100 ml	E 752 cm^{-1}	Concentration g/100 ml	E 776 cm^{-1}	Concentration g/100 ml	E 815 cm^{-1}
0·25	0·120	0·60	0·115	0·50	0·09
0·50	0·235	1·15	0·220	1·00	0·20
1·00	0·465	2·35	0·460	2·10	0·405
2·00	0·820			3·15	0·60

An unknown mixture of the three cresols in cyclohexane was analysed and the percentage light absorption at 752, 776 and 815 cm^{-1} was 14·5, 50 and 41 respectively. Determine the concentration of each cresol and the percentage composition of the mixture.

(After WHIFFEN & THOMPSON (1945), *J. chem. Soc.*, 268.)

7.10. The determination of the proportion of carbon monoxide haemoglobin and oxyhaemoglobin in mixtures of the two by means of characteristic extinction coefficient quotients was worked out by Heilmeyer and Krebs. The wavelengths selected were 576 and 560 nm, corresponding respectively to the first absorption maximum and minimum of oxyhaemoglobin. Their results are tabulated below.

Use these data to construct a calibration curve for the determination. What percentage of oxyhaemoglobin is present in mixtures giving extinction coefficient quotients (576/560) of 1·580, 1·220 and 0·930 ? Suggest how it might prove possible to determine the absolute concentrations of oxyhaemoglobin and carbon monoxide haemoglobin in such mixtures and indicate what additional information would be required.

(After HEILMEYER & KREBS, quoted by HEILMEYER (1943), in *Spectrophotometry in Medicine* Trs. Jordan & Tippell. London : Adam Hilger Ltd.)

Percentage of carbon monoxide haemoglobin	Quotient K_{576}/K_{560}	Percentage of carbon monoxide haemoglobin	Quotient K_{576}/K_{560}
0	1·725	55	1·190
5	1·666	60	1·153
10	1·611	65	1·115
15	1·558	70	1·078
20	1·507	75	1·042
25	1·457	80	1·007
30	1·410	85	0·974
35	1·363	90	0·940
40	1·318	95	0·908
45	1·275	100	0·877
50	1·233		

7.11. The following data are taken from experiments to determine the extinction coefficients of reduced nicotinamide-adenine dinucleotide and nicotinamide-adenine dinucleotide phosphate. The sample of dinucleotide used need not be pure if an enzyme reaction is used with pure substrate and the dinucleotide is in excess. Under suitable conditions the change in absorption is due to the reaction of a quantity of dinucleotide equivalent to the added substrate. The reactions selected were :

Pyruvic acid + $NADH_2$ $\rightleftharpoons$ Lactic acid + NAD
Acetaldehyde + $NADH_2$ $\rightleftharpoons$ Ethanol + NAD
D-Isocitric acid + NADP $\rightleftharpoons$ α-Oxoglutaric acid + $NADPH_2$ + CO_2

and under the conditions selected proceeded from left to right virtually to completion. The lactic dehydrogenase of rabbit muscle also reacts with NADP, although at a much slower rate, and this dinucleotide was also used in the system. The cells were of 1 cm light path and the observed extinctions at 340 nm were as follows :

System	Extinction		Concentration of substrate $10^3 \times$ μmoles/cm³
	Initial*	Final	
1. Pyruvate − $NADH_2$	0·648	0·190	73·3
2. Pyruvate − $NADPH_2$	0·494	0·212	45·0
3. Acetaldehyde − $NADH_2$	0·620	0·485	22·1
4. Isocitrate − NADP	0·167	0·526	60·6

* Corrected for dilution due to substrate addition.

Calculate the molar extinction coefficient for each experiment.

(After HORECKER & KORNBERG (1948), *J. biol. Chem.*, **175**, 385.)

7.12. The following data were obtained in an experiment to calibrate a nephelometer for bacterial turbidity measurements. A suspension of *Aerobacter aerogenes* containing 610×10^6 cells per ml was diluted over a tenfold range and the original suspension used as a reference standard. With this suspension in the nephelometer the light intensity was adjusted to give a scale reading of 100 and the other suspensions were in turn placed in the instrument and the readings taken. These are given below.

Cell numbers millions/ml	610	549	488	427	366	305	264	183	122	61
Galvanometer reading	100	91·5	81·5	73·0	66·0	58·5	47·0	42·0	32·0	23·5

Plot the resultant calibration curve and comment on the behaviour of this curve at low bacterial densities.

7.13. Absorption spectra of solutions (in 0·75 N-sodium hydroxide) of the 2,4-dinitrophenylhydrazones derived from equimolar amounts of pyruvic and α-oxoglutaric acids display an isosbestic point at 431 nm. This property may be used to analyse mixtures of the two keto acids. The 2,4-dinitrophenylhydrazones are prepared in alkaline solution and the extinctions measured at 431 nm, and also at 390 nm where the spectra of the two acids have widely different extinction values.

The method was calibrated by investigating the relationship between extinction at 431 nm and the concentration of either acid (Table 1), and by measuring extinctions at 390 and 431 nm for carefully prepared mixtures of the two keto acids (Table 2). The data obtained are given below.

TABLE 1

Keto acid per determination μmole	E_{431}	
	Pyruvic acid	α-Oxoglutaric acid
0·10	0·095	0·100
0·25	0·260	0·255
0·50	0·550	0·545
0·75	0·812	0·805
1·00	1·060	1·058

TABLE 2

Percentage composition of mixture		E_{390}	E_{431}
Pyruvic acid	α-Oxoglutaric acid		
100·0	0·0	0·141	0·321
83·3	16·7	0·168	0·320
67·7	33·3	0·196	0·323
50·0	50·0	0·223	0·323
33·3	67·7	0·252	0·323
16·7	83·3	0·279	0·326
0·0	100·0	0·301	0·320

(a) Construct the necessary calibration curves and determine the total quantity of keto acid used in the experiment of Table 2.

(b) Three solutions, containing pyruvic and α-oxoglutaric acids as the sole keto acids, were analysed by this technique. 2 ml of solution was used in each case. The relevant data obtained were:

Solution	E_{390}	E_{431}
A	0·344	0·430
B	0·563	0·915
C	0·476	0·645

Determine the molar concentration of α-oxoglutaric acid present in each solution.

7.14. 3 ml of a solution of partially reduced nicotinamide-adenine dinucleotide were placed in a 1 cm cell and the extinction determined at 340 and 260 nm. The values obtained were 0·200 and 0·800 respectively. Calculate the molar concentrations of oxidized and reduced forms of the dinucleotide.

The molar extinction coefficient of NADH at 340 nm is $6·22 \times 10^3$ and the molar extinction coefficients of NAD^+ and NADH have the same value of 18×10^3 at 260 nm.

(Leeds Honours Course Finals, 1959. *Note:* this was only part of the question set ; see also Problem 5.20.)

7.15. Give a brief account of the following enzymes :

(a) tryptophan peroxidase (pyrollase) ;
(b) homogentisate oxidase ;
(c) maleylacetoacetate isomerase.

The molar extinction coefficients of tryptophan and tyrosine in 0·1N-NaOH are :

Wavelength (nm)	Molar extinction coefficient Tryptophan	Tyrosine
294·5	2375	2375
280	5225	1576

A solution of tryptophan and tyrosine in 0·1N-NaOH had an optical density in a 1 cm cell of 0·500 at 294·5 nm and 0·700 at 280 nm. What are the concentrations of these two amino acids ?

(Leeds Honours Course Finals, 1962.)

7.16. The extinction at 400 nm of 0·5 mM-p-nitrophenol was measured at a series of different pH values with the following results :

pH	3·0	4·0	5·0	6·0	6·5	7·0	7·5	8·0	9·0	10·0
E_{400}	0·009	0·009	0·014	0·040	0·088	0·179	0·263	0·328	0·370	0·371

Write an equation for the reaction involved and determine the pK of p-nitrophenol.

7.17. The extinction at 295 nm of glycyl-L-tyrosine was measured at various pH values at 25° and ionic strength 0·16 with the results appended below :

pH	1·09	4·6	9·7	10·2	10·5	13·0	14·0
ε molar (295 nm)	0	35	748	1370	1710	2280	2280

Comment on the structural features of the dipeptide which are responsible for this behaviour and calculate pK_{a_3} for glycyl-L-tyrosine.

7.18. The effect of temperature on the extinction of the DNA from bacteriophage φX-174 has been compared with that on native thymus DNA in 0·1 M-NaCl and heat-denatured thymus DNA in 0·1 M-NaCl. The φX DNA was in 0·2 M-NaCl + 1 mM-phosphate buffer, pH 7·5. The results are recorded as

$E(P)$ values, which term Chargaff & Zamenhof introduced for the extinction per g atom of phosphorus ; for ϕX DNA in 0·2 M-NaCl at 37°, $E(P)$ is 8700.

Temperature °C	$E(P)$ at 260 nm of DNA samples		
	ϕX	Native thymus	Heat-denatured thymus
0		6500	7000
20	8050	6590	7500
40	9050		8150
55	9505	6550	8450
60	9700		
70	9950	6550	
80	10100		8750
90	10150		

The ϕX DNA reacts with formaldehyde as does the denatured DNA, but the native DNA does not react with this reagent.

What can you deduce concerning the structure of the ϕX DNA ?

(After SINSHEIMER (1959), *J. mol. Biol.*, **1**, 37.)

7.19.   (a) State the Beer-Lambert Laws. Define the terms *extinction* and *transmittance* and derive a relationship between them.

The extinction at 260 nm of 4 ml of a solution of adenine in 0·1 N-HCl is 0·630. What is the extinction at the same light path of :

  (i) 8 ml of the solution ;

  (ii) 4 ml of the solution to which has been added 4 ml of a similar solution of adenine having an extinction at 260 nm of 0·200?

(b) Given the following information, devise a procedure for the assay of compound A in the presence of compound B.

Compound A has molar extinction coefficients in 0·1 N-HCl of a_1 and a_2 at wavelengths λ_1 and λ_2, respectively. A sample of compound B in 0·1 N-HCl gave extinctions of b_1 and b_2 at wavelengths λ_1 and λ_2 respectively.

Mention any factors limiting the accuracy of the method.

(Hull Honours Course Finals, Part I, 1965.)

7.20.   Three different extracts of barley leaves were prepared for the determination of ratio of chlorophyll a to chlorophyll b in this tissue. A different spectrophotometer was used in each case. Absorption measurements were made at 663 nm and 645 nm, where the specific absorptivities (k, concentration expressed in g/l) of the chlorophylls are

Chlorophyll a	663 nm, k 82·0	645 nm, k 16·8
Chlorophyll b	663 nm, k 9·27	645 nm, k 45·6

The following results were obtained :

1. Measurements made in a Unicam S.P. 600 spectrophotometer :

$$A_{663} \quad 0·341 \qquad A_{645} \quad 0·131$$

2. Measurements made in a Hilger Uvispek :

$$A_{663} \quad 1·040 \qquad A_{645} \quad 0·385$$

3. Measurements made in an Optica recording spectrophotometer and the following values read from the chart :

$$A_{663} \quad 0\cdot174 \qquad A_{645} \quad 0\cdot070$$

Use these figures to calculate the ratio of chlorophyll a to chlorophyll b in the barley leaf. Assume that the only significant errors in these determinations occur in the spectrophotometric measurements and comment on the differences, if any, in the values of the ratio obtained with the three instruments.

(University of Wales, Cardiff, Honours Course Finals, 1966.)

7.21. What do you understand by the following terms used in spectrophotometry : per cent. transmission, optical density, extinction, extinction coefficient, molar extinction coefficient, band width ?

The molar extinction coefficient of adenine in 0·1 N-HCl is $13\cdot4\times10^3$ ($\lambda = 262\cdot5$ nm ; path length $= 1$ cm). What would the extinction and per cent. transmission of a 5 μg/ml solution of adenine be when measured under the same conditions ? Molecular weight of adenine $= 135$.

(Hull Biochemistry Special I Course, 1968.)

7.22. Briefly outline the fundamental origins of ultraviolet, infrared and nuclear magnetic resonance spectra.

A globular protein containing residues of all the commonly occurring amino acids has 60-70 per cent. α-helical content. The tertiary structure is a heavily folded conformation, stabilized by several disulphide bridges. Large regions of the peptide chain are buried in the centre of the molecule. Discuss the expected spectral properties of this molecule and how they may be affected by denaturation with urea, or denaturation with urea followed by treatment with performic acid.

(University of Manchester, Biological Chemistry Honours Finals, 1968.)

7.23. Account for the following :

(a) The extinction of ribonuclease in aqueous urea at constant pH (7) at 2870 Å shows a linear increase with increasing urea concentration up to 4 M but thereafter the experimental curve changes slope and rapidly drops to a minimum at 8 M-urea. (Native ribonuclease contains six tyrosyl residues ; three of these are accessible to the surrounding solvent and the other three are buried inside the non-polar region of the molecule.)

(b) The dye, Biebrich scarlet (I) has a strong red colouration

I

On addition of a solution of this dye to poly-L-arginine solution in a 1 : 2 ratio of dye to amino acid residues, the colour disappears almost completely.

(c) The 100 Mc nuclear magnetic resonance spectrum of native ribonuclease presents a picture of a limited number of broad smooth peaks with little fine structure. Treatment of the protein with performic acid transforms this spectrum into one with sharp peaks and much detailed fine structure.

Discuss how the above observations might be usefully applied in structural studies of proteins.

(University of Manchester, Biological Chemistry Honours Course Finals, 1969.)

7.24.   Proton magnetic resonance was used to determine the nature of the interactions between hen egg white lysosyme and β-methyl-N-acetyl-D-glucosamine.   The chemical shift of the acetamido protons on the small molecule was observed as a function of concentration at an enzyme concentration of 3×10^{-3} M and at $20°$ C.   As the substrate concentration was lowered there was a progressive upfield shift of the resonance of these acetamido protons in comparison to the resonance of the acetamido protons on the free substrate.   From the data given calculate the equilibrium constant K_{diss} for the dissociation and chemical shift of the bound substrate using the formula

$$S_0 = \frac{E_0\,\varDelta}{\delta\text{obs}} - (K_{diss} + E_0) \qquad . \qquad . \qquad . \qquad (1)$$

Where S_0 = Total substrate concentration.
$\quad\quad E_0$ = Enzyme concentration.
$\quad\quad \delta$obs = Observed chemical shift relative to the free substrate.
$\quad\quad \varDelta$ = Chemical shift of the bound substrate.

Substrate concentration $\times 10^2$ (M)	Observed chemical shift from substrate resonance (cycles/second)
1·4	3·571
2·3	2·703
3·1	2·564
3·7	2·273
4·7	1·961
6·2	1·695

(i) Can you deduce anything about the enzyme binding site from this information?

(ii) State the assumption in the derivation of equation (1) which makes it valid. Is the above equation a valid relationship for treating the results of this experiment?

(iii) Given the information that the experiment was performed on a 100 megacycle NMR spectrometer, express $\varDelta$ in parts per million rather than cycles/second.

(iv) Can you suggest another way to obtain similar information from this type of experiment?

(v) If the K_{diss} for $30°$ C is $1·7 \times 10^{-2}$ M, what is the $\varDelta H°$ of this reaction?

$\ln X = 2·303 \log X \qquad R = 1·987$ cal mole $^{-1} T^{-1}$

(University of Glasgow, Honours Course Finals, 1970.)

7.25.   Distinguish briefly between optical rotatory dispersion and circular dichroism.

The variation of optical rotation of protein solutions with wavelength may be expressed quantitatively by a relationship due to Moffitt.   One form of this relationship is the equation :

$$[m'] \lambda = a_0\lambda_0^2/(\lambda^2 - \lambda_0^2) + b_0\lambda_0^4/(\lambda^2 - \lambda_0^2)^2$$

where λ_0 is assumed to be 212 nm.
Define the terms in the equation.

An optical rotatory dispersion study of a protein from bovine adrenal chromaffin granules yielded the following data in the range 300-500 nm :

$[m'](\lambda^2 - \lambda_0^2)/\lambda_0^2$	$\lambda_0^2/(\lambda^2 - \lambda_0^2)$
−750	1·10
−740	1·00
−730	0·85
−720	0·65
−700	0·55
−690	0·45
−685	0·35
−680	0·30
−675	0·35

Use these results to make an estimate of the α-helical content of the protein.
A characteristic feature of the protein was its unusual amino acid composition : glutamic acid, 26 per cent. (w/w) and proline, 8·6 per cent. (w/w) with low proportions of all other amino acids.   How can this amino acid composition be correlated with the result obtained for the helical content ?

(University of Manchester, Biological Chemistry Honours Finals, 1968.)

CHAPTER VIII

MANOMETRY

MANOMETRIC techniques are of considerable importance for the investigation of biochemical reactions in which gas exchanges are directly or indirectly involved. For instance the progress of oxidation, decarboxylation and photosynthetic reactions may be followed directly by the consumption or evolution of the gas concerned, whereas fermentation reactions and hydrolyses, during the course of which acid is produced, can also be measured by carrying out the reaction in bicarbonate buffer and recording the carbon dioxide evolution.

The advantage of manometric techniques lies in their extreme versatility and they have been put to many varied uses in furthering biochemical knowledge. At the present time the Warburg constant volume manometer and the Barcroft-Haldane differential instrument find most favour, and of these the former is more widely used. In this chapter attention will be confined to the Warburg apparatus, and the reader is referred to the excellent books by Dixon (1951) and Umbreit, Burris and Stauffer (1964) for details concerning other types of manometer.

The Warburg Constant Volume Manometer

The principle involved is that if the volume of a gas is held constant, at constant temperature, any changes in the quantity of gas may be measured by changes in pressure. The instrument is illustrated in Fig. 8.1 and consists of a flask attached to the manometer by means of a ground glass joint. The flask may have one or more side bulbs which permit the addition of the substrate or reagents at intervals as required ; it has also a centre well to which alkali is added when carbon dioxide is to be absorbed. The manometer fluid is contained in a reservoir and its level can be adjusted by means of the screw clamp. The tap permits the flask to be opened to the air. When assembled, the apparatus is fitted on a shaking device attached to a thermostatic water bath and so arranged that the flask is completely submerged. Accurate temperature control is necessary. In operation the level of fluid in the closed limb of the manometer is always adjusted to the zero mark and the level in the open limb recorded. This observed pressure difference (in millimetres) when multiplied by a constant, which must be determined for each flask and manometer, gives the quantity of gas evolved or absorbed. To enable the constant to be calculated it is necessary to know the gas volume of the flask, the volume of fluid in it, the gas being exchanged, the temperature, and the density of the manometer fluid. The way in which the constant is derived is given below.

The customary convention is to regard all quantities of gas evolved as positive and all quantities absorbed as negative. Gas evolution is indicated by a rise of liquid in the open limb of the manometer when readings will therefore be positive. In the following treatment all quantities of gas are expressed in μl of dry gas at N.T.P.

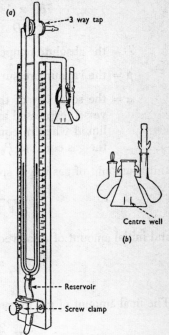

FIG. 8.1

(a) Warburg constant volume manometer with single side arm flask attached.

(b) Enlarged view of double side arm flask.

Let x = the quantity of gas evolved in μl at N.T.P. (if the gas is absorbed x will be negative).

V_g = the volume of the gas space in the vessel (including the connecting and manometer tubes down to the zero mark).

h = the observed manometer reading in mm.

V_f = the volume of fluid in the vessel.

P = the initial pressure in the vessel of the gas being determined.

P_0 = normal pressure (760 mm Hg) in mm of manometric fluid. If D is the density of the fluid,

$$P_0 = \frac{760 \times 13 \cdot 60}{D} \text{ (density of Hg} = 13 \cdot 60 \text{ at } 0° \text{ C).}$$

T = the absolute temperature of the thermostatic bath.

p = the vapour pressure of water at temperature T.

a = the solubility of the evolved gas in the liquid in the vessel (expressed as μl gas at N.T.P. dissolved in 1 μl liquid when in equilibrium with a partial pressure of the gas equal to P_0).

Initial amount of gas in gas space

$$= V_g \frac{273(P - p)}{T P_0}$$

and initial amount of dissolved gas

$$= V_f \frac{a(P - p)}{P_0}.$$

The final amount of gas in gas space

$$= V_g \frac{273(P - p + h)}{T P_0}$$

and the final amount of dissolved gas

$$= V_f \frac{a(P - p + h)}{P_0}.$$

But the final total amount of gas present will be the sum of the amount initially present and the amount x evolved.

Thus

$$\left(V_g \frac{273}{T} + a_f V \right) \frac{P - p + h}{P_0} = \left(\frac{V_g 273}{T} + V_f a \right) \frac{P - p}{P_0} + x,$$

whence

$$x = h \left[\frac{\dfrac{V_g 273}{T} + V_f a}{P_0} \right] . \qquad . \qquad (8.1)$$

From this equation it will be appreciated that the expression within the brackets remains constant for a given gas with a given manometer and flask provided the liquid volume and the temperature are unchanged. This term is known as the flask 'constant' (k) of the apparatus and, as already mentioned, multiplication of the pressure difference h by it gives the quantity in μl of dry gas at N.T.P. evolved. Thus

$$x = hk \quad . \quad . \quad . \quad . \quad (8.2)$$

CALIBRATION.—The value of the constant may be determined, i.e. the instrument calibrated, by one of three methods :

1. Calculation by use of equation 8.1.

2. Addition or withdrawal of a measured amount of gas from the flask by means of a graduated pipette and noting the change in manometer reading (Münzer and Neumann method).

3. By liberating or absorbing a known amount of gas in the manometer vessel by means of a chemical reaction, e.g. liberation of CO_2 from bicarbonate by acid.

THERMOBAROMETER.—Since the Warburg manometer has one end of the manometer tube open to the air, readings will be affected by slight changes in barometric pressure or in temperature of the thermostat which may occur during the course of an experiment. To eliminate errors arising in this way, an additional manometer containing water is always set up to function as a thermobarometer. Whenever readings are taken, the thermobarometer is also read and its value subtracted from those of the other manometers ; changes in external conditions are thereby compensated.

Example 8.1.—A Warburg manometer was calibrated by the bicarbonate method. 2 ml of 2 N-hydrochloric acid were placed in the flask and 1 ml of 10^{-2} M-sodium bicarbonate added from the side arm after equilibration, when the manometer and thermobarometer readings were respectively $+4$ mm and -0.5 mm. When gas evolution ceased the manometer reading was $+133.5$ mm and the thermobarometer registered -3.5 mm. Determine the flask constant.

The reaction employed is

$$NaHCO_3 + HCl = NaCl + CO_2 + H_2O$$

and each mole of bicarbonate yields 1 mole of CO_2.

$$1 \text{ ml of } 10^{-2} \text{ M NaHCO}_3 \equiv \frac{0.01}{1000} \text{ moles}$$

$$\equiv 10 \text{ } \mu\text{moles.}$$

Hence 10 μmoles of CO_2 are evolved by the excess acid and

$$x_{CO_2} = 22 \cdot 4 \times 10 = 224 \ \mu l.$$

The true manometer reading is $[133 \cdot 5 - 4] - [-3 \cdot 5 - (-0 \cdot 5)]$

$$= 129 \cdot 5 + 3 \cdot 0 = 132 \cdot 5 \text{ mm}.$$

Hence the flask constant $k_{CO_2} = \dfrac{224}{132 \cdot 5} = 1 \cdot 62.$

Units

The basis of all manometric measurements is the gram-molecular volume (G.M.V.), which is the volume occupied by one gram-molecule of any gas at normal temperature and pressure and has the value of 22·4 litres. In manometry very much smaller quantities are usually dealt with, and the following table defines the units employed :

1 mole of any gas at N.T.P. occupies   22·4 litres

1 millimole (m-mole) $= 10^{-3}$ mole occupies   22·4 millilitres (ml)

1 micromole (μmole) $= 10^{-6}$ mole occupies   22·4 microlitres (μl)

For example, the decarboxylation of 2 μmoles of aspartic acid results in the formation of 2 μmoles of CO_2 and β-alanine respectively

$$2 \ \begin{array}{c} \text{COOH} \\ | \\ \text{CHNH}_2 \\ | \\ \text{CH}_2 \\ | \\ \text{COOH} \end{array} \rightarrow 2 \ \begin{array}{c} \text{CH}_2\text{NH}_2 \\ | \\ \text{CH}_2 \\ | \\ \text{COOH} \end{array} + 2\text{CO}_2$$

Hence the volume of CO_2 evolved will be $2 \times 22 \cdot 4 = 44 \cdot 8 \ \mu l.$ Specific bacterial amino acid decarboxylases have been used by Gale for the quantitative analysis of amino acids in amounts that are far too small to be analysed by classical chemical methods.

RESPIRATORY QUOTIENTS

The respiratory quotient or R.Q. is defined as the relationship between the volume of carbon dioxide produced and the volume of oxygen consumed in respiration

$$\text{i.e. R.Q.} = \frac{CO_2 \text{ produced}}{O_2 \text{ consumed}}.$$

The R.Q. gives an indication of the nature of the metabolism which is taking place. Complete oxidation of carbohydrate gives a value of 1·0 whereas an average fat yields a value of approximately 0·7 and an average protein a value of 0·8. Oxidation of a mixture of substrates results in an intermediate value.

Example 8.2.—What is the R.Q. when (*a*) glucose and (*b*) triolein ($C_{57}H_{104}O_6$) is completely oxidized ?

(*a*) $C_6H_{12}O_6 + 6O_2 = 6CO_2 + 6H_2O$ R.Q. $= 6/6$ $= 1·0.$

(*b*) $C_{57}H_{104}O_6 + 80O_2 = 57CO_2 + 52H_2O$ R.Q. $= 57/80 = 0·71.$

METABOLIC QUOTIENTS

To express the rate of oxygen uptake by a particular tissue or micro-organism values known as *metabolic quotients* are used. They take various forms and are denoted by the symbol Q_X, where X is the metabolite being measured. Thus Q_{O_2} is defined as the μl oxygen taken up per milligram dry weight of biological material per hour. It has a negative value. Where the metabolite is a solid or liquid it is for the purpose of this definition regarded as a gas at N.T.P. so that 1 μmole is equivalent to 22·4 μl. Other examples, where a different tissue basis is employed, are as follows and it will be noted that the tissue basis is given in brackets :

$$Q_{O_2}(N) = \mu l \ O_2/mg. \ \text{tissue nitrogen/hour}$$

$$Q_{O_2}(C) = \mu l \ O_2/mg. \ \text{tissue carbon/hour}$$

If the atmosphere used in the manometer is other than air, it is also specified as a suffix to the quotient and the general form of the quotient then becomes

$$Q \ ^{\text{gas atmosphere}}_{\text{metabolite measured}} \ (\text{tissue basis})$$

For instance, if oxygen uptake is measured in an atmosphere of pure oxygen and expressed on the basis of tissue nitrogen, the metabolic quotient will be denoted by $Q_{O_2}^{O_2}(N)$ whereas a fermentation reaction, carried out in an atmosphere of nitrogen and followed by CO_2 evolution from bicarbonate buffer, on the basis of tissue carbon will be given as $Q_{CO_2}^{N_2}(C)$. It is important, of course, that the weight basis used should represent as nearly as possible 'active' cell material, since inclusion of inert material will make very misleading the comparison of Q_X values with those of other tissues or organisms.

If metabolic quotients are given in any units other than μl of gas X per milligram dry weight of biological material per hour, e.g. μmoles of X/mg dry wt./h, the symbol used is q_X. Production and removal of metabolites are indicated by quotients with $+$ or $-$ signs respectively, although these may be omitted if no confusion can arise.

Example 8.3.—A bacterial suspension, respiring in the absence of substrate, absorbed $32\,\mu l$ of oxygen in the initial 15 minutes. If the Warburg flask contained 12 mg dry weight of organisms what is the Q_{O_2} value based on this initial rate ?

$$Q_{O_2} = \mu l \ O_2/\text{mg dry weight/hour}$$
$$= \frac{32 \times 60}{15 \times 12} = \underline{10 \cdot 7}.$$

Measurement of Oxygen Consumption; the 'Direct Method' of Warburg

The respiration of most living cells, as opposed to many enzyme preparations, results in the consumption of oxygen and the evolution of carbon dioxide. Where oxygen and carbon dioxide are the only gases involved, oxygen consumption can be measured by absorbing the CO_2 in alkali placed in the centre well. This keeps the CO_2 pressure zero in the gas phase and the net gaseous exchange is then due to the oxygen consumed. This is the basis of the 'Direct Method' of Warburg for determining oxygen uptake.

One of the major disadvantages of the method lies in the fact that the atmosphere must be free from CO_2, and it has been shown for certain tissues (brain, liver, etc.) that oxygen uptake is stimulated by the presence of CO_2. Consequently, results obtained by the 'Direct Method' may not be a true reflection of metabolism under normal conditions. This difficulty may be circumvented by using Warburg's 'Indirect Method', details of which may be found in the recommended books.

The 'Direct Method' is used in two main ways, for the determination of (*a*) the rate of oxidation of a substrate, i.e. the Q_{O_2} value, and (*b*) the total oxygen consumption in the presence of a measured quantity of substrate. It is customary to determine both in the same experiment.

In determinations of both rate and total amount of oxygen uptake the status of the endogenous respiration (i.e. respiration of the tissue in the absence of external substrate) must be assessed and often leads to uncertainty. This uncertainty arises because the presence of the added substrate may or may not suppress the endogenous respiration, and the problem presented is whether a constant rate of endogenous respiration should be assumed, and the values deducted from the oxygen uptake observed in the presence of the substrate (see Dawes & Ribbons, 1962). It is customary in published work to report the endogenous respiration and to state whether or not it has been subtracted from the values recorded for the oxidation of the substrate.

OXIDATIVE ASSIMILATION

When measurements of total oxygen consumption were made with non-proliferating suspensions of certain micro-organisms metabolizing various substrates, it was discovered that oxidation ceased at levels corresponding to 30-60 per cent. oxidation of the substrate. Analysis showed that no substrate remained in the manometer flask and addition of a further quantity of substrate permitted resumption of oxidation. This led to the concept of *oxidative assimilation* and it has been demonstrated, mainly by Barker and Clifton and their colleagues, that energy derived from oxidation of certain substrates is utilized by the micro-organisms to enable a portion of the substrate to be synthesized into cellular components, as indicated by an increase in the carbon content of the cells. Thus where total oxygen uptake corresponds to 40 per cent. oxidation the remaining 60 per cent. has been assimilated by the cells.

Example 8.4.—Oxidation of ethanol by the colourless alga *Prototheca zopfii* was investigated manometrically and 5 μmoles of ethanol found to give a total oxygen consumption of 187 μl. Over the same period the endogenous respiration was 36 μl. What was the percentage of ethanol synthesized into cellular material ?

Assuming no suppression of endogenous respiration, the true oxygen uptake is 187 − 36 = 151 μl.

$$= \frac{151}{22 \cdot 4} = 6 \cdot 74 \ \mu\text{moles.}$$

For complete oxidation of ethanol to carbon dioxide and water

$$C_2H_5OH + 3O_2 = 2CO_2 + 3H_2O$$

3 moles of oxygen are required per mole of ethanol.

Thus 5 μmoles ethanol require 15 μmoles O_2 for complete oxidation.

$$\text{Percentage oxidation} = \frac{6 \cdot 74}{15} \times 100 = 44 \cdot 9.$$

Hence percentage synthesized into cellular material = 55·1.

Diffusion of Oxygen in Tissue Slices

The rate of diffusion of gases into tissue slices is dependent on their thickness, and clearly it is of importance to know the limiting thickness which ensures that the oxygen concentration within the tissue is adequate to support respiration at the maximum rate. Warburg has shown that the maximum permissible thickness of tissue slices can be calculated from the formula

$$d = \sqrt{8c_0 \frac{D}{A}}$$

where D = the diffusion coefficient of O_2 in the tissue in ml O_2 at N.T.P./cm²/min when the pressure gradient is 1 atmosphere/cm.

A = the rate of respiration of the tissue in ml/min/ml tissue

c_0 = the concentration of oxygen in atmospheres

d = the thickness of the tissue slice in cm at which the oxygen concentration at the centre of the slice is just zero.

According to Krogh, D at 38° is $1 \cdot 4 \times 10^{-5}$, A for liver tissue may be taken as 5×10^{-2}, and c_0 for air is $0 \cdot 2$.

Hence

$$d = \sqrt{\left(\frac{8 \times 0 \cdot 2 \times 1 \cdot 4 \times 10^{-5}}{5 \times 10^{-2}} \right)} = 2 \cdot 1 \times 10^{-2} \text{ cm.}$$

If an atmosphere of pure oxygen is used ($c_0 = 1 \cdot 0$) then $d = 4 \cdot 7 \times 10^{-2}$ cm. These results indicate that liver slices should be no thicker than $0 \cdot 2$ mm for experiments conducted in air, and no more than about $0 \cdot 4$ mm for those in pure oxygen. As it is difficult to prepare slices $0 \cdot 2$ mm thick whereas thicker ones of about $0 \cdot 3$ mm (which are much less fragile) can be obtained without trouble, it is customary to use the thicker slices in an atmosphere of pure oxygen.

Carbon Dioxide Production by the 'Direct Method'

Subject to the possible errors already mentioned, one is able to determine carbon dioxide evolution by the 'Direct Method' and hence obtain the R.Q. Two flasks are set up with the same contents and carbon dioxide is absorbed by alkali in one but not in the other. The former gives the oxygen uptake, whereas the gas exchange in the latter is due to the combined effect of oxygen consumption and carbon dioxide evolution; the carbon dioxide output can therefore be determined. Flask 1 contains alkali. Let h_{O_2} be the observed manometer change, x_{O_2} the amount of oxygen absorbed and $k_{1_{O_2}}$ the flask constant for oxygen. Flask 2 has no alkali. Let h be the observed manometer change, h_{CO_2} the movement due to the carbon dioxide, x_{CO_2} the amount of carbon dioxide evolved and $k_{2_{O_2}}$ and $k_{2_{CO_2}}$ the flask constants for oxygen and carbon dioxide respectively.

Now, the amount of oxygen absorbed is obtained from flask 1 and is given by

$$x_{O_2} = h_{O_2} k_{1_{O_2}}$$

For flask 2 the observed manometer change is

$$h = h_{O_2} + h_{CO_2}$$

$$= \frac{x_{O_2}}{k_{2_{O_2}}} + \frac{x_{CO_2}}{k_{2_{CO_2}}}$$

Whence $$x_{CO_2} = \left(h - \frac{x_{O_2}}{k_{2_{O_2}}} \right) k_{2_{CO_2}} \quad . \quad . \quad . \quad (8.3)$$

Thus, obtaining x_{O_2} from flask 1, x_{CO_2} may be obtained from the observed manometer change in flask 2.

Example 8.5.—Equal weights of rat liver slices were placed in two Warburg flasks together with saline. Alkali was added to the centre well of one flask which had an oxygen constant of 1·55. The other flask had carbon dioxide and oxygen constants of 1·77 and 1·42 respectively. After 60 minutes the alkali-containing flask showed a manometer change of 105 mm and the other manometer recorded a fall in level of 45·5 mm. Determine the respiratory quotient of the tissue slices.

The oxygen absorbed is given by the alkali-containing flask

$$x_{O_2} = -1 \cdot 55 \times 105 = 162 \cdot 8 \ \mu l.$$

The carbon dioxide evolved is given by the expression

$$x_{CO_2} = \left[-45 \cdot 5 - \left(\frac{-162 \cdot 8}{1 \cdot 42} \right) \right] 1 \cdot 77$$

$$= [-45 \cdot 5 + 114 \cdot 7] \ 1 \cdot 77$$

$$= +122 \cdot 5 \ \mu l.$$

Hence R.Q. $$= \frac{122 \cdot 5}{162 \cdot 8} = 0 \cdot 75.$$

Note that this example represents an ideal case where the weights of tissue are equal in the two flasks. In actual practice this is very difficult to achieve and therefore results must be reduced to comparable values by dividing the observed manometer changes by the weight of tissue present in the flasks. Then h_{O_2} and h are in terms of mm per mg, x_{O_2} in terms of μl per mg and substitution of these values in equation 8.3 gives x_{CO_2} in terms of μl per mg. The R.Q. is then obtained from the x_{CO_2}/x_{O_2} ratio.

RETENTION OF CARBON DIOXIDE BY BUFFERS

At physiological pH values carbon dioxide reacts with buffer solutions and is converted into bicarbonate. This means that in the presence of buffers the observed manometer change will not include all the CO_2 evolved and the value for x_{CO_2} will be decreased. Phosphates and proteins have this effect and cause retention of CO_2.

$$Na_2HPO_4 + CO_2 + H_2O \rightleftharpoons NaH_2PO_4 + NaHCO_3.$$

Consequently manometric determinations employing phosphate buffer or serum as media are subject to error and the retained CO_2 must be liberated from the bicarbonate at the end of the experiment in order to obtain the total CO_2 evolution. Addition of acid from a side bulb of the flask liberates the CO_2, but since the original medium and tissue probably contained a small quantity of CO_2 initially, a third manometer must be set up in which acid is added at the beginning of the experiment. This gives the bound CO_2 initially present and must be subtracted from the amount given by the second manometer to obtain the true amount of CO_2 evolved during the experiment. Consequently for experiments using phosphate buffer solutions three manometers are necessary, one with alkali to record oxygen uptake and two without alkali, one of which has acid added at the beginning and the other at the end of the experiment, in order to determine the CO_2 evolved.

Example 8.6.—Washed suspensions of *Escherichia coli* consume oxygen and evolve carbon dioxide in the absence of exogenous substrates. By determination of the R.Q. of endogenous respiration, an indication of the nature of intracellular reserve material undergoing oxidation was obtained. Three Warburg flasks were employed, one containing alkali in the centre well, and the other two containing acid in the side arm to determine the CO_2 evolved. Each flask contained : bacterial suspension (1 ml) ; phosphate buffer (1·8 ml) ; and acid or alkali (0·2 ml) as indicated. The acid was tipped in one flask at zero time, and after 30 min in the other; these manometers were read 10 min after the acid was tipped.

Flask	Flask constants		Manometer readings (mm) at times (min)			
	k_{O_2}	k_{CO_2}	0	10	30	40
1. KOH	1·71		+187		+31	
2. Acid	1·89	2·05	−7			+25·5
3. Acid		1·99	−5	+40		

The oxygen uptake is obtained from flask 1

$$x_{O_2} = h_1 k_{1O_2}$$
$$= (31 - 187) \times 1·71 = -266·8 \ \mu l.$$

The carbon dioxide output is derived from flasks 2 and 3

$$x_{CO_2} = \left(h_2 - \frac{x_{O_2}}{k_{2O_2}} \right) k_{2CO_2} - h_3 k_{3CO_2}$$

$$= \left(32·5 + \frac{266·8}{1·89} \right) 2·05 - 45 \times 1·99$$

$$= (32·5 + 141·1)2·05 - 89·6 = 266·3$$

Hence R.Q. $= \dfrac{266·3}{266·8} = \underline{0·998}.$

A value of 0·998 indicates that carbohydrate is the substrate of endogenous respiration in *Escherichia coli.*

Carbon Dioxide Production by other Methods

The 'Direct Method' is not widely used for CO_2 output determinations and is replaced either by the 'Indirect Method' of Warburg or the First or Second Methods of Dickens and Simer or by the method of Dixon and Keilin. Discussion of these methods lies outside the scope of a book of this size and the student is referred to the prescribed texts for details.

The Use of Bicarbonate-Carbon Dioxide Buffers

In manometry bicarbonate-CO_2 buffers find great application and the relationship between the bicarbonate and CO_2 concentrations and the pH is of considerable importance.

When CO_2 dissolves in aqueous solution, the following equilibria may be envisaged:

$$CO_2(gas) \rightleftharpoons CO_2(\text{dissolved in solution}) \rightleftharpoons H_2CO_3 \rightleftharpoons H^+ + HCO_3^-.$$

The dissociation constant for the first dissociation of carbonic acid is given by the expression

$$K_{a1} = \frac{(H^+)(HCO_3^-)}{(H_2CO_3)} \quad . \quad . \quad . \quad (8.4)$$

but since (H_2CO_3) is dependent on the concentration of dissolved CO_2 (which in turn is governed by the partial pressure of CO_2 in the gas phase) equation 8.4 may be rewritten as

$$K_{a1} = \frac{(H^+)(HCO_3^-)}{(CO_2)} \quad . \quad . \quad . \quad (8.4a)$$

where (CO_2) is the concentration of dissolved gas. The Henderson-Hasselbalch equation (2.10) enables the following relationship to be written :

$$pH = pK_{a1} + \log \frac{(HCO_3^-)}{(CO_2)} \quad . \quad . \quad (8.5)$$

As discussed in Chapter II, the activity of the bicarbonate ion must be used for strict accuracy. However, to make for ease in calculation, the activity coefficient of the bicarbonate ion may be included in the pK_a term when the concentrations of HCO_3^- and CO_2 can then be used. Thus

$$pH = pK_{a1}' + \log \frac{[HCO_3^-]}{[CO_2]} \quad . \quad . \quad (8.6)$$

where $\qquad pK_{a1}' = pK_{a1} + \log f_{HCO_3^-} \quad . \quad . \quad (8.7)$

pK_{a1}' has the value of 6·317 at infinite dilution and 38°. The value is corrected for concentrations other than infinite dilution by the expression

$$6·317 - 0·5\sqrt{I} \quad . \quad . \quad . \quad (8.8)$$

where I is the ionic strength. Manometric experiments rarely employ bicarbonate concentrations higher than 0·1 M and it has been shown that the correction for this concentration is small enough to be neglected for most purposes. Although equation 8.6 does not take into account the second dissociation constant of carbonic acid, its effect may be ignored provided the pH is less than 8.

A problem arises in connexion with the value taken for the bicarbonate concentration, for it represents the contribution of both the added bicarbonate and the bicarbonate arising from the dissociation of carbonic acid. However, certain working rules have been established, and providing the hydrogen ion concentration is one-hundredth of the $NaHCO_3$ concentration, the bicarbonate concentration for use in equation 8.6 may be taken as equal to the concentration of $NaHCO_3$ added. The error involved in neglecting the carbonic acid contribution is less than 0·1 per cent. For example, at pH 6, $[H^+] = 10^{-6}$ M and the lowest bicarbonate concentration permissible would be 10^{-4} M; at pH 7 the bicarbonate concentration should be greater than 10^{-5} M. There is, therefore, over the physiological range of pH, no practical difficulty in providing sufficient bicarbonate to eliminate errors due to neglect of the second dissociation constant of carbonic acid. For many purposes it is even possible to cut the above relationship by one-tenth without introducing large errors, e.g. providing the bicarbonate concentration is ten times the hydrogen ion concentration, $[HCO_3^-]$ may be taken as equal to the added bicarbonate concentration.

The CO_2 concentration in equation 8.6 is in terms of moles per litre of solution whereas the value usually known is the percentage by volume or partial pressure in mm of mercury. Conversion of these units to the former is effected as follows.

If $(CO_2\%)$ is the percentage by volume of CO_2 at atmospheric pressure P, and α is the solubility of CO_2 (in ml gas per ml liquid), then the partial pressure of CO_2 will be

$$p_{CO_2} = \frac{P(CO_2\%)}{100},$$

and under standard conditions (760 mm) $p_{CO_2} = \dfrac{P(CO_2\%)}{760 \times 100}$

atmospheres. The quantity of CO_2 (in ml) dissolved per litre of solution will be p_{CO_2} multiplied by 1000α, and to convert this to moles per litre it must be divided by 22400. Hence

$$[CO_2] \text{ in moles/litre} = \frac{P(CO_2\%)1000\alpha}{760 \times 100 \times 22400} \qquad (8.9)$$

The Henderson-Hasselbalch equation then becomes

$$pH = pK_{a1}' + \log [HCO_3^-] - \log P(CO_2\%) - \log \frac{\alpha}{760 \times 2240}$$

$$\qquad\qquad\qquad\qquad\qquad\qquad\qquad\qquad (8.10)$$

$$\text{or pH} = pK_{a1}' + \log [HCO_3^-] - \log p_{CO_2} - \log \frac{a}{760 \times 22 \cdot 4}$$

$$\qquad \qquad \qquad \cdot \qquad \cdot \qquad \cdot \qquad \cdot \qquad (8.11)$$

Inspection of this equation, after substituting the value of $6\cdot317$ for pK_{a1}' at $38°$, reveals the fact that it is impossible to measure CO_2 at pH 7 unless there is CO_2 in the atmosphere, for as $[CO_2]$ tends to zero, the $[HCO_3^-]/[CO_2]$ ratio increases and hence the pH also increases. Under normal conditions the concentration of CO_2 in the atmosphere is very small and in order to maintain a physiological pH value of 7 it is necessary either to decrease appreciably the bicarbonate concentration or to supply CO_2 in the atmosphere of the manometer flask. The former procedure is impracticable because, as already mentioned, retention of CO_2 by tissue buffers increases the bicarbonate concentration, and hence addition of CO_2 to the gas phase is the only practicable method. By taking a fixed bicarbonate concentration and varying the partial pressure of CO_2, or vice versa, it is possible to adjust the pH to any value in the physiological range of 6 to 8. Bicarbonate ion cannot exist in solution at pH 5 or below, therefore all CO_2 liberated remains in the gas phase under these conditions.

Example 8.7.—Bicarbonate–Ringer solution contains $0\cdot025$ M-bicarbonate and is in equilibrium with an atmosphere containing 5 per cent. CO_2 at $38°$. The solubility coefficient of CO_2 is $0\cdot537$ and pK_{a1}' is $6\cdot317$ at $38°$. What is the initial pH value of the solution ? Atmospheric pressure is 740 mm.

$$\text{pH} = pK_{a1}' + \log [HCO_3^-] - \log \frac{P(CO_2\%)\alpha}{760 \times 2240}$$

$$= 6\cdot317 + \log 0\cdot025 - \log \frac{740 \times 5 \times 0\cdot537}{760 \times 2240}$$

$$= 6\cdot317 + (\overline{2}\cdot3979) - (\overline{3}\cdot0672)$$

$$= 6\cdot317 - 1\cdot6021 + 2\cdot9328$$

$$= 7\cdot65.$$

REFERENCES AND SUGGESTED READING

DAWES, E. A. & RIBBONS. D. W. (1962). The endogenous metabolism of micro-organisms. *A. Rev. Microbiol.*, **16**, 241.

DIXON, M. (1951). *Manometric Methods*, 3rd ed. London : Cambridge University Press.

UMBREIT, W. W., BURRIS, R. H. & STAUFFER, J. F. (1964). *Manometric Techniques*, 4th ed. Minneapolis : Burgess.

PROBLEMS

8.1. The following data were obtained in experiments to calibrate five Warburg manometers by the mercury method. The volumes (in ml) of the flasks to the zero point of the manometers were respectively :

(a) $21\cdot24$;   (b) $21\cdot00$;   (c) $19\cdot22$;   (d) $22\cdot21$;   (e) $20\cdot54$

The solubility of oxygen at 37° is 0·0239 and the density of the manometer fluid 1·030.

Determine the oxygen constants of these manometers at 37° for 3 ml fluid content of the flasks.

8.2. The carbon dioxide constants of four Warburg manometers were determined by means of the bicarbonate method.

2 ml of a solution containing 0·3361 g sodium bicarbonate per litre were pipetted into the main compartments and 1 ml of 2 N-sulphuric acid was placed in the side bulbs. When the reaction had ceased, the observed manometer changes (in mm) were (a) +93 ; (b) +102·5 ; (c) +88 ; (d) +98·5 and the thermobarometer showed a difference of +3·0 mm.

Determine the constants.

8.3. Oxygen constants for Warburg manometers may be obtained by measuring the nitrogen evolved when excess of alkaline hydrazine solution is allowed to react with potassium iodate. The nitrogen constant thus obtained is practically identical with the oxygen constant and may be used as such. The reaction involved is :

$$2KIO_3 + 3N_2H_4 = 2KI + 3N_2 + 6H_2O.$$

1 ml of a solution containing 0·3210 g potassium iodate per 250 ml was pipetted into the side bulbs of a series of manometers and 2 ml of an alkaline hydrazine solution were placed in the main compartment. After the reaction had taken place the following manometer changes (in mm) were recorded : (a) +128 ; (b) +109 ; (c) +136 ; (d) +145, and the thermobarometer had changed by −2 mm.

What are the constants for these manometers ?

8.4. Calculate the respiratory quotients for the complete oxidation of the following compounds : (a) glycerol, (b) pyruvic acid, (c) oxaloacetic acid, (d) α-oxoglutaric acid, (e) citric acid, (f) glutamic acid, (g) lysine, (h) capric acid, (j) stearic acid.

8.5. Air freed from moisture and carbon dioxide contains 21·00 per cent. oxygen and 78·06 per cent. nitrogen by volume. What are the partial pressures of these gases when the atmospheric pressure is 76 cm Hg ?

8.6. Using a gas phase containing 5 per cent. carbon dioxide, what concentration of bicarbonate must be employed in order to obtain media of pH values 6·0, 6·5, 7·0 and 7·5 ? At 38°, pK_{a_1}' for carbonic acid is 6·317 and α_{CO_2} is 0·537. Assume atmospheric pressure to be 760 mm.

8.7. The following rates of evolution of carbon dioxide, produced by the action of a preparation of a decarboxylase on L- (+) -ornithine, were obtained using a Warburg respirometer :

μl. CO_2 evolved per 5 minutes	Ornithine concentration, mM.
10·0	0·556
36·4	3·33
52·7	8·34
66·6	25·0

Determine by an accurate graphical method the maximum rate of evolution attainable under these conditions, and derive any relationship you employ.

(Leeds Honours Course Finals, 1951.)

8.8. A solution of oxaloacetic acid of unknown concentration was analysed manometrically by the aniline-citrate method which catalyses the decarboxyla-

L

tion of the keto acid. After deduction of the appropriate control the carbon dioxide evolution from 1 ml of the solution was 168 μl.

Determine the molarity of the oxaloacetic acid solution and also express the concentration in terms of percentage (w/v).

8.9. Briefly outline how you would determine *three* of the following : (*a*) L-histidine ; (*b*) pyruvic acid ; (*c*) succinic acid ; (*d*) hydrogen peroxide, using manometric methods, including in your answer the exact constituents of the media used in the flasks ; details of operation of the manometers are not required.

Urea can be determined manometrically using a urease preparation which catalyses the reaction :

$$CO(NH_2)_2 + H_2O = CO_2 + 2NH_3.$$

Using 1 ml of a urea solution and a buffer of pH 5, the gas evolutions recorded on control and reaction manometers were respectively 8·7 and 83·3 μl (N.T.P.). Calculate the molarity of the urea solution employed.

(Leeds Honours Course Finals, 1949.)

8.10. The oxidation of glycerol by the alga *Prototheca zopfii* has been investigated by Barker and the effect of glycerol concentration on total oxygen consumption is recorded below :

Glycerol present µmoles	Total oxygen consumed (μl)	
	Control (*no substrate*)	Observed with substrate (*no correction applied*)
6·90	53	170
13·8	75	329
27·6	106	659
55·2	161	1292

Determine the extent to which oxidative assimilation has occurred and ascertain whether these data are consistent with the supposition that an increasing substrate concentration suppresses endogenous respiration of the alga.

(After BARKER (1936), *J. cell. comp. Physiol.*, 8, 231.)

8.11. The effect of variation of bacterial concentration on the oxidation of glucose by *Pseudomonas aeruginosa* was studied and the following results obtained :

Relative concentration of bacteria	Oxygen consumption (μl)	
	Endogenous	5 µmoles glucose
1	45	491
2	95	540
4	200	645

Endogenous respiration has not been subtracted from the observed uptake in the presence of substrate.

Is there any evidence for oxidative assimilation by these cells and, if so, to what extent ? What conclusions may be drawn about the effect of the substrate on the endogenous respiration of the organism ?

(Adapted from NORRIS, CAMPBELL & NEY (1949), *Canad. J. Res.*, C, 27, 157.)

8.12. *Aerobacter aerogenes* ferments citric acid in accordance with the following scheme :

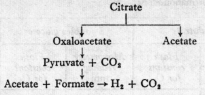

Arsenite, at a concentration of $2·5 \times 10^{-3}$ M, inhibits the fermentation, and to investigate its locus of action manometers were set up containing 2·5 μmoles of citrate with and without arsenite. Total gas evolution was measured in each case in an atmosphere of nitrogen and 5 per cent. carbon dioxide, and after deduction of the respective controls, values of 48μl and 139 μl respectively were obtained in the presence and absence of arsenite. From these data deduce the probable point of action of arsenite on the fermentation. What further experiment would you carry out to confirm your conclusions ?

(After DAGLEY & DAWES (1953), *J. Bact.*, 66, 259.)

8.13. The effect of sodium azide and 2,4-dinitrophenol (DNP) on the aerobic utilization of glucose by a strain of baker's yeast has been investigated.

Each Warburg vessel contained 0·1 ml of 0·1 M-glucose, 0·1 ml of NaN_3, DNP or water and 1·8 ml of a 0·66 M-phosphate buffer (pH 6·0) suspension of yeast. Results obtained were as follows :

Reagent	Oxygen consumed μl	CO_2 evolved μl
—	466	481
10^{-3} M DNP	495	779
10^{-4} M NaN_3	543	691
5×10^{-4} M NaN_3	104	520

Determine the respiratory quotient in the presence and absence of these reagents. What explanation can you offer for the observed results ?

(After PICKETT & CLIFTON (1941), *Proc. Soc. exp. Biol. Med.*, 46, 443.)

8.14. The metabolism of glucose by rat liver slices was investigated manometrically. Equal weights of slices were placed in three Warburg flasks containing phosphate buffer of pH 7·0 and alkali was added to the centre well of one of these. 5 N-Sulphuric acid was added to the side arms of the other two flasks which were used to determine carbon dioxide evolution. At the start of the experiment acid was tipped in one of the flasks, which had a carbon dioxide constant of 1·85, and the manometer change was +21 mm.

At the end of the experiment the alkali-containing flask recorded −121 mm pressure (k_{O_2}, 1·65 for this manometer) and the other manometer, after tipping the acid, gave a reading of +8·0 mm. The constants for this latter flask were k_{O_2}, 1·59 and k_{CO_2}, 1·80.

Determine the oxygen consumption and carbon dioxide evolution and hence the respiratory quotient.

8.15. The effect of aeration on the endogenous and glucose metabolism of non-proliferating suspensions of *Sarcina lutea* has been investigated. The cells, suspended in water, were incubated at 37° and a vigorous stream of sterile air passed through the culture. At intervals samples of cells were withdrawn, centrifuged, washed and set up in Warburg manometers to determine the oxygen consumption in the presence and absence of 5 μmoles of glucose. The following data refer to manometer readings at 60 minutes, obtained with 1 ml bacterial suspension per manometer :

Period of aeration hours	No substrate		5 μmoles glucose		Bacterial dry weight mg/ml
	Manometer reading mm	Flask constant for O_2	Manometer reading mm	Flask constant for O_2	
0	−272	1·64	−329	1·69	31·0
1	−158	1·47	−245	1·55	31·2
2	−70	1·57	−183	1·67	32·3
3·5	−34·5	1·56	−191	1·60	32·6
4·5	−14·0	1·43	−213	1·52	33·5

Determine the Q_{O_2} values for endogenous and glucose metabolism and comment on the effect of aeration on these quotients.

(After DAWES & HOLMS, unpublished data.)

8.16.   Cell-free extracts of a phthalate-grown pseudomonad oxidize both 4,5-dihydroxyphthalate and protocatechuate (3,4-dihydroxybenzoate) to 3-oxoadipate. The gaseous exchanges occurring during these transformations were determined after 60 min incubation at 30°, when the reaction had gone to completion. The following data were obtained.

Substrate	KOH present		KOH absent Acid tip after 60 min.			KOH absent Acid tip at 0 min.	
	mm	k_{O_2}	mm	k_{O_2}	k_{CO_2}	mm	k_{CO_2}
4,5-Dihydroxy-phthalate (6 μmoles)	−91	1·45	+84	1·54	1·74	+20	1·70
Protocatechuate (6 μmoles)	−97	1·38	+10·5	1·49	1·69	+18	1·83

4,5-Dihydroxyphthalate is also decarboxylated anaerobically to protocate-chuate. Determine the stoicheiometry of the reactions and formulate a reaction sequence based on these observations.

(After RIBBONS & EVANS (1960). *Biochem. J.*, 76, 310.)

8.17.   After anaerobic growth on glucose, washed cells of an unidentified *Clostridium* were found to decarboxylate acetoacetic acid at the following rates (corrected for non-enzymic reaction) :

pH	6·5	6·0	5·5	5·0	4·5	4·0
Q_{CO_2}	7	14	23	42	36	25

Steam distillation of samples from the growth medium acidified with $2N\text{-}H_2SO_4$ showed that volatile acids were formed and amounts determined by titration were compared with the quantity of glucose remaining

Acid (equivalents)	0	25	58	42	30
Glucose (moles)	100	70	50	30	0

In the light of the above information what can you say about the fermentation of glucose by this organism, what end-products would you expect to find in the medium and how are they formed ? How would you identify the organism by analysis of these products ?

(Hull Special Biochemistry I Course, 1966.)

8.18. What do you understand by the term oxidative assimilation ? The complete oxidation of ethanol in the dark by the green alga *Chlamydomonas reinhardii* was investigated manometrically and 5 μmoles of ethanol were found to give a total oxygen consumption of 187 μl. Over the same period the endogenous respiration was 36 μl. The same experiment carried out with a mutant strain of *Chlamydomonas reinhardii* gave a total oxygen uptake of 336 μl, while the endogenous respiration was 40 μl. What was the percentage of ethanol synthesized into cellular material in the parent strain and the mutant ? What differences in the metabolism of the mutant compared with the parent strain would give the observed result ?

(University of Bradford, Applied Biology Honours B. Tech. Course, Fourth Year, 1967.)

8.19. How would you use a manometric method to assay the following :

(a) The L-glutamic acid content of a protein hydrolysate ?

(b) The cytochrome oxidase activity of a mitochondrial preparation ?

(c) The urea produced by a slice of liver using an ammonium salt as substrate ?

In a Warburg manometer at 37° and in 3 ml liquid a reaction starting at pH 7·4 ($14·5 \times 10^{-3}$M bicarbonate and a gas phase of 5 per cent. CO_2) evolves from acid production 200 μl of CO_2 and then stops. What is the final pH ? You may assume the following form of the Henderson-Hasselbalch equation applies under the conditions of your experiment (37° and 740 mm Hg barometric pressure) :

$$\log[HCO_3^-] = pH - pK' + \log(2·46 \times 10^{-4} \times CO_2)$$

where $pK' = 6·32$ and CO_2 is the percentage of CO_2 in the gas phase.

(St. Andrews, Queen's College Honours Finals, 1967.)

CHAPTER IX

BACTERIAL GROWTH

The Bacterial Growth Cycle

If a number of living bacteria are inoculated into a given volume of a suitable nutrient medium contained in a closed system such as a flask (a so-called *batch culture*), their subsequent growth activities may be conveniently divided into three well-defined phases:

(*a*) the lag phase, before cell division commences,

(*b*) the exponential or logarithmic growth phase, when the cells are dividing at constant rate, and

(*c*) the stationary phase, when cell division ceases.

The overall effect of these phases, referred to as the bacterial growth cycle, is shown in Fig. 9.1. Other phases may be distinguished, such as the acceleration phase between (*a*) and (*b*) and

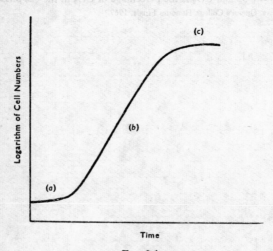

FIG. 9.1

Typical bacterial growth cycle.

the deceleration phase between (b) and (c), but quite often the transition from lag to exponential and from exponential to stationary phase is so sharp that these other phases may be non-existent.

It is worth emphasizing that the growth cycle, as described, is not a fundamental property of the bacterial cells but is a consequence of their interaction with the environment *in a closed system*. As we shall see subsequently, when growth occurs in a suitable open system it is possible to maintain cells in the exponential phase of growth over long periods of time.

The Exponential Growth Phase

If an individual organism in the medium is observed it will be seen to increase in size until, finally, division into two daughter cells occurs, each of which in turn increases in size and divides. The period between one cell division and the next is called the 'generation time', and when observations are made on a single cell this value is found to fluctuate quite considerably. However, in most bacterial cultures we are dealing with populations of many millions of cells and under these circumstances the average generation time for all cells in the culture is fairly constant, and is referred to as the *mean generation time*.

If n_0 cells are inoculated into a nutrient medium, then at the end of one generation time there will be $2n_0$ cells, at the end of two generations $4n_0$ cells (or 2^2n_0) and at the end of z generations $2^z n_0$ cells. If n cells are present at the end of z generations, then

$$n = n_0 . 2^z \quad . \quad . \quad . \quad . \quad (9.1)$$

Taking logarithms, $\log n = \log n_0 + z \log 2$

and $$\frac{\log n - \log n_0}{\log 2} = z.$$

Let T be the mean generation time and t the time taken for the population to increase exponentially from n_0 to n, then

$$\frac{t}{T} = z$$

and substituting in the above equation

$$\frac{\log n - \log n_0}{t} = \frac{\log 2}{T} \quad . \quad . \quad . \quad (9.2)$$

When bacterial growth is followed experimentally, results are usually expressed by a graph in which the logarithm of the cell count is plotted against the time (semi-logarithmic graph paper is useful for this purpose). For the exponential phase of growth this results in a linear plot for log n against time. From such graphical representation the above derived relationship (equation 9.2) follows by similar triangles (Fig. 9.2).

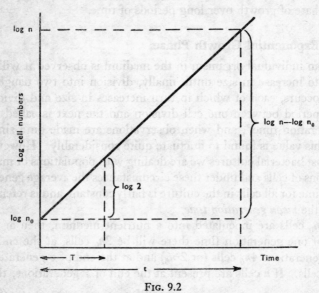

FIG. 9.2

Logarithmic plot of bacterial cell numbers against time.

Monod (1949) has pointed out the advantages of using logarithms to the base 2 for plotting bacterial growth curves, when an increase of one unit in $\log_2 n$ corresponds to one generation time, i.e. a doubling of the population. (Because of fluctuations in the generation times of individual cells, to which reference has already been made, some microbiologists prefer the term *doubling time*, t_d, to mean generation time, T.) Thus the number of generations of cells is given by the difference of ordinates at any two intervals of time. Logarithmic tables to the base 2 ($\log_2 = 3 \cdot 322 \log_{10}$) are given by Finney, Hazlewood & Smith (1955).

An alternative method of recording growth rates is by means of the specific growth rate k. The rate of increase of a bacterial population is given by the expression

$$\frac{dn}{dt} = kn \qquad . \qquad . \qquad . \qquad (9.3)$$

where dn/dt measures the growth rate, and the rate of increase per unit of organism concentration $\left(\dfrac{1}{n}\dfrac{dn}{dt}\right)$ is termed the *specific growth rate*, k, sometimes denoted by μ.[1]

Integration of the above expression yields $n = n_0 e^{kt}$ or

$$\ln \frac{n}{n_0} = kt$$

Therefore the slope of the line in the plot of $\log n$ against t (Fig. 9.2) is $k/2 \cdot 303$. With k evaluated, the mean generation time is obtained from the equation

$$\ln 2 = kT$$

whence

$$T = \frac{0 \cdot 693}{k} \qquad . \qquad . \qquad . \qquad (9.4)$$

Example 9.1.—An inoculum of 4×10^5 bacteria grew in a medium without lag to a population of $3 \cdot 68 \times 10^7$ cells in 6 hours and had not then reached the stationary phase. What was the mean generation time of the organism in this medium ?

Data given : $n_0 = 4 \times 10^5$ and $n = 3 \cdot 68 \times 10^7$. $t = 360$ minutes.

Where T is the mean generation time, since there was no lag period :

$$\frac{\log n - \log n_0}{t} = \frac{\log 2}{T}$$

$$\frac{7 \cdot 5658 - 5 \cdot 6021}{360} = \frac{0 \cdot 3010}{T}$$

$$T = \frac{0 \cdot 3010 \times 360}{1 \cdot 9637} = 55 \cdot 2 \approx 55 \text{ minutes.}$$

Example 9.2.—25 ml of peptone medium were inoculated with 4×10^6 cells of *Escherichia coli* and incubated at $37°$ C. The stationary phase of 3×10^9 cells per ml was reached after 284 minutes and no lag phase occurred. What was the mean generation time in the peptone medium ?

Data given : $n_0 = 4 \times 10^6$ cells per 25 ml, $n = 3 \times 10^9$ cells per ml and $t = 284$ minutes.

[1] It will be noted that in the theoretical treatment of continuous cultivation (p. 320) μ is defined in terms of bacterial mass, $x\left(\mu = \dfrac{1}{x}\dfrac{dx}{dt}\right)$, rather than bacterial number, n. In a steady state system in a chemostat k and μ, defined in terms of bacterial numbers and mass respectively, will be identical, but this is not necessarily so in batch culture where, at different stages of the growth cycle, bacterial size varies considerably.

Note particularly that n_0 and n are not expressed in the same units, and therefore n_0 must be converted to cells per ml before the calculation can be carried out.

$$n_0 = \frac{4 \times 10^6}{25} = 1\cdot6 \times 10^5 \text{ cells per ml.}$$

$$T = \frac{284 \times 0\cdot3010}{9\cdot4771 - 5\cdot2041} = \frac{85\cdot48}{4\cdot2730} = \underline{20 \text{ minutes.}}$$

The exponential growth rate remains constant over wide ranges of concentration of a given nutrient and it is only when the nutrient concentration has fallen to a very low value that the growth rate decreases. The relationship between the growth rate and the nutrient concentration is similar to that for the saturation of an enzyme with its substrate (see, for example, Fig. 6.2), the maximum growth rate being achieved at very low concentrations of the nutrient. For this reason the exponential phase represents a steady state system in which the relative concentrations of all metabolites and all enzymes within the cell are constant. A steady state culture therefore offers an ideal system for the study of various biochemical problems, such as protein and induced enzyme synthesis, and attention has been directed to the possibility of maintaining cells in the exponential phase for extended periods. This interest has resulted in the development of techniques for *continuous culture* in which, unlike 'batch' culture (i.e. growth in a closed system with a given volume of medium), sterile medium is fed at a steady rate of flow into the culture vessel and culture emerges from it at the same rate. By this means it is not only possible to maintain the cells in exponential growth for long periods but also to achieve this at reduced rates of growth, which are controlled by the concentration of substrate and the rate of flow of the replenishing medium. Continuous culture methods are now widely used for biochemical investigations of micro-organisms and the theory of the process is treated on p. 320.

The Lag Phase

The examples cited are ones where the inoculum of n_0 cells commences to divide immediately without a lag phase. If, as frequently happens, there is a period of lag before exponential growth begins, the time t measured for the increase in population to n is not occupied exclusively by cell division but includes also the time L—the length of the lag phase (Fig. 9.3).

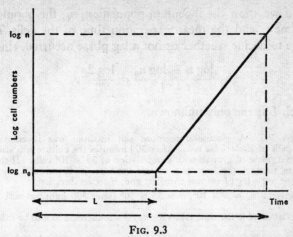

FIG. 9.3

Lag and exponential phases of growth cycle.

In this connexion one point remains to be made. The transition from lag to exponential phase is not always sharp, and difficulty may be experienced in deciding exactly where the lag phase ends. To overcome this difficulty, the length of the lag phase is

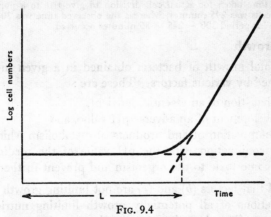

FIG. 9.4

Graphical determination of the length of the lag phase.

conventionally defined as the intercept on the line drawn through the logarithm of the inoculum population, parallel to the time axis, obtained by extrapolating the growth curve backwards. This is made clear in Fig. 9.4.

Thus, if we know the inoculum population, n_0, the population n after time t, and the mean generation time of the cells T, it is possible to decide whether or not a lag phase occurred, since

$$\frac{\log n - \log n_0}{t - L} = \frac{\log 2}{T} . \qquad . \qquad . \qquad (9.5)$$

of which L is the only unknown.

Example 9.3.—A glucose-ammonium salt medium was inoculated with 5×10^5 cells of *Escherichia coli*. After 300 minutes the culture was still in the exponential phase of growth with a population of 35×10^6 cells. If the mean generation time of the organism in this medium is 40 minutes, determine whether or not a lag phase was manifest and, if so, its duration.

Data given : $n_0 = 5 \times 10^5$, $n = 35 \times 10^6$ (after 300 minutes) and $T = 40$ minutes.

These data enable the time taken for n_0 cells to increase to n to be calculated. Thus

$$\frac{\log n - \log n_0}{t} = \frac{\log 2}{T}$$

Substituting the known values

$$\frac{7 \cdot 5441 - 5 \cdot 6990}{t} = \frac{0 \cdot 3010}{40} .$$

Therefore $t = \dfrac{1 \cdot 8451 \times 40}{0 \cdot 3010} = \dfrac{73 \cdot 804}{0 \cdot 3010} = 245$ minutes.

Thus the time taken for actual cell division to give rise to a population of 35×10^6 cells was 245 minutes, whereas the observed time was 300 minutes. Hence the lag period $300 - 245 = 55$ minutes occurred.

Total Growth

The total growth of bacteria obtained in a given medium is determined by various factors. These are :

(a) exhaustion of an essential nutrient,

(b) development of an adverse pH value, and

(c) accumulation of end products of metabolism which, taken in conjunction with the pH value of the medium, may become toxic to the organism and prevent further growth.

Provided that factors (b) and (c) are not limiting growth and that concentrations of all potentially growth limiting nutrient substances are adjusted so that they are in excess compared to one given nutrient, then a linear relationship exists between the total growth (i.e. stationary population) and the concentration of that latter substance. The graph of such a relationship is shown in Fig. 9.5, and where linearity breaks down, factors other than nutrient concentration are limiting growth.

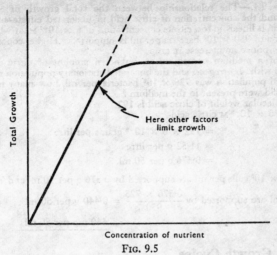

Concentration of nutrient

FIG. 9.5

Relationship between total bacterial growth and concentration of a given nutrient.

The determination of total growth as a function of the concentration of a given nutrient finds extensive application in the quantitative determination of amino acids, vitamins of the B group, purines and pyrimidines. Such methods of microbiological assay are specific and extremely sensitive.

The linear relationship observed means that

$$\frac{\text{Weight of bacteria formed}}{\text{Weight of limiting material utilized}} = Y$$

where Y is the *yield constant* or *coefficient*. During the last decade there has been considerable interest in the determination of molar growth yields of micro-organisms, i.e. g dry weight of cells obtained per mole of substrate utilized, since it is possible then to relate the growth yield to the energy (as ATP) made available by the substrate. Clearly if the substrate is used as a source of cellular material as well as a source of energy, the quantity of substrate assimilated must be deducted from the total substrate utilized before the growth can be related to the ATP yield. These important aspects of biological energetics are discussed by Bauchop & Elsden (1960), Gunsalus & Shuster (1961) and Payne (1970).

Example 9.4.—The relationship between the total growth of *Aerobacter aerogenes* and the concentration of citric acid in unaerated citrate-ammonium salt medium is linear up to a citrate concentration of 6×10^{-2} M, at which value a population of 950×10^6 bacteria per ml is supported. Higher concentrations of citrate produce no increase in crop.

50 ml of a medium containing an unknown amount of citric acid were inoculated with *A. aerogenes* and the unaerated stationary population measured. If this final population was 725×10^6 bacteria per ml, how many milligrams of citric acid were present in the medium ?

The molecular weight of citric acid is 192.

Therefore 6×10^{-2} M citric acid

$$= 192 \times 6 \times 10^{-2} \text{ grams per litre}$$
$$= 11 \cdot 52 \text{ g per litre}$$
$$= 0 \cdot 576 \text{ g per 50 ml.}$$

Thus 950×10^6 cells per ml are supported by $0 \cdot 576$ g per 50 ml and 725×10^6

cells per ml are supported by $\dfrac{0 \cdot 576 \times 725}{950} = 0 \cdot 440$ g per 50 ml

$$= \underline{440 \text{ mg per 50 ml.}}$$

Diauxic Growth Cycles

In certain circumstances more complex growth cycles may be encountered, the most familiar being two consecutive exponential growth phases separated by a lag phase. This phenomenon of biphasic growth or 'diauxie', i.e. double growth, is observed when certain bacteria are grown in media containing either limiting amounts of two different carbohydrates or of a carbohydrate and an organic acid. The first growth phase corresponds to the exclusive utilization of one of the compounds followed by a period of adaptation before the other substrate is metabolized and growth recommences. A typical diauxic growth curve is shown in Fig. 9.6, where the substrates are glucose and citric acid respectively. In this particular instance it has been shown that glucose inhibits the formation of the citrate transport mechanism ('permease') so that during growth in a medium containing both substrates *Aerobacter aerogenes* is unable to take citrate into the cell, even though the tricarboxylic acid cycle is fully operative during growth on glucose. In other examples it is usually found that the preferentially utilized substrate represses either or both the permease and the enzyme(s) of initial attack on the second substrate.

Continuous Cultivation of Micro-organisms

The theoretical aspects of continuous cultivation were first considered by Monod (1950) and Novick & Szilard (1950) and sub-

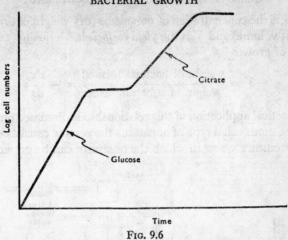

FIG. 9.6

Diauxic growth curve of *Aerobacter aerogenes* in citrate medium containing a limiting concentration of glucose.

sequently discussed thoroughly by Herbert, Elsworth & Telling (1956), on whose paper the following treatment is based.

The specific growth rate $\mu\left(= \dfrac{1}{x}\dfrac{dx}{dt}\right)$ is determined by the substrate concentration and may be described by the expression

$$\mu = \mu_m\left(\frac{s}{K_s + s}\right) \qquad . \qquad . \qquad . \qquad (9.6)$$

where μ_m is the maximum value of μ at saturation concentrations of substrate, s is the substrate concentration and K_s is a *saturation constant* numerically equal to the substrate concentration at which $\mu = \mu_m/2$ (compare this equation with equation 6.3). It follows from equation 9.6 that exponential growth can occur at specific growth rates having any value between zero and μ_m, provided the substrate concentration can be held constant at a given value. Apparatus for the continuous cultivation of micro-organisms is designed to enable this condition to be met.

There is a simple relationship between growth and utilization of substrate such that the growth rate is a constant fraction of the rate of substrate utilization. Thus

$$\frac{dx}{dt} = - Y \frac{ds}{dt} \qquad . \qquad . \qquad . \qquad (9.7)$$

where x is the concentration of organisms (dry weight of organisms per unit volume) and Y is the *yield coefficient*. Thus for any finite period of growth

$$Y = \frac{\text{weight of bacteria formed}}{\text{weight of substrate used}} = -\frac{dx}{ds}$$

The practical application of this relationship is discussed on p. 319.

In the more usual type of apparatus the reactor consists of some form of culture vessel in which the organism can be grown under

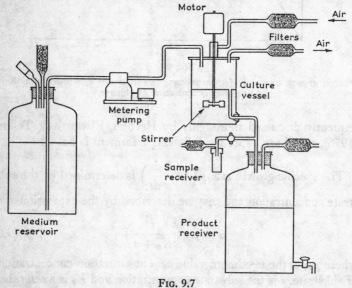

Fig. 9.7

Diagram of a continuous culture apparatus (schematic).

suitable conditions. Sterile growth medium is fed into the vessel at a flow rate f and culture emerges at the same rate, a constant level device keeping the volume of the culture, v, constant. Efficient stirring is necessary so that the entering medium approaches the ideal condition of instantaneous mixing and uniform distribution throughout the vessel—the so-called 'completely-mixed' vessel (Fig. 9.7).

The period that a particle remains in the culture vessel, referred to as its *residence time*, will be determined not by the absolute values of flow rate and culture volume, but by their ratio, f/v, which is termed the *dilution rate, D,* i.e. the number of complete volume changes per unit time. The mean residence time of a particle in the culture vessel is thus equal to $1/D$.

Now if it is assumed that the bacteria in the culture vessel are neither growing nor dividing, then provided that mixing is complete, every organism in the vessel has an equal probability of leaving it within a given period of time. In consequence, it can be shown that the fraction of the total organisms in the culture vessel having a residence time $\geqslant t$ is e^{-Dt}. The *wash-out rate,* i.e. the rate at which organisms initially present in the vessel would be washed out if growth ceased but the flow of medium continued, is therefore

$$- \frac{dx}{dt} = Dx \qquad . \qquad . \qquad . \qquad (9.8)$$

where x, as before, is the concentration of organisms in the vessel. Thus, it is possible to describe, in fairly simple terms, the distribution of residence-times and the wash-out rate for a completely-mixed continuous culture vessel. This expression, in conjunction with the kinetics of the growth process, can now be applied to the problems of continuous culture.

Consider bacteria growing in a completely-mixed type of vessel, as previously described, with a constant culture volume, with the inflowing medium containing a single organic substrate (e.g. glucose) at a concentration s_R and with all other substrates in excess. It is assumed that the degree of aeration is sufficient to prevent the oxygen supply becoming a limiting factor and therefore the sole growth-limiting factor is the supply of the organic substrate. The experimenter is able to control two variables in such a system, namely the substrate concentration and the flow rate of the incoming culture medium, and the theory must therefore be able to describe how variation of these parameters affects the growth rate and the concentrations of organisms and of substrate in the culture vessel.

In the growth vessel the organisms are growing at a rate described by equation 9.3 and simultaneously being washed away at a rate determined by equation 9.8. The net rate of change of concentration of organisms is expressed by the following balance equation, where the individual terms refer to rates in each case

increase = growth − output

or
$$\frac{dx}{dt} = \mu x - Dx \qquad . \qquad . \qquad . \qquad (9.9)$$

When $\mu = D$, $dx/dt = 0$ and x is constant and independent of time, i.e. there is a steady state concentration of organisms in the vessel. If $\mu > D$, then dx/dt is positive and the concentration of organisms will increase, while if $\mu < D$, dx/dt is negative and the concentration of organisms will decrease, eventually to zero, corresponding to the washing out of the culture vessel.

Under steady state conditions, therefore, the specific growth rate, μ, of the organisms in the culture vessel is exactly equal to the dilution rate, D. It is necessary to know, however, what values of D make a steady state possible, and to find this the effect of dilution rate on the concentration of substrate in the culture vessel must also be known, since the value of μ is dependent on s (equation 9.6).

In the growth vessel, substrate enters at a concentration s_R, is consumed by the organisms, and flows out at concentration s. The net rate of change of substrate concentration is thus given by another balance equation, namely

increase = input − output − consumption

$$= \text{input} - \text{output} - \frac{\text{growth}}{\text{yield coefficient}} \quad \begin{array}{l}\text{(from} \\ \text{equation 9.7)}\end{array}$$

or
$$\frac{ds}{dt} = Ds_R - Ds - \frac{\mu x}{Y} \quad . \qquad . \qquad (9.10)$$

Now equations 9.9 and 9.10 both contain μ, which is itself a function of s (equation 9.6), so substituting for μ we obtain:

from equation 9.9
$$\frac{dx}{dt} = x\left[\mu_m\left(\frac{s}{K_s + s}\right) - D\right] \quad . \qquad (9.11)$$

and from equation 9.10
$$\frac{ds}{dt} = D(s_R - s) - \frac{\mu_m x}{Y}\left(\frac{s}{K_s + s}\right)$$
$$. \qquad . \qquad . \qquad (9.12)$$

These two equations define completely the behaviour of a continuous culture in which the fundamental growth relationships are those described by equations 9.3, 9.6 and 9.7.

From a consideration of equations 9.11 and 9.12 it will be apparent that if s_R and D are held constant and D does not exceed a certain critical value (see equation 9.15), then unique values of x and s exist for which both dx/dt and ds/dt are zero, i.e. the system is in a steady state. If equations 9.11 and 9.12 are solved for $dx/dt = 0$ and $ds/dt = 0$, the steady state values of x and s, designated $\tilde{x}$ and $\tilde{s}$, are given by the expressions

$$\tilde{s} = K_s\left(\frac{D}{\mu_m - D}\right) \qquad \qquad (9.13)$$

$$\tilde{x} = Y(s_R - \tilde{s})$$

$$= Y\left[s_R - K_s\left(\frac{D}{\mu_m - D}\right)\right] \qquad (9.14)$$

With these equations, and knowing the values of the constants μ_m, K_s and Y for a given organism and growth medium, the steady-state concentrations of organisms and substrate in the culture vessel can be predicted for any value of the dilution rate and concentration of inflowing substrate. If these predicted values for $\tilde{x}$ and $\tilde{s}$ are then plotted as a function of the dilution rate, D, a graphical representation of the operation of the system is obtained, as shown in Fig. 9.8.

A consideration of Fig. 9.8 reveals that over most of the range of possible dilution rates the steady state concentration of substrate in the culture vessel is very low, i.e. the substrate is almost completely consumed by the organisms. Only when the dilution rate approaches a critical value, designated D_c, corresponding to that at which wash-out occurs, does unused substrate appear in the culture; at D_c the substrate concentration becomes equal to s_R. The concentration of organisms has a maximum value when the dilution rate is zero, the substrate concentration then also being zero; this situation corresponds to the final stage of a batch culture in which substrate exhaustion has limited growth. As the dilution rate increases the substrate concentration increases in the manner previously noted, and the concentration of cells falls, eventually becoming zero at dilution rate D_c. The critical dilution rate, D_c, is therefore of considerable practical importance and, from equation 9.9, will be seen to be equal to the highest possible value of μ, which is attained when $\tilde{s}$ is at its maximum value s_R. Thus

$$D_c = \mu_m\left(\frac{s_R}{K_s + s_R}\right) \qquad . \qquad . \qquad (9.15)$$

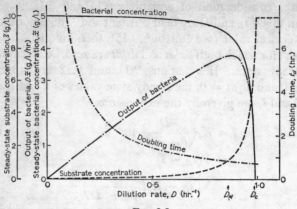

FIG. 9.8

Steady-state relationships in a continuous culture (theoretical). The steady-state values of substrate concentration, bacterial concentration and output at different dilution rates are calculated from equations 9.13 and 9.14 for an organism with the following growth constants:

$$\mu_m = 1 \cdot 0 \text{ h}^{-1}, \quad Y = 0 \cdot 5 \quad \text{and} \quad K_s = 0 \cdot 2 \text{ g/l};$$

and a substrate concentration in the inflowing medium of $s_R = 10$ g/l.

(HERBERT, ELSWORTH & TELLING, 1956)

and when $s_R \gg K_s$, which is usually the case, $D_c \sim \mu_m$. Whenever the dilution rate exceeds D_c, dx/dt will be negative (see equation 9.9) and the attainment of a steady state with $\tilde{x} > 0$ is impossible since the organisms will be washed out of the culture vessel faster than they can grow.

Fig. 9.8 also shows the theoretical *output* of organisms, Dx (i.e. g cells/litre/hour), as a function of dilution rate. It will be evident that this curve attains a maximum at a dilution rate, designated D_M, which is the optimum for the production of cells or products. Also plotted is the doubling time of the organisms, t_d, i.e. the time required for the mass concentration of cells to double ($t_d = \ln 2/\mu$).

The foregoing treatment has been based on the type of continuous culture apparatus known as a *chemostat* or *bactogen*, which is operated by controlling the flow rate. There is, however, another type of apparatus termed a *turbidostat*, which is operated by controlling the cell density. It should be noted that the equations derived for the chemostat apply equally to the turbidostat. The essential difference between the two types of apparatus is that in the chemostat the dilution rate is set at a pre-determined value and the cell concentration allowed to find its own level, whereas the cell concentration is held at a fixed value and the dilution rate allowed to find its own level in the turbidostat.

Example 9.5.—*Aerobacter cloacae* was grown in continuous culture in a small pilot plant having a culture vessel of 20 litres capacity. The carbon source in an ammonium-salts medium was glycerol at a concentration of 2·5 g per litre. What bacterial densities would be obtained by adjusting the flow rate of the incoming medium to values of 5·0, 10·0 and 20·0 litres per hour and what would be the corresponding outputs of the systems?

In a number of batch culture experiments it was ascertained that one litre of this medium yielded 1·325 g dry weight of bacteria and that the maximum specific growth rate was 0·85 h⁻¹. A value for K_s was obtained with some difficulty because its value was so small that over most of the lower range of dilution rates the substrate concentration was too low to measure with accuracy; the value of $1·35 \times 10^{-4}$ M may therefore be subject to considerable error.

(Data from HERBERT, ELSWORTH & TELLING (1956).)

The dilution rates corresponding to flow rates of 5·0, 10·0 and 20·0 l/h will be given by the expression $D = f/v$, where $v = 20$ l. Hence the values are 0·25, 0·50 and 1·0 h⁻¹.

The critical dilution rate, D_c, at which washout occurs, will be reached when the substrate concentration of the effluent is equal to that of the incoming medium, i.e. when $s_R = \tilde{s} = 2·5$ g/l.

Thus
$$\tilde{s} = K_s\left(\frac{D_c}{\mu_m - D_c}\right) = 2·5$$

K_s must be expressed in the same concentration units as the other terms, i.e. g/l. The molecular weight of glycerol ($C_3H_8O_3$) is 92, hence $1·35 \times 10^{-4}$ M is equivalent to $92 \times 1·35 \times 10^{-4}$ g/l, i.e. 0·0124 g/l.

Therefore
$$0·0124\left(\frac{D_c}{0·85 - D_c}\right) = 2·5$$

$$0·0124\, D_c = 2·5 \times 0·85 - 2·5\, D_c$$

$$D_c = \frac{2·5 \times 0·85}{2·512} = \underline{0·846 \text{ h}^{-1}}$$

Thus wash-out will occur with dilution rates equal to or greater than 0·846 h⁻¹, as for example with the highest flow rate used (20 l/h ≡ D of 1·0 h⁻¹). For the other two flow rates the steady state bacterial concentration $\tilde{x}$ is obtained from equation 9.14, namely

$$\tilde{x} = Y(s_R - \tilde{s}) = Y\left[s_R - K_s\left(\frac{D}{\mu_m - D}\right)\right]$$

The yield coefficient Y (g dry weight of bacteria produced per g of glycerol consumed) will be equal to $1 \cdot 325/2 \cdot 5 = 0 \cdot 53$, since $s_R = 2 \cdot 5$ g/l.

The steady state bacterial concentration for $D = 0 \cdot 25$ h^{-1} is thus

$$\tilde{x} = 0 \cdot 53 \left[2 \cdot 5 - 0 \cdot 0124 \left(\frac{0 \cdot 25}{0 \cdot 85 - 0 \cdot 25} \right) \right]$$

$$= 0 \cdot 53 \left[2 \cdot 5 - \frac{0 \cdot 0124 \times 0 \cdot 25}{0 \cdot 60} \right]$$

$$= 0 \cdot 53 [2 \cdot 5 - 0 \cdot 005]$$

$$= \underline{1 \cdot 323 \text{ g/l.}}$$

For $D = 0 \cdot 50$ h^{-1}

$$\tilde{x} = 0 \cdot 53 \left[2 \cdot 5 - \frac{0 \cdot 0124 \times 0 \cdot 50}{0 \cdot 35} \right]$$

$$= 0 \cdot 53 [2 \cdot 5 - 0 \cdot 018]$$

$$= \underline{1 \cdot 316 \text{ g/l.}}$$

The output of bacteria at each flow rate will be $D\tilde{x}$ g dry weight per litre per hour, thus:

For $D = 0 \cdot 25$ h^{-1} Output $= 0 \cdot 25 \times 1 \cdot 323$
                              $= \underline{0 \cdot 331 \text{ g/l/h.}}$

For $D = 0 \cdot 5$ h^{-1} Output $= 0 \cdot 5 \times 1 \cdot 316$
                              $= \underline{0 \cdot 658 \text{ g/l/h.}}$

As the culture vessel is of 20 litres capacity, the total output of the apparatus per hour will be 6·62 and 13·16 g dry weight of cells respectively.

REFERENCES AND SUGGESTED READING

BAUCHOP, T. & ELSDEN, S. R. (1960). The growth of micro-organisms in relation to their energy supply. *J. gen. Microbiol.*, 23, 457.

FINNEY, D. J., HAZLEWOOD, T. & SMITH, M. J. (1955). Logarithms to base 2. *J. gen. Microbiol.*, 12, 222.

GUNSALUS, I. C. & SHUSTER, C. W. (1961). *The Bacteria*, 2, 1, ed. I. C. Gunsalus and R. Y. Stanier. New York: Academic Press.

HERBERT, D., ELSWORTH, R. & TELLING, R. C. (1956). The continuous culture of bacteria ; a theoretical and experimental study. *J. gen. Microbiol.*, 14, 601.

HINSHELWOOD, C. N. (1946). *Chemical Kinetics of the Bacterial Cell*, Chaps II and III. London : Oxford University Press.

MONOD, J. (1949). The growth of bacterial cultures. *A. Rev. Microbiol.*, **3**, 371.

MONOD, J. (1950). La technique de culture continue ; théorie et applications. *Annls. Inst. Pasteur*, **79**, 390.

NOVICK, A. & SZILARD, K. (1950). *Proc. natn. Acad. Sci. U.S.A.*, **36**, 708.

PAYNE, W. J. (1970). Energy yields and growth of heterotrophs. *A. Rev. Microbiol.* **24**, 17.

PROBLEMS

9.1. *Aerobacter aerogenes* was inoculated simultaneously into three different media : peptone and two chemically defined media containing respectively glucose and citric acid as the sole sources of carbon. The inoculum in each case was 3×10^5 cells per ml. After 4·5 hours measurements of the cell population in each culture revealed values of 9732×10^6, $41·11 \times 10^6$ and $1·427 \times 10^6$ cells per ml for peptone, glucose and citrate respectively. The citrate culture was the only one to display a lag period, the duration of which was 90 minutes, and the cells in each culture were still growing logarithmically.

What were the mean generation times for the organism in these media ? Comment on the results.

9.2. The following data were obtained in an experiment in which anaerobically grown *Aerobacter aerogenes* was inoculated into a glucose-ammonium sulphate medium and grown with gentle aeration at 37° C. The inoculum used was $2·0 \times 10^6$ bacteria per ml.

Time (minutes)	150	200	250	280	310	340	370	400
Bacterial population (millions per ml)	14·1	38·9	104·7	190·6	346·7	616·5	794·2	812·7

Determine graphically the mean generation time of the organism under these conditions, whether or not a lag period occurs and, if so, its duration.

9.3. A glucose-ammonium salt medium was inoculated with 2·5 million cells per ml of *Aerobacter aerogenes* and grown initially without aeration. After a period, a gentle stream of air was passed through the culture until growth ceased. From the protocol given below, determine the mean generation time of the organism for unaerated and aerated growth and also the time after inoculation that the air stream was turned on.

Time (minutes)	Bacterial population (millions per ml)
100	13·80
130	23·44
160	39·81
190	66·07
220	117·5
250	223·9
280	436·5
310	812·8
340	1122·0
370	1148·0

9.4. *Escherichia coli* was grown in a medium containing 0·5 g per litre fructose as the sole carbon source, the sugar being completely utilized within 538 minutes. The inoculum population used was 5×10^4 bacteria per ml. If the relationship between stationary population and fructose concentration is linear up to 0·80 g per litre sugar, at which concentration the population supported is $3·2 \times 10^8$ bacteria per ml, determine the mean generation time of the organism in this medium. What length of time would elapse before growth ceased in the presence of excess fructose if the same size inoculum were used ? It may be assumed that the breakdown in linearity of the total growth and fructose concentration relationship is quite sharp.

9.5. The addition of certain compounds to cultures of *Aerobacter aerogenes* which are displaying very long lag periods in glucose-ammonium sulphate media has the effect of either shortening or prolonging the length of this lag phase. In an experiment to determine the effect of various organic acids, all added to a concentration of 0·06 g per litre, each tube received an inoculum of 2×10^4 bacteria per ml. Comparison with a control culture containing no added organic acid revealed the fact that all the cultures were growing at the same rate with a mean generation time of 40 minutes. 700 minutes after inoculation the populations in the various tubes were as follows :

	Bacterial population (millions per ml)
Control	3·467
Succinic acid	10·00
Malic acid	1·413
α-Oxoglutaric acid	4·677
DL-Aspartic acid	37·15
DL-Glutamic acid	53·70

Determine the effect of the various compounds on the length of the lag phase.

(After DAGLEY, DAWES & MORRISON (1950), *J. gen. Microbiol.*, 4, 437.)

9.6. An inoculum of 10^6 cells of *Aerobacter aerogenes* in a glucose-ammonium salt medium containing phenol to a concentration of $3·4 \times 10^{-3}$ M displayed a lag phase of 100 minutes before logarithmic growth began. After 230 minutes the culture (in the logarithmic phase) had a population of $73·55 \times 10^5$ cells. A control tube (no phenol) was given an identical inoculum and grew without lag. At 180 minutes its population was $15·85 \times 10^6$ cells and it was still growing exponentially.

From these data, determine whether the concentration of phenol employed exerted any effect on the rate of division of the organism.

9.7. A facultatively anaerobic organism was inoculated into three identical 25 ml quantities of medium and grown (*a*) with aeration, (*b*) without aeration, and (*c*) strictly anaerobically in a Fildes-McIntosh jar. The aerated culture lagged for 55 minutes before active cell division began, but the other two grew immediately and after 250 minutes the populations in the three tubes were (*a*) 446·7, (*b*) 281·8 and (*c*) 63·10 millions of bacteria per ml. If the inoculum per tube was 5×10^7 cells, determine the mean generation times for growth under the three conditions of aerobiosis.

9.8. A medium containing 0·25 g per litre glucose and an unknown amount of galactose was inoculated with 5×10^5 cells of *Escherichia coli* per ml medium. No lag was exhibited and the cells grew with a mean generation time of 40 minutes utilizing the glucose. When the glucose was exhausted, the cells adapted themselves to galactose and then grew with a mean generation time of 45 minutes to attain a stationary population of $3·08 \times 10^8$ cells per ml, 6 hours 30 minutes after inoculation.

Determine the length of the phase of adaptation between the two growth cycles and also the amount of galactose present in the medium, given that the relationship between stationary population and sugar concentration is linear up to 0·9 g per litre of either sugar, at which concentration the stationary population supported is $3·7 \times 10^8$ bacteria per ml.

(Glasgow Double Science Course, 1952.)

9.9. What factors influence the size of the bacterial crop that a given medium will support ?

After light inoculation with *Escherichia coli* crops of 120 and 510 millions of cells/ml grew respectively in media containing 240 and 1,020 mg/litre of either glucose or mannose. A medium containing a mixture of the sugars was similarly inoculated with the following results :

Time after inoculation (minutes)	220	260	280	310	340	372	400	440
Cell count (millions/ml)	69	141	200	224	251	355	490	500

Using a graphical method, find the approximate concentration of each sugar present.

(Leeds Honours Course Finals, 1953.)

9.10. A medium containing $0·4 \times 10^{-2}$ M glucose and $7·1 \times 10^{-2}$ M citrate was inoculated with cells of *Aerobacter aerogenes*, which had been grown anaerobically on citrate, to give an inoculum population of $6·31 \times 10^6$ cells per ml. The cells displayed no lag period and grew exponentially from the time of inoculation. Determine the length of the lag phase between the phases of growth on glucose and on citrate if, after 400 minutes, the cells were growing exponentially and utilizing citrate, and the population at this time was 5495×10^5 cells per ml.

The mean generation times of the organism for citrate and glucose under these conditions of unaerated growth are respectively 80 and 38 minutes. The relationship between the concentration of glucose supplied as the sole carbon source and stationary population is linear up to $8·6 \times 10^{-3}$ M, at which level the population supported is 518×10^6 cells per ml.

(After DAGLEY & DAWES (1953), *J. Bact.*, 66, 259.)

9.11. The following data were obtained in a study of the relationship between the stationary population of *Aerobacter aerogenes* and the concentration of glucose in a simple ammonium salt medium :

Glucose concentration (g per litre)	0·125	0·30	0·50	0·80	1·00	1·25
Stationary population (millions per ml)	51·5	123·3	205·5	329	371	372

Three media containing unknown quantities of glucose were given light inocula and growth ceased at respective populations of 112, 348 and 372 millions of organisms per ml. What were the millimolar concentrations of glucose present in these media ?

(Glasgow Double Science Course, 1957.)

9.12. Discuss the value to the microbiological biochemist of accurate quantitative measurements of bacterial growth.

A bacterium was inoculated from a broth medium into a glycerol-ammonium

salts medium containing limited amounts of L-cysteine and DL-methionine.
Growth was measured turbidimetrically and the following data were obtained:

Time (min)	Cell population (millions per ml)	Time (min)	Cell population (millions per ml)
0	0·156	590	3·236
90	0·156	650	5·495
180	0·158	740	12·02
245	0·269	860	19·95
290	0·457	950	21·88
350	0·912	1040	44·67
410	1·906	1130	89·13
470	2·512	1220	125·90
530	2·514	1410	126·20

Derive the growth constants which completely characterize the growth of this
culture, and interpret the pattern of growth observed under these particular
nutritional conditions.

(Hull Special Biochemistry I Course, 1964.)

9.13. Discuss the value to the biochemist of determinations of the molar
growth yield of a micro-organism.

A facultative bacterium was found to metabolize both glucose and pyruvate
with the production of gas ; the volume of gas evolved was halved when KOH
was placed in the centre well of the Warburg vessel. Acetate was found to be a
product of the fermentation.

Several determinations gave an average total crop of 5·9 mg dry weight of
cells per 50 ml of medium containing 36 mg of glucose (limiting glucose).
What is the molar growth yield?

How can you explain the value of the molar growth yield obtained?

[The molar growth yield of yeast under similar conditions is 21 g dry weight
per mole of glucose.]

(Hull Special Biochemistry I Course, 1965)

9.14. What measurements would you make to enable the growth of a
bacterium in a given medium to be completely characterized ? (Experimental
details are *not* required.) Discuss the value of total growth measurements to the
biochemist.

A facultative bacterium produced gas from both glucose and pyruvate and the
gas volume yielded was halved when KOH was added to the centre well of the
Warburg vessel. Acetate was a product of the fermentation.

Several determinations of total growth gave an average value of 2·95 mg dry
weight of cells per 25 ml of medium containing 72 mg of glucose per 100 ml
(glucose exhaustion limited growth).

Calculate the molar growth yield. Offer an explanation for the value of the
molar growth yield obtained. [The molar growth yield of yeast under similar
conditions is 21 g dry weight of yeast per mole of glucose.]

(Hull Special Biochemistry I Course, 1967.)

CHAPTER X

OXIDATION-REDUCTION POTENTIALS

OXIDATION may be defined as the loss of electrons and reduction as the gain of electrons. The two processes are complementary, so that whenever one substance is oxidized another must be reduced; these are, respectively, the electron donor and the electron acceptor. A typical oxidation-reduction system (usually referred to as a redox system) may be written as

$$\text{Reduced form} \rightleftharpoons \text{Oxidized form}^{n+} + ne$$

or, in abbreviated form,

$$\text{red} \rightleftharpoons \text{ox}^{n+} + ne$$

where n is the number of electrons involved in the reaction. The formal analogy with an acid and its conjugate base will be apparent (p. 71). The electrons furnished by this system may now be accepted by an appropriate second redox system, the oxidized form of which has a suitable affinity for the electrons. The individual and overall reactions can then be written as

$$\text{red}_1 \rightleftharpoons \text{ox}_1{}^{n+} + ne$$
$$\underline{\text{ox}_2{}^{n+} + ne \rightleftharpoons \text{red}_2}$$
$$\text{red}_1 + \text{ox}_2{}^{n+} \rightleftharpoons \text{ox}_1{}^{n+} + \text{red}_2$$

The reaction is reversible and will therefore be characterized by an equilibrium constant for any given temperature and from which the free energy change may be determined. The overall reaction thus comprises the two individual redox systems which are usually referred to either as *half-cell reactions*, as will be subsequently discussed, or as *redox couples*.

If an inert metal electrode, such as platinum or silver, is immersed in a solution of a redox system, a potential difference is set up between the electrons in solution and those in the metal. This gives rise to an *electrode potential* and it can be shown that the value of this potential, designated by E, is given by the Nernst equation

$$E = E^\circ + \frac{RT}{nF} \ln \frac{(\text{ox})}{(\text{red})} \quad . \quad . \quad . \quad (10.1)$$

where E° is a special constant for a given system known as the *standard electrode potential*, R is the gas constant equal to 8·314 absolute joules per degree per mole, T the absolute temperature, n the number of equivalents involved in the reaction, F the faraday (equal to 96,494 coulombs) which is necessary to convert one equivalent of an element to an equivalent of ions, and (ox) and (red) are the activities of the oxidized and reduced forms of the redox system. The standard electrode potential, E°, is the potential of the system when all the solutes are in their standard states of unit activity. The terms emf and potential are often used interchangeably but the term potential should be reserved exclusively for half-cell reactions and junction potentials, and the term emf restricted to complete cells. The distinction then is that the emf is a directly measurable quantity whereas the potential can only be measured by reference to an arbitrary standard, as we shall now discuss.

An electrode immersed in a solution of redox system constitutes a half-cell, the potential difference of which it is impossible to measure since it is a single electrode ; to do so it must be combined with another half-cell by means of a KCl-agar bridge, thereby making a complete cell, and the potential difference of the two half-cells measured by connecting them to a potentiometer. Preferably a reference standard half-cell should be used. The hydrogen electrode is such a standard and its potential is arbitrarily taken as zero. The cell reaction is

$$\tfrac{1}{2}H_2 \rightleftharpoons H^+ + e$$

and, by definition, the potential of this system is zero at all temperatures when an inert metal electrode is placed in a solution of unit activity with respect to hydrogen ions, i.e. pH = 0, in equilibrium with gaseous hydrogen at a pressure of 1 atmosphere.

To determine the potential difference of a system it is not necessary to use a hydrogen electrode. Any other half-cell, the potential difference of which has been accurately determined with respect to the hydrogen half-cell, may be used, and in fact the calomel cell is usually more convenient in practice than the hydrogen electrode. The symbol E_h is used for electrode potential to show that the hydrogen electrode is taken as the reference standard. Consequently $E_h = E - E_H$, where E_H is the potential of the hydrogen electrode which, by definition, is zero. Hence the Nernst general equation for the potential of a redox system becomes

$$E_h = E° + \frac{RT}{nF} \ln \frac{(ox)}{(red)} \qquad . \qquad . \quad (10.1a)$$

At 30°, which is the temperature frequently employed for electrode measurements, the factor $2.303RT/nF$ (converting natural logarithms to the base 10) has a value of 0.06 for $n = 1$ and 0.03 for $n = 2$. Thus, for $n = 1$

$$E_h = E° + 0.06 \log \frac{(ox)}{(red)} \qquad . \qquad . \quad (10.2)$$

The general equation reveals the fact that the electrode potential of a redox system is dependent on a constant, the standard electrode potential, and upon the logarithm of the ratio of the activities of oxidized to reduced forms. With dilute solutions the molar concentrations may be used in place of the activities, i.e. the activity coefficients are assumed to be unity. The standard electrode potential under these conditions is usually denoted by E_o to distinguish it from $E°$. The potential will be increased, or become more positive, if the proportion of oxidized to reduced form is increased and, conversely, become lower or more negative if the proportion of reduced form becomes greater. As the logarithmic term in equation 10.1a involves the *ratio* of the activities of the oxidized and reduced forms, it follows that the electrode potential is independent of the total concentration of the system, if it is assumed that dilution has no effect on activity coefficients or affects them all in a like manner. For example, any system that is 40 per cent. reduced will have the same E_h value irrespective of whether its total concentration is, say, 1.0, 0.5 or 0.02 M. For this reason it will be apparent that the electrode potential is not a measure of the *capacity* of a system for oxidation or reduction but only of its oxidation or reduction intensity, in the same way that pH is a measure of the acidity or alkalinity of a system but not of its buffering power.

When the system is half-reduced and the ratio of oxidized to reduced activities becomes unity, $E_h = E°$ and hence the standard electrode potential is that of a half-reduced system. $E°$ is of considerable importance because, knowing its value, the electrode potential at any degree of oxidation or reduction of the system can be calculated. Conversely, the degree of oxidation can be obtained from the measured electrode potential. $E°$ is a measure of the oxidation or reduction intensity of a system and enables a list of redox systems to be drawn up in order of their standard electrode potentials. Any given system will be capable of being oxidized by a system more positive, i.e. above it on the scale, and in its turn will oxidize any system more negative than itself (below it on

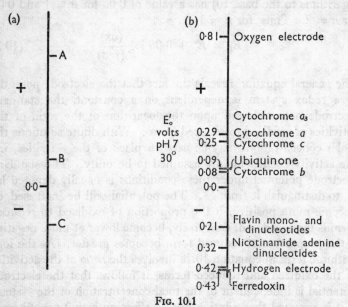

FIG. 10.1
(a) Standard electrode potential scale of oxidation-reduction systems.
(b) Biological oxidation-reduction systems.

the scale). Thus in Fig. 10.1 (a) system A will oxidize systems B and C, and B will oxidize C, whereas system C will reduce systems B and A, and B will reduce A. It should be mentioned that a catalyst may be needed in order to bring about the reaction between two such systems. In Fig. 10.1 (b) are drawn up some systems of biological interest and, when considering them, it should be borne in mind that the values shown represent half-reduced systems, i.e. the mid-point of the range of electrode potential which may be manifest under biological conditions. It should also be emphasized that electrode potentials determined *in vitro* may not necessarily have the same value *in vivo*, due, for example, to the effect of protein binding. Table 10.1 lists the standard electrode potentials of some oxidation-reduction half-reactions of biological importance.

TABLE 10.1

Standard Electrode Potentials of some Oxidation-Reduction Half Reactions of Biological Importance

System	E (pH 7·0, 30°) volt
$\frac{1}{2}O_2 + 2H^+ + 2e \rightarrow H_2O$	+0·816
$Fe^{3+} + e \rightarrow Fe^{2+}$	+0·771
$\frac{1}{2}O_2 + H_2O + 2e \rightarrow H_2O_2$	+0·30
Cytochrome a $Fe^{3+} + e \rightarrow$ cytochrome a Fe^{2+}	+0·29
Cytochrome c $Fe^{3+} + e \rightarrow$ cytochrome c Fe^{2+}	+0·25
2,6-*Dichlorophenolindophenol* (ox) + 2H$^+$ + 2e → 2,6-*dichlorophenolindophenol* (red)	+0·22
Crotonyl-CoA + 2H$^+$ + 2e → butyryl-CoA	+0·19
Ubiquinone + 2H$^+$ + 2e → dihydro-ubiquinone	+0·10
Cytochrome b $Fe^{3+} + e \rightarrow$ cytochrome b Fe^{2+}	+0·08
Phenazine methosulphate (ox) + e →*phenazine methosulphate* (red)	+0·08
Dehydroascorbate + 2H$^+$ + 2e → ascorbate	+0·06
Fumarate + 2H$^+$ + 2e → succinate	+0·031
Methylene blue + 2H$^+$ + 2e →*leuco-methylene blue*	+0·011
FAD + 2H$^+$ + 2e → FADH$_2$	−0·06
Oxaloacetate + 2H$^+$ + 2e → malate	−0·102
Acetaldehyde + 2H$^+$ + 2e → ethanol	−0·163
Pyruvate + 2H$^+$ + 2e → lactate	−0·190
Riboflavin + 2H$^+$ + 2e → riboflavin H$_2$	−0·200
1,3-Diphosphoglycerate + 2H$^+$ + 2e → 3-phospho-glyceraldehyde + P$_i$	−0·290
NAD$^+$ + 2H$^+$ + 2e → NADH + H$^+$	−0·320
NADP$^+$ + 2H$^+$ + 2e → NADPH + H$^+$	−0·320
Benzylviologen + H$^+$ + e →*benzylviologen*-H	−0·359
Acetyl-CoA + 2H$^+$ + 2e → acetaldehyde + CoA	−0·410
CO$_2$ + 2H$^+$ + 2e → formate	−0·420
H$^+$ + e → $\frac{1}{2}$H$_2$	−0·420
Ferredoxin-Fe^{3+} + e → ferredoxin-Fe^{2+}	−0·432
Methylviologen + H$^+$ + e →*methylviologen*-H	−0·440
Acetate + 2H$^+$ + 2e → acetaldehyde	−0·60

All components are at unit activity with the exception of H$^+$ which is at 10^{-7} M. Gases are at 1 atm pressure.
Redox indicators are italicized.

Note that $E°$ refers to measurements at zero pH and, since it is often impracticable to measure certain systems at this pH value, the term $E°'$ is introduced to indicate the standard electrode potential at a *stated* pH other than zero (see p. 344). When molar concentrations are used instead of activities the corresponding value is denoted by E_o'.

Example 10.1.—At pH 7·0 E_o' for the methylene blue-leucomethylene blue system is 0·011 volt at 30°. If the measured electrode potential of a solution of this system is 0·065 volt, calculate the percentage of the reduced (leuco) form present.

The reaction for the reduction of methylene blue may be written as

$$MB + H_2 \rightleftharpoons MBH_2.$$

Two hydrogen atoms (equivalent to 2 protons plus 2 electrons) are involved so that $n = 2$ for this system and

$$E_h = E_o' + \frac{2 \cdot 303RT}{2F} \log \frac{[MB]}{[MBH_2]}$$

$$0 \cdot 065 = 0 \cdot 011 + 0 \cdot 03 \log \frac{[MB]}{[MBH_2]}$$

$$\log \frac{[MB]}{[MBH_2]} = \frac{0 \cdot 054}{0 \cdot 03} = 1 \cdot 8,$$

whence

$$\frac{[MB]}{[MBH_2]} = 63 \cdot 1.$$

Let x be the percentage of MBH_2, then

$$\frac{100 - x}{x} = 63 \cdot 1 \text{ and } x = \underline{1 \cdot 56}.$$

The system, therefore, contains 1·56 per cent. of leucomethylene blue.

Example 10.2.—The riboflavin-leucoriboflavin system has an E_o' value of −0·186 volt at pH 7·0 and 20°. Calculate the oxidation-reduction potential of this system when it contains (a) 10 per cent. and (b) 80 per cent. of the oxidized form.

The reaction may be written

$$X + H_2 \rightleftharpoons XH_2,$$

where X represents riboflavin and XH_2 its reduced form. Two hydrogen atoms or, in principle, two electrons are involved, hence $n = 2$.

$$\begin{aligned}
E_h &= E_o' + \frac{2 \cdot 303RT}{nF} \log \frac{[X]}{[XH_2]} \\
&= -0 \cdot 186 + \frac{2 \cdot 303 \times 8 \cdot 314 \times 293}{2 \times 96494} \log \frac{[X]}{[XH_2]} \\
&= -0 \cdot 186 + 0 \cdot 029 \log \frac{[X]}{[XH_2]} \, .
\end{aligned}$$

Therefore in (a)

$$\begin{aligned}
E_h &= -0 \cdot 186 + 0 \cdot 029 \log \frac{10}{90} \\
&= -0 \cdot 186 + 0 \cdot 029 \log 0 \cdot 1111 \\
&= -0 \cdot 186 + 0 \cdot 029 \, (\bar{1} \cdot 0457) \\
&= -0 \cdot 186 - 0 \cdot 029 \times 0 \cdot 9543 \\
&= -0 \cdot 186 - 0 \cdot 028 \\
&= \underline{-0 \cdot 214 \text{ volt.}}
\end{aligned}$$

(b)
$$E_h = -0.186 + 0.029 \log \frac{80}{20}$$
$$= -0.186 + 0.029 \times 0.6021$$
$$= -0.186 + 0.017$$
$$= \underline{-0.169 \text{ volt.}}$$

Consider now the reactions between two different redox systems. If two systems at the same pH but with different potentials are mixed, reaction will occur between them until equilibrium is reached, i.e. until both systems achieve the same potential. To illustrate this, the succinate dehydrogenase and methylene blue systems, as used in the Thunberg tube technique, may be considered. The reactions involved are

$$succinate^= \rightleftharpoons fumarate^= + 2H$$

$$MB + 2H \rightleftharpoons MBH_2$$

Net reaction : succinate$^=$ + MB $\rightleftharpoons$ fumarate$^=$ + MBH$_2$

The equilibrium constant for this reaction is given by the expression

$$K_c = \frac{[\text{fum}][\text{MBH}_2]}{[\text{succ}][\text{MB}]} .$$

If 1 ml of a solution containing succinate and fumarate, both at 0 001 M, together with the enzyme succinate dehydrogenase, are in the evacuated Thunberg tube, and 1 ml of a methylene blue solution is in the hollow stopper, also 0.001 M with respect to both oxidized and reduced forms, at pH 7.0 and 30°, then for each system $E_h = E_o'$, because both are half-reduced. E_o' for the succinate system is +0.005 volt and for the MB-MBH$_2$ system +0.011 volt.

When the solutions are mixed by tipping the tube the succinate-fumarate system, having the lower potential, reduces the methylene blue system and is itself oxidized. Consequently the ratio [fum]/[succ] increases and the ratio [MB]/[MBH$_2$] decreases, causing an increase in the potential of the former and a decrease in the potential of the latter until equilibrium is attained, when both are at the same potential, i.e. $E_{h_1} = E_{h_2}$. Therefore

$$E_{o \atop \text{fum-succ}}' + 0.03 \log \frac{[\text{fum}]}{[\text{succ}]} = E_{o \atop \text{MB-MBH}_2}' + 0.03 \log \frac{[\text{MB}]}{[\text{MBH}_2]}$$

M

and $\quad \Delta E_0' = 0.011 - 0.005 = 0.03 \log \dfrac{[\text{fum}][\text{MBH}_2]}{[\text{succ}][\text{MB}]}$.

That is, $\qquad 0.006 = 0.03 \log \dfrac{[\text{fum}][\text{MBH}_2]}{[\text{succ}][\text{MB}]}$.

But in the example considered it follows that, since the reaction started with equal concentrations of all four components, [fum] = [MBH$_2$] and [succ] = [MB] at all stages of the reaction. Therefore the above relationship reduces to

$$0.006 = 0.03 \times 2 \log \dfrac{[\text{fum}]}{[\text{succ}]} ,$$

whence $\qquad \log \dfrac{[\text{fum}]}{[\text{succ}]} = 0.10$

and $\qquad \dfrac{[\text{fum}]}{[\text{succ}]} = \dfrac{[\text{MBH}_2]}{[\text{MB}]} = 1.26.$

Let x be the percentage of MBH$_2$, then

$$\dfrac{x}{100 - x} = 1.26 \text{ and } x = 55.8.$$

Thus, at equilibrium, the dye would be 55·8 per cent. reduced, showing that under these conditions the amount of reduction occurring is quite small, namely from 50 to 55·8 per cent.

Free Energy of Oxidation-Reduction Reactions

In the example just discussed it was seen that

$$\Delta E_0' = 2.303 \, \frac{RT}{nF} \log \frac{[\text{fum}][\text{MBH}_2]}{[\text{succ}][\text{MB}]}.$$

But the equilibrium constant, K_c, of this reaction is equal to [fum][MBH$_2$]/[succ][MB], and if it is assumed that when the oxidized and reduced components of the system are at equilibrium the ratio of their activity coefficients can be taken as unity, then we can equate the thermodynamic equilibrium constant K with K_c. This assumption enables the following general expression to be derived :

$$\Delta E_0' = \frac{RT}{nF} \ln K$$

or
$$nF \Delta E_o' = RT \ln K.$$

But from thermodynamics we have the relationship between the standard free energy change of a reaction, $\Delta G°$, and the equilibrium constant (equation 3.9):

$$\Delta G° = -RT \ln K,$$

from whence it follows that

$$\Delta G° = -nF \Delta E_o' \qquad . \qquad . \qquad . \quad (10.3)$$

This means that the standard free energy change of the reaction between two redox systems may be calculated from the difference in E_o' values. In similar manner it is possible to calculate the free energy change for the two individual systems:

succ-fum $\qquad \Delta G_1° = -nF \Delta E_{o(\text{succ-fum})}'$

MB-MBH$_2$ $\qquad \Delta G_2° = -nF \Delta E_{o(\text{MB-MBH}_2)}'$

and the free energy change for the combined systems is then

$$\Delta G° = \Delta G_2° - \Delta G_1°.$$

It will be noted that although the subtraction may be carried out in two ways, e.g. $\Delta G_2° - \Delta G_1°$ or $\Delta G_1° - \Delta G_2°$, only one of these will give the negative $\Delta G°$ value characteristic of a spontaneous reaction. The value of $\Delta G°$ in this calculation will be in joules, therefore to express it in calories it is necessary to divide by 4·185. Hence the free energy change of a reaction is given by the general equation

$$\Delta G = -nF \Delta E \qquad . \qquad . \qquad . \quad (10.3a)$$

where ΔE is the potential difference between the participating systems, provided n is the same for these two systems.

Example 10.3.—Calculate the standard free energy change and the equilibrium constant of the reaction

lactate + acetaldehyde $\rightleftharpoons$ pyruvate + ethanol.

At pH 7·0 and 25°, E_o' for the pyruvate-lactate system is $-0·18$ volt and for the acetaldehyde-ethanol system is $-0·16$ volt. NAD is the coenzyme for both systems.

The individual reactions coupled by NAD are :

$$\text{lactate} + NAD^+ \rightleftharpoons \text{pyruvate} + NADH + H^+ \qquad E_o' = -0 \cdot 18 \text{ volt}$$

$$\text{acetaldehyde} + NADH + H^+ \rightleftharpoons \text{ethanol} + NAD^+ \qquad E_o' = 0.16 \text{ volt}$$

Sum :   lactate + acetaldehyde $\rightleftharpoons$ pyruvate + ethanol

Since the redox potential of the acetaldehyde/ethanol system is slightly more positive than that of the lactate/pyruvate system, under the conditions used it will be expected that some acetaldehyde will be reduced and some lactate oxidized. Thus :

$$\Delta E_o' = -0 \cdot 16 - (-0 \cdot 18) = +0 \cdot 02 \text{ volt}$$

and $n = 2$ for this system.

Since $\Delta G^\circ = -nF\Delta E_o'$

$$\Delta G^\circ = \frac{-2 \times 96494 \times 0 \cdot 02}{4 \cdot 185} \text{ calories}$$

$$= -923 \text{ calories.}$$

The equilibrium constant for the reaction is given by

$$\Delta G^\circ = -RT \ln K$$

Thus, at 25° $-923 = -2 \cdot 303 \times 1 \cdot 987 \times 298 \log K$

and $\log K = \dfrac{923}{2 \cdot 303 \times 1 \cdot 987 \times 298}$

$$\log K = 0 \cdot 676$$

$$K = 4 \cdot 75$$

This value of K indicates that the equilibrium mixture contains significant concentrations of all the components. To ascertain what these concentrations are, if we assume the initial concentrations of all reactants are unity and we let x be the concentrations of pyruvate and ethanol and $(1-x)$ the concentrations of lactate and acetaldehyde at equilibrium, then

$$K = \frac{x^2}{(1-x)^2} = \frac{x^2}{x^2 - 2x + 1}$$

Thus $4 \cdot 75x^2 - 9 \cdot 5x + 4 \cdot 75 = x^2$

and $3 \cdot 75x^2 - 9 \cdot 5x + 4 \cdot 75 = 0$

Solving the quadratic equation

$$x = \frac{-(-9 \cdot 5) \pm \sqrt{((-9 \cdot 5)^2 - 4 \times 3 \cdot 75 \times 4 \cdot 75)}}{2 \times 3 \cdot 75}$$

$$= \frac{9 \cdot 5 \pm \sqrt{(90 \cdot 25 - 71 \cdot 25)}}{7 \cdot 50}$$

$$= \frac{9 \cdot 5 \pm \sqrt{19 \cdot 0}}{7 \cdot 50} = \frac{9 \cdot 5 \pm 4 \cdot 352}{7 \cdot 50}$$

$$= \frac{5 \cdot 141}{7 \cdot 50} \text{ or } \frac{13 \cdot 859}{7 \cdot 50}$$

Of these, only the first is permissible, hence

$$x = 0.685$$

Thus, at equilibrium lactate would be 68·5 per cent. oxidized and acetaldehyde 68·5 per cent. reduced.

Potentiometric Titration

It is possible to follow the course of a reaction between two redox systems potentiometrically, provided they are reversible and electromotively active, i.e. not sluggish, by placing a platinum electrode into the solution of the reducing agent, coupling with a reference electrode and measuring the potential difference as the solution of oxidizing agent is added. A curve relating the potential difference to the percentage oxidation or reduction can be constructed (Fig. 10.2(a) and (b).)

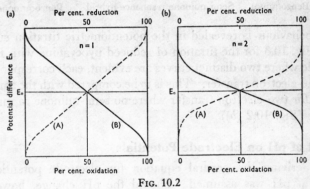

FIG. 10.2

Potentiometric titration curves for oxidation-reduction systems (a) where $n = 1$ and (b) where $n = 2$.

Curve A shows the effect of adding a strong oxidizing agent to the reduced form of a redox system, and curve B the reverse process, reduction of the oxidized form of the redox system. At the mid-point of the titration, i.e. at 50 per cent. oxidation, (ox) = (red), therefore $E_h = E°$ and there is a point of inflexion in the curve. (Compare with the mid-point of an acid-base titration where $pH = pK_a$.) The slope of the curve at the inflexion point is directly related to the number of electrons involved in the process and is therefore useful in determining the type of reaction under investigation. Fig. 10.2(a) shows the titration curve for a system where $n = 1$ and it will be noticed that the curve is steeper than in Fig. 10.2(b), where $n = 2$. Some biological redox systems which involve an overall transfer of two electrons may actually occur in two distinct one-electron steps, e.g. the reduction of flavins via semi-quinone formation. The model for this system is the oxidation of benzoquinol (hydroquinone) to benzoquinone (quinone) via the intermediate resonance-stabilized semi-quinone structure.

Benzoquinol   Semi-quinone resonance hybrid   Benzoquinone

This behaviour is revealed in the potentiometric titration curve, as in Fig. 10.3 for the titration of reduced pyocyanine with ferricyanide where two distinct curves are evident, each corresponding to a one-electron transfer. This is to be compared with the smooth curve for two-electron transfer where no semi-quinone formation occurs (Fig. 10.2 (b)).

Effect of pH on Electrode Potentials

In deriving the general equation for electrode potential, a constant pH was assumed. Should the pH change, however, complications may arise due to changes in the ionization of components of the system ; for example, either or both of oxidized and reduced forms may be ionized. This is frequently the case with biological systems.

Consider the general case of the reduced form being a weak dibasic acid (as, for example, succinic acid). The reaction will be

$$red^= \rightleftharpoons ox + 2e$$

and $$E_h = E_b + \frac{RT}{2F} \ln \frac{(ox)}{(red^=)} \qquad . \qquad . \qquad (10.4)$$

Now the dibasic acid ionizes in two stages as follows

$$H_2red \rightleftharpoons H^+ + Hred^-$$

$$Hred^- \rightleftharpoons H^+ + red^=$$

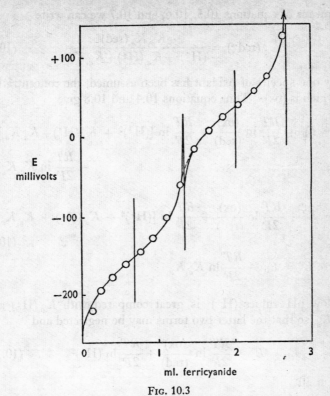

FIG. 10.3

Potentiometric titration of reduced pyocyanine with ferricyanide ion at pH 1·82. Note the two oxidation steps corresponding to one-electron transfers. At 50 per cent. oxidation the semi-quinone form of pyocyanine is present.

(FRIEDHEIM & MICHAELIS, 1931.)

and the corresponding primary and secondary dissociation constants are given by

$$K_{a_1} = \frac{(H^+)(Hred^-)}{(H_2red)} \qquad . \qquad . \qquad . \qquad (10.5)$$

$$K_{a_2} = \frac{(H^+)(red^=)}{(Hred^-)} \qquad . \qquad . \qquad . \qquad (10.6)$$

The total stoicheiometric concentrations of the reduced form (red) will be the sum of the ionized and unionized fractions, i.e.

$$(red) = (H_2red) + (Hred^-) + (red^=) \qquad . \qquad . \qquad (10.7)$$

By means of equations 10.5, 10.6, and 10.7 we can write

$$(\text{red}^=) = \frac{K_{a_1}K_{a_2}(\text{red})}{(\text{H}^+)^2 + K_{a_1}(\text{H}^+) + K_{a_1}K_{a_2}} \qquad . \quad (10.8)$$

Only one species of oxidant has been assumed, the concentration of which is (ox). Thus equations 10.4 and 10.8 give

$$E_h = E_b + \frac{RT}{2F} \ln \frac{(\text{ox})}{(\text{red})} + \frac{RT}{2F} \ln [(\text{H}^+)^2 + K_{a_1}(\text{H}^+) + K_{a_1}K_{a_2}]$$

$$- \frac{RT}{2F} \ln K_{a_1}K_{a_2}$$

or

$$E_h = E^\circ + \frac{RT}{2F} \ln \frac{(\text{ox})}{(\text{red})} + \frac{RT}{2F} \ln [(\text{H}^+)^2 + K_{a_1}(\text{H}^+) + K_{a_1}K_{a_2}]$$

$$. \quad . \quad . \quad (10.9)$$

where $E^\circ = E_b - \dfrac{RT}{2F} \ln K_{a_1}K_{a_2}$

At low pH values $(\text{H}^+)^2$ is great compared with $K_{a_1}(\text{H}^+)$ and $K_{a_1}K_{a_2}$ so that the latter two terms may be neglected and

$$E_h = E^\circ + \frac{RT}{2F} \ln \frac{(\text{ox})}{(\text{red})} + \frac{RT}{2F} \ln (\text{H}^+)^2 \qquad . \quad (10.10)$$

and at 30°

$$E_h = E^\circ + 0{\cdot}03 \log \frac{(\text{ox})}{(\text{red})} - 0{\cdot}06 \text{ pH} \qquad . \qquad . \quad (10.11)$$

When (ox) = (red), E_h varies with pH in a linear manner and the slope of the E_h (or, in effect, E'_o) versus pH curve will be $-0{\cdot}06$. At higher pH values, where $K_{a_1}(\text{H}^+)$ is much greater than $(\text{H}^+)^2$ or $K_{a_1}K_{a_2}$ and the latter two terms may be neglected, the slope of the line will be $-0{\cdot}03$. Extrapolation of the lines of slope $-0{\cdot}06$ and $-0{\cdot}03$ to their point of intersection gives a value for pK_{a_1} of the dibasic acid. At high pH values $K_{a_1}K_{a_2}$ is the only significant term in the last term of equation 10.9 and the pH curve then has zero slope. This enables the value of pK_{a_2} to be obtained by extrapolation of the lines of slope $-0{\cdot}03$ and 0 to their point of intersection. Fig. 10.4 illustrates the pH dependence curve for a dibasic ionizing system. It will be obvious that this is not a very exact method for obtaining pK_a values.

When (H^+) is constant the last term in equation 10.9 is a constant at constant temperature and may be combined with $E^°$ to give $E^{°\prime}$, so that

$$E_h = E^{°\prime} + \frac{RT}{2F} \ln \frac{(ox)}{(red)} \qquad . \qquad . \quad (10.12)$$

where $E^{°\prime}$ is the standard electrode potential at the given constant pH value.

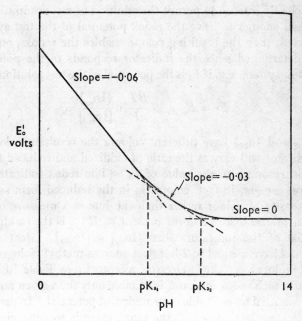

FIG. 10.4

Graph of E_0' values for a dibasic ionizing oxidation-reduction system plotted against pH and showing method of obtaining values for pK_{a_1} and pK_{a_2} by extrapolation of lines of different slope to their points of intersection.

The variation of E_h with pH in the linear portions of the pH dependence curve may be expressed, at $30°$, by the equation

$$E_h = E^° + \frac{0\cdot06}{n} \log \frac{(ox)}{(red)} - \frac{0\cdot06a}{n} pH \qquad . \quad (10.13)$$

where n is the number of electrons involved and a the number of hydrogen ions released in the oxidation process. The symbol

$-\Delta E_h / \Delta \text{pH}$ is used to denote the pH dependence of E_h and is often encountered in tables of oxidation-reduction potentials.

Redox Indicators

Certain dyestuffs which can function as reversible redox half-reactions, and which possess oxidized and reduced forms of different colours, can be used as redox indicators in an analogous fashion to the use of pH indicators (p. 83). Provided the colour is sufficiently intense to permit the addition of a concentration of dye small enough to leave the redox potential of the test system unaffected, then the resulting colour enables the redox potential to be determined since the indicator responds to the potential of the test system, e.g. if E_t is the potential of the test solution then

$$E_t = E'_{o(\text{In})} + \frac{RT}{nF} \ln \frac{[\text{In}_{ox}]}{[\text{In}_{red}]}$$

As In_{ox} and In_{red} have different colours the resultant colour of the indicator will vary as the ratio of oxidized and reduced forms alters in response to the value of E_t. (The redox indicators in common use are, in fact, colourless in the reduced form so that the intensity of colour rather than the hue is a measure of the fraction of the oxidized form present.) $E'_{o(\text{In})}$ is the (midpoint) potential of the indicator when $[\text{In}_{ox}] = [\text{In}_{red}]$. Most redox indicators have n equal to 2 but dyes such as methyl viologen and benzyl viologen are single-electron acceptors (see Table 10.1).

Clearly each redox indicator is suitable only for a given range of redox potential on each side of its midpoint potential. In practice, for indicators having $n = 2$, the range extends to approximately 0·06 volt more positive and 0·06 volt more negative than $E'_{o(\text{In})}$ at 25°, beyond which values the indicator displays its fully oxidized and fully reduced colours respectively. With single-electron indicators the range is about twice as great, i.e. about 0·12 volt on each side of the midpoint potential.

Thiol Compounds

The oxidation of a thiol (sulphydryl) group under appropriate conditions produces a disulphide rather than a sulphonic acid, i.e. instead of

$$\text{RSH} + 1\tfrac{1}{2}\text{O}_2 \longrightarrow \text{RSO}_3\text{H}$$

we have the possibility of

$$2RSH \longrightarrow RS-SR + 2H^+ + 2e$$

$$\begin{matrix} SH & HS & S——S \\ | & | \longrightarrow & | & | \\ R_1C——CR_2 & & R_1C——CR_2 + 2H^+ + 2e \end{matrix}$$

and

These reactions are exemplified by the amino acid cysteine which undergoes oxidation to cystine either as the free molecule or when combined as a residue in a peptide chain. In the latter case the disulphide bridge formed may be either between cysteine residues of the same peptide chain, as in pancreatic ribonuclease, or between residues of different chains which are thereby held together, e.g. the A and B chains of insulin. These types of linkage are usually referred to as intrachain and interchain bonds respectively.

Thiol-disulphide redox systems are generally 'sluggish' in their electromotive activity, that is they attain electronic equilibrium with a platinum electrode only slowly. In consequence there are difficulties in measuring their electrode potentials with accuracy and this, in some measure, accounts for the conflicting results recorded in the literature for many such systems. The problem of ionization must also be taken into account, especially in the case of cysteine and cystine where amino and carboxyl groups may be ionized in addition to the thiol group. The resulting equation for the electrode potential may therefore be quite complicated.

If we consider a system of the type

$$2RSH \rightleftharpoons RSSR + 2H^+ + 2e$$

Then

$$E_h = E_o + \frac{RT}{2F} \ln \frac{[RSSR][H^+]^2}{[RSH]^2}$$

$$= E_o + \frac{RT}{2F} \ln \frac{[RSSR]}{[RSH]^2} + \frac{RT}{F} \ln [H^+]$$

Where, however, the possibility exists of various ionized forms, e.g. ^+RSH, $^\pm RSH$, ^-RSH, $^-RS^-$, as for cysteine, and $^+RSSR^+$, $^\pm RSSR^+$, $^\pm RSSR^\pm$, $^-RSSR^\pm$, and $^-RSSR^-$ as for cystine, an appropriate reaction must be selected for the electrode equation. If, say, the reaction

$$2^+RSH \rightleftharpoons {}^+RSSR^+ + 2H^+ + 2e$$

is chosen, then

$$E_h = E_o + \frac{RT}{2F} \ln \frac{[^+RSSR^+][H^+]^2}{[^+RSH]^2}$$

By analogy with the case discussed in the previous section (p. 344), the total stoicheiometric concentrations of the reduced forms [red] will be given by

$$[red] = [^+RSH] + [^\pm RSH] + [^-RSH] + [^-RS^-]$$

and the oxidized by

$$[ox] = [^+RSSR^+] + [^\pm RSSR^+] + [^\pm RSSR^\pm] + [^-RSSR^\pm] +$$
$$[^-RSSR^-]$$

The resulting electrode equation (compare with equation 10.9) will therefore have a complicated third term involving the dissociation constants of all the ionizing groups and $[H^+]$. If, however, the pH is constant the expression reduces simply to

$$E_h = E_o'' + \frac{RT}{2F} \ln \frac{[ox]}{[red]^2} \qquad . \qquad . \qquad (10.14)$$

where E_o'', the potential at the midpoint of the titration curve, obviously does not have the significance it has in other cases. For the characterization of such systems the condition $[ox] = [red]^2$ is usually adopted, when equation 10.14 becomes $E_h = E_o''$.

A number of enzymes are known to possess thiol groups which must be maintained in the reduced state for enzymic activity to be manifest. It is common practice in these instances to add reagents which will achieve this objective whenever purification procedures and kinetic experiments are carried out. Compounds such as 2-mercaptoethanol and glutathione have been widely used in this way. Thioglycollate is often added to the media used for the growth of anaerobic bacteria to secure the low redox potential essential for their proliferation ; the reaction involved is

$$2 \begin{array}{c} CH_2SH \\ | \\ COO^- \end{array} \rightleftharpoons \begin{array}{c} CH_2S-SCH_2 \\ | \qquad\qquad | \\ COO^- \qquad COO^- \end{array} + 2H^+ + 2e$$

which has E_o' (pH 7) of -0.34 volt.

Cleland (1964) introduced the use of dithiothreitol as a reagent for maintaining monothiol compounds completely in the reduced state, and for the quantitative reduction of disulphides. The compound has a low redox potential ($E_0' = -0.33$ volt at pH 7·0), is highly water-soluble and has the desirable characteristic that, unlike other thiol compounds, it is relatively free of odour.   The reaction of dithiothreitol with a disulphide is as follows :

$$R\text{--}S\text{--}S\text{--}R + HS.CH_2(CHOH)_2CH_2SH \longrightarrow$$
$$\text{Dithiothreitol}$$

$$R\text{--}SH + R\text{--}S\text{--}SCH_2(CHOH)_2CH_2SH$$

$$R\text{--}S\text{--}SCH_2(CHOH)_2CH_2SH \longrightarrow$$

$$
\begin{array}{c}
S\text{---}CH_2 \\
\diagup \qquad \diagdown \\
S \qquad\qquad CHOH + RSH \\
\diagdown \qquad \diagup \\
CH_2\text{---}CHOH
\end{array}
$$

The cyclization step involving the formation of the dithiane ring occurs more readily than the formation of the dithiolane ring of lipoamide.   Dithiothreitol has the advantage over other thiol reagents in that the second stage of the reaction with a disulphide is intramolecular, so that one molecule of the reagent per molecule of disulphide suffices.

The redox potential of the dithiothreitol (DTT) system was measured by equilibrating the DTT-oxidized DTT system with the $NAD^+ - NADH$ system in the presence of lipoamide and dihydrolipoate dehydrogenase and measuring the amount of NADH at equilibrium at 340 nm (making suitable corrections for the extinction of lipoamide and DTT_{ox} at this wavelength).   The equilibrium constant for the reaction of DTT with NAD^+ to give DTT_{ox} and NADH, i.e.

$$K_{app} = \frac{[DTT_{ox}][NADH]}{[DTT][NAD^+]} \simeq 2\cdot5 \text{ at pH } 7\cdot0, \text{ and } 35 \text{ at pH } 8\cdot1$$

Assuming the redox potential of NAD^+ to be -0.330 volt at pH 7·0, then the redox potential of dithiothreitol is -0.332 volt at pH 7·0 and -0.336 volt at pH 8·1.

Table 10.2 records the standard electrode potentials for some thiol systems.

TABLE 10.2

Electrode Potentials of some Thiol Systems

System	Temperature °C	E_o', pH 7·0 volt
Lipoic acid	23	−0·286
3-Thiolactic acid	30	−0·32
Cysteine	25	−0·33
Dithiothreitol	25	−0·33
Thioglycollic acid	25	−0·34
Glutathione	25	−0·35
Thiourea	30	−0·48

E_o' is calculated for the conditions that [ox] = [red]^2, where the appropriate reaction is

$$2RSH \rightleftharpoons RSSR + 2H^+ + 2e$$

The rH Scale

An alternative method of expressing an oxidation-reduction reaction, and one which has obvious advantages when considering biochemical redox systems, is

$$AH_2 \rightleftharpoons A + H_2.$$

This enables a hydrogen scale of oxidation-reductions to be used since the oxidized and reduced forms are related to hydrogen by the above equation. rH is defined as

$$rH = -\log (H_2) \qquad . \qquad . \qquad . \quad (10.15)$$

where (H_2) is the hydrogen pressure in atmospheres (compare with pH = $-\log (H^+)$). The value of rH is obtained from the general electrode equation (10.1a) and the electrode reaction of the hydrogen half-cell:

$$\tfrac{1}{2}H_2 \rightleftharpoons H^+ + e.$$

Thus
$$E_h = E^\circ + \frac{RT}{F} \ln \frac{(H^+)}{\sqrt{(H_2)}} \qquad . \qquad . \quad (10.16)$$

But for the hydrogen electrode $E_h = 0$ when (H^+) and (H_2) are both unity, hence E° must also be zero. Therefore

$$E_h = 2·303 \frac{RT}{F} \log \frac{(H^+)}{\sqrt{(H_2)}} \qquad . \qquad . \quad (10.16a)$$

and since pH $= -\log (H^+)$ and rH $= -\log (H_2)$

$$E_h = 2 \cdot 303 \frac{RT}{F} \left(\frac{rH}{2} - pH \right)$$

or

$$E_h = 2 \cdot 303 \frac{RT}{2F} (rH - 2pH). \qquad . \qquad (10.17)$$

At 30° this becomes

$$E_h = 0 \cdot 03(rH - 2pH) \qquad . \qquad . \qquad (10.17a)$$

When the pressure of hydrogen gas is 1 atmosphere, rH $= 0$ and

$$E_h = -0 \cdot 06pH \qquad . \qquad . \qquad . \qquad (10.18)$$

that is the potential varies only when there is an alteration in pH, and therefore at pH 7 $E_h = -0 \cdot 42$ volt. This is the basis for the use of the hydrogen electrode in the determination of pH values.

Dixon (1949) has advocated the use of the rH scale rather than the scale of electrode potentials since it confers certain advantages. He has developed a series of useful rules which has no counterpart on any other scale. The rH of hydrogen gas at 1 atmosphere pressure is zero and this is taken as the datum line for the rH scale. The free energy of hydrogen at this pressure is also zero ; this follows from the relationship $\Delta G = -nFE_h$, since E_h is zero for the standard hydrogen electrode. The scale extends from the hydrogen datum line to oxygen at 1 atmosphere pressure, which is at rH $= 41$. All other O-R systems are placed at such positions on the scale that their distance is proportional to the free energy change of the O-R reaction. Most biological systems lie within the range rH 0-25. For convenience, a calorie scale is used in conjunction with the rH one ; rH is related to the free energy by the equation

$$\Delta G = -2 \cdot 303RT \times rH \text{ calories} . \qquad . \qquad (10.19)$$

At 30° this becomes

$$\Delta G = -1380 \text{ rH calories} \qquad . \qquad . \qquad (10.19a)$$

Here the gas constant R has the value of $1 \cdot 987$ calories per degree per mole. Note particularly that equation 10.19 applies to systems involving two electrons, since they are being balanced against $H_2 \rightleftharpoons 2H^+ + 2e$ and $rH = -\log(H_2)$. Where $n = 1$, however, the system is being balanced against $\frac{1}{2}H_2 \rightleftharpoons H^+ + e$ and the free energy is related to the work done in expanding a *half* mole of hydrogen from 1 atmosphere to the required pressure. Hence for cases where $n = 1$

$$\varDelta G = -2 \cdot 303 RT \times \frac{rH}{2} \qquad \qquad .(10.19b)$$

The free energy change for the reaction between any two O-R systems is given by

$$\varDelta G = -2 \cdot 303 RT \varDelta rH . \qquad \qquad .(10.19c)$$

where $\varDelta rH$ is the difference between the rH values of the two systems. This applies whatever the percentage reduction of the systems. At equilibrium $\varDelta G$ is, of course, zero. The standard free energy change of the reaction, $\varDelta G°$, is a property of the reaction itself and, as seen in Chapter III, is defined as the free energy change when all the reactants are at unit activity. If we assume the activity coefficients of all reactants to be unity (an assumption often made with biological systems but by no means always justified), then $\varDelta G° = \varDelta G$ when the concentrations are 1 M. Under these conditions (ox) = (red) for each system and this corresponds to the midpoint of the rH versus percentage reduction curves, designated by rH° (compare with $E°$). Hence :

$$\varDelta G° = -2 \cdot 303 \ RT \varDelta rH° \qquad \qquad .(10.19d)$$

and comparing with equation 3.9 it will be seen that

$$\log K = \varDelta rH° \qquad \qquad . \qquad . \qquad . (10.20)$$

where K is the equilibrium constant of the reversible O-R reaction. This affords an extremely simple method of obtaining the equilibrium constant of the reaction between two O-R systems and is very useful in the study of linked dehydrogenase systems.

Applying the van't Hoff Isochore (equation 4.8), to rH° values, we have

$$\varDelta H = 2 \cdot 303 \ RT^2 \frac{d(rH°)}{dT} \qquad \qquad . \qquad . (10.21)$$

$$= 418100 \frac{d(rH°)}{dT} \text{ calories at } 30°,$$

so that the heat of reaction, ΔH, can be determined provided the rate of change of rH° with temperature is known.

As previously discussed in connexion with electrode potentials, the rH expressions become more complex if ionization and pH changes occur. The reader is referred to Dixon's book for full treatment of these cases. Dixon denotes the half-reduced ionizing system by the symbol r′H. r′H is related to E_0' by the equation

$$r'H = \frac{2E_o'F}{2 \cdot 303RT} + 2pH \qquad . \qquad . \quad (10.22)$$

and rH° is related to $E°$ by

$$rH° = \frac{2E°F}{2 \cdot 303RT} \cdot \qquad . \qquad . \qquad . \quad (10.23)$$

which are derived from equation 10.17 and the fact that $E°$ refers to pH = 0.

Example 10.4.—Determine the electrode potential of the hydrogen electrode at 30° in equilibrium with a partial pressure of 10^{-5} atmosphere of hydrogen at (*a*) pH 6·0 and (*b*) pH 7·5.

Using equation 10.17, we have :

$$E_h = \frac{RT}{2F} \times 2 \cdot 303(rH - 2pH)$$

$$rH = -\log (H_2) = -\log 10^{-5} = 5.$$

Therefore at pH 6·0 and 30°

$$E_h = \frac{8 \cdot 314 \times 303 \times 2 \cdot 303}{2 \times 96494} (5 - 2 \times 6)$$

$$= -0 \cdot 03 \times 7$$

$$= \underline{-0 \cdot 21 \text{ volt.}}$$

At pH 7·5, $\qquad E_h = 0 \cdot 03(5 - 2 \times 7 \cdot 5)$

$$= \underline{-0 \cdot 30 \text{ volt.}}$$

Example 10.5.—The lactate dehydrogenase system of muscle requires NAD as coenzyme and has an r′H value of 8 at pH 7·5. The reduced coenzyme is oxidized by flavoprotein and the $NADH_2$-NAD system has r′H = 4 at the same pH value. All measurements are at 30°. Determine the free energy change on interaction of the two systems if all reactants are at the same initial concentration.

$$\Delta r'H = 8 - 4 = 4$$

and $\qquad \Delta G = -2 \cdot 303 \, RT\Delta r'H$

$$= -1380 \times 4$$

$$= \underline{-5520 \text{ calories.}}$$

Example 10.6.—The E'_0 values for the malate-oxaloacetate and methylene blue-leucomethylene blue systems are -0.102 and $+0.011$ volt respectively at pH 7 and 30°. Determine the r'H values for these systems and the free energy change and equilibrium constant of the reaction between them. Assume that all reactants are present at the same initial concentration.

$$r'H = \frac{2E'F}{2.303RT} + 2pH.$$

For the malate-oxaloacetate system

$$r'H = -\frac{2 \times 0.102 \times 96494}{2.303 \times 8.314 \times 303} + 14$$

$$= \frac{-0.102}{0.03} + 14 = -3.4 + 14$$

$$\underline{r'H = +10.6.}$$

For the MB-MBH₂ system

$$r'H = \frac{0.011}{0.03} + 14 = 0.36 + 14$$

$$\underline{r'H = +14.36.}$$

The free energy change is given by equation 10.19c.

$$\Delta G = -2.303RT\Delta r'H$$

$$= -2.303 \times 1.987 \times 303 \times (14.36 - 10.6)$$

$$= -1380 \times 3.76$$

$$\underline{\Delta G = -5189 \text{ calories.}}$$

The equilibrium constant of the reaction is obtained from equation 10.20 when, as already seen,

$$\log K = \Delta r'H = 3.76$$

and

$$K = \frac{(\text{oxaloacetate})(\text{MBH}_2)}{(\text{malate})(\text{MB})} = \underline{5754.}$$

This question specifically asked for the calculation of r'H values, but if these had not been required, ΔG for the reaction could have been obtained in an alternative manner by use of equation 10.3a.

$$\Delta G = -nF\Delta E.$$

In this case $n = 2$ and $\Delta E = 0.011 - (-0.102) = +0.113$ volt.

Therefore $\Delta G = -2 \times 96494 \times 0.113$ joules

$$= \frac{-2 \times 96494 \times 0.113}{4.184} \text{ calories}$$

$$= \underline{-5212 \text{ calories.}}$$

The agreement between this and the previously derived value is fairly reasonable.

Oxidative Phosphorylation

Aerobic organisms obtain their energy by the oxidation of suitable substrates. Protons and electrons, which are removed from substrates by the appropriate dehydrogenases, combine with oxygen to form water. Electrons are transported by a system of carriers and at appropriate points along the electron transport chain synthesis of ATP from inorganic orthophosphate (P_1) and ADP occurs in the process known as *respiratory chain phosphorylation*. This distinguishes the process from substrate phosphorylation and photophosphorylation which are associated with anaerobic (though oxidative) and photosynthetic energy-yielding processes respectively.

The major energy-yielding reactions in aerobic organisms are those concerned with fatty acid oxidation and with oxidations via the tricarboxylic acid cycle; these are catalysed by multi-enzyme systems located in the mitochondria. By measurements of the oxygen consumed and the inorganic orthophosphate converted to organic phosphate, evidence has been obtained that 3 moles of ATP are synthesized for each atom of oxygen consumed, i.e. a P/O ratio of 3. (Other work suggests that, in some cases, this ratio may be higher; the reader is referred to the review by Griffiths (1965) for further details.) Spectrophotometric techniques, allied to the judicious use of inhibitors, have indicated the three regions of the electron transport chain where energy coupling takes place, as shown in Fig. 10.5. At each of these three sites ATP synthesis occurs and the process may be formulated thus:

1. $AH_2 + B + I \rightleftharpoons A \sim I + BH_2$
2. $A \sim I + B \rightleftharpoons A + X \sim I$
3. $X \sim I + P_1 \rightleftharpoons X \sim P + I$
4. $X \sim P + ADP \rightleftharpoons X + ATP$

where AH_2 and B are electron transport carriers of the respiratory chain, such as $NADH_2$ and FAD, and X and I are unidentified intermediates which permit oxidation to be linked to phosphorylation. In this formulation the oxidative reaction precedes the phosphorylation step but other models have been proposed in which phosphate is incorporated prior to the oxidative reaction.

Recently evidence has been presented which suggests that there are two species of cytochrome b present in mitochondria with very different standard electrode potentials (E'_o, pH 7·0, of +0·245v and +0·020v respectively). It has been proposed that energy conservation in the cytochrome b region of the electron transport chain is associated with the reduction of the high potential cytochrome b by the low potential species.

Mitchell (1966) has proposed a mechanism for oxidative phosphorylation which eliminates the requirement for an intermediate

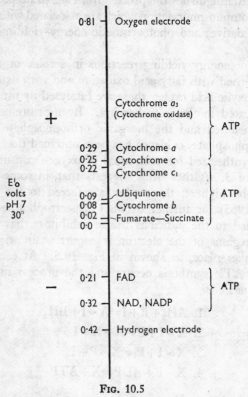

FIG. 10.5

The biological electron transport chain showing the three sites at which ATP synthesis is coupled to electron transport.

complex between one of the electron carriers and an intermediate of the phosphorylation sequence. Briefly, Mitchell suggests that electron transport results in the extrusion of protons to the outside of the mitochondrial membrane. This membrane is impermeable to protons, but within the membrane is located an ATPase system which catalyses the reaction :

$$2H^+_{inside} + ATP \rightleftharpoons ADP + POH + 2H^+_{outside}$$

Since this reaction is reversible the build up of protons on the outside of the membrane will result in a synthesis of ATP with a concomitant removal of protons to the interior of the mitochondrion.

Mitochondria and sub-mitochondrial preparations carry out exchange reactions between ADP-ATP, P_i-ATP and $H_2{}^{18}O$-ATP which are believed to be involved in oxidative phosphorylation. Thus the ADP-ATP exchange would be represented by reaction 4 in the above scheme, i.e.

$$X{\sim}P+ADP \rightleftharpoons X+ATP$$

and the P_i-ATP exchange as the sum of reactions 3 and 4

$$I{\sim}X+P_i \rightleftharpoons I+X{\sim}P$$
$$\underline{X{\sim}P+ADP \rightleftharpoons X+ATP}$$
$$P_i+ADP \rightleftharpoons ATP$$

Methods for the measurement of oxidative phosphorylation may be direct or indirect.

Direct Methods.—First, it is essential that the oxygen consumption be measured and this can be achieved either with the Warburg apparatus (p. 293) or by means of an oxygen electrode, which enables the oxygen concentration in solution to be measured. Second, it is necessary to measure the conversion of inorganic phosphate and ADP to ATP in accordance with the equation

$$ADP+P_i \rightleftharpoons ATP$$

This is possible by measuring the amount of inorganic orthophosphate utilized, which is equivalent to the ATP formed. It is common practice to include in the system a suitable trap for the ATP formed, as for example by the inclusion of glucose and hexokinase. In this system glucose is converted to glucose 6-phosphate:

$$\text{Glucose}+ATP \xrightarrow{\text{Hexokinase}} \text{Glucose 6-phosphate}+ADP$$

and the glucose 6-phosphate formed is thus a measure of the ATP synthesized. Glucose 6-phosphate is then assayed spectrophotometrically by measuring the reduction of $NADP^+$ in the glucose 6-phosphate dehydrogenase reaction

$$\text{Glucose 6-phosphate}+NADP^+ \rightleftharpoons \text{Gluconate 6-phosphate}$$
$$+NADPH+H^+.$$

Indirect Methods.—If the tissue contains phosphatases which hydrolyse the organic phosphate formed to an appreciable extent, or if the organic phosphate undergoes further metabolism with the release of inorganic phosphate, indirect assays must be used. These may take the form of measurements of the ability of an oxidation process to maintain an ATP-organic phosphate pool at a constant level, or the rate of incorporation of ^{32}P into ATP or other organic phosphates. Carefully controlled conditions are essential since the presence of phosphatases and of side-reactions, particularly in crude tissue preparations, make the quantitative evaluation difficult. Fluoride is usually added to preparations to inhibit phosphatases but the inhibition may be incomplete.

Example 10.7.—Sub-mitochondrial preparations oxidize D-β-hydroxybutyric acid quantitatively to acetoacetic acid with coupled production of ATP. The P/O ratio for such preparations can be determined by carrying out the reaction in the presence of [^{32}P]orthophosphate and a glucose-hexokinase trapping system. The acetoacetate formed can be determined colorimetrically by reaction with diazotized p-nitroaniline, and ATP by measuring the production of radioactive glucose 6-phosphate. Residual inorganic phosphate is removed by treatment with ammonium molybdate followed by extraction of the phosphomolybdate complex into a mixture of isobutanol and benzene. The labelled glucose 6-phosphate remains in the aqueous layer and can then be assayed.

The P/O ratio will be given by the ratio of the amount of labelled glucose 6-phosphate formed to that of acetoacetate produced, since the oxidation step is equivalent to the uptake of 1 atom of oxygen.

An experiment was carried out in which 2 m-moles of DL-β-hydroxybutyrate were incubated with 3 m-moles of potassium [^{32}P]orthophosphate (registering 500,000 counts/min.), 0·25 m-mole of ADP, 2·5 m-moles of glucose, hexokinase and histidine buffer, pH 6·5. 0·2 ml of submitochondrial particles, obtained by digitonin treatment, was added and the total volume of the reaction mixture was 1·0 ml.

After incubation for 20 minutes at 20°, with shaking in air, the reaction was terminated by the addition of 3·5 ml of 0·19 M-trichloroacetic acid. The precipitated protein was removed by centrifuging and the supernatant liquid used for analysis of glucose 6-[^{32}P] phosphate and acetoacetate.

A sample (0·5 ml) of the deproteinized reaction mixture was treated with ammonium molybdate and extracted with isobutanol-benzene reagent. The total volume of the aqueous phase was 4 ml and 1 ml was taken for counting. The sample registered 100 counts/min (corrected).

The dilution factors involved are $\dfrac{4 \times (1·0 + 3·5)}{0·5} = 36$. Hence the labelled

ATP formed is equivalent to 3600 counts/min. But the specific activity of the P_i in the reaction mixture was 500,000/3 counts/min/m-mole or 166·6 counts/min/μmole.

Therefore $\dfrac{3600}{166·6} = 21·6$ μmoles of ATP were formed.

Acetoacetate analysis revealed the total production of 8·8 μmoles.

Hence P/O ratio $= \dfrac{21·6}{8·8} = \underline{2·45}$

The Oxygen Electrode.—The Clark oxygen electrode in its original or in modified form has been fairly widely used for investigations of oxidative phosphorylation. Essentially it consists of a platinum disk electrode separated from a calomel half-cell by a very thin film of potassium chloride solution, the platinum electrode being 0·6—0·8 volt negative with respect to the calomel half-cell. The film of potassium chloride solution is separated from the reaction mixture by a thin polyethylene or teflon membrane, which is freely permeable to oxygen. The current which flows through this electrode is directly proportional to the oxygen activity. The electrode is attached to a recorder so that direct traces of the electrode current can be obtained. The oxygen electrode measures the activity and not the concentration of oxygen in solution. The following relation holds

$$\frac{\log f}{I} = K$$

where f is the activity coefficient, I is the ionic strength and K is a constant which, to a certain extent, depends on the nature of the electrolyte. Although the solubility of oxygen in water is known and that in the medium can be calculated, the effective ionic strength of a complex medium is not readily determined and, additionally, non-electrolytes present affect the solubility of oxygen in water. For these reasons the electrode is usually calibrated directly by measuring the oxidation of a known amount of NADH by a sub-mitochondrial particle preparation. (The NADH is determined accurately, either spectrophotometrically or by measuring the change in extinction at 340 nm when an enzyme such as alcohol dehydrogenase is added.) In this way the number of recorder scale divisions corresponding to the oxidation of a given quantity of NADH is found. As the reaction involved is

$$NADH + H^+ + \tfrac{1}{2}O_2 \longrightarrow NAD^+ + H_2O$$

1 μmole of NADH utilizes 0·5 μmole of oxygen and therefore the oxygen concentration corresponding to the NADH concentration added may be determined. A linear response is found between electrode current and the concentration of dissolved oxygen.

The course of such a calibration is shown in Fig. 10.6. The reaction medium is bubbled with water-saturated air at a given temperature, say 25°. A measured volume of the medium is then

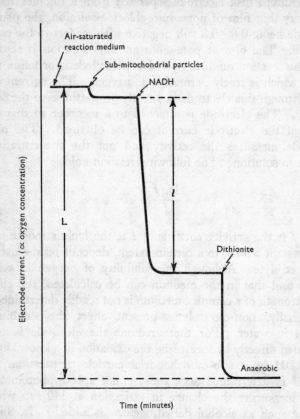

FIG. 10.6

Typical profile of an experiment to determine the concentration of oxygen in an air-saturated medium using an oxygen electrode. The oxidation of a known amount of NADH by sub-mitochondrial particles is used for the measurement, the oxidation of 1 μmole of NADH being equivalent to the utilization of 1 μatom of oxygen.

introduced into the electrode cell, taking care that air bubbles are absent. This gives the initial electrode reading corresponding to the air-saturated medium. Sub-mitochondrial particles (0·1 ml) are now added and this causes a slight fall in the trace if the particles are in an anaerobic state. Addition of NADH (say 1

μmole) causes an immediate oxygen consumption, as seen by the rapid and linear decrease of electrode current, until all the added NADH is oxidized. This decrease corresponds to l. Dithionite is next added to bring the oxygen concentration to zero. The difference between this last value and the initial reading for the air-saturated medium gives the current equivalent to the air-saturated oxygen concentration (L).

Since 1 μmole of NADH, equivalent to 0·5 μmole of oxygen, corresponds to an electrode current l, then in the total volume of the electrode cell contents (v ml) there will be

$$\frac{L \times 0.5 \times 10^3}{l \times v} \; \mu\text{moles/litre}$$

In a typical experiment of this sort values of about 240 μM are obtained for the oxygen concentration of the air-saturated medium.

Example 10.8.—The P/O ratio for rat-heart mitochondria oxidizing α-oxoglutarate may be determined with the oxygen electrode in the following manner.

The reaction mixture containing 10 mM inorganic orthophosphate, 0·25 M-sucrose, 5 mM-MgCl$_2$ and 10 mM-α-oxoglutarate, at pH 7·4, is bubbled with water-saturated air and then 4 ml introduced into the electrode chamber. A mitochondrial preparation, suspended in 0·25 mM-sucrose-10 mM-tris buffer, pH 7·4, is then added (0·1 ml, containing 1-2 mg protein). Fig. 10.7a shows the fall in electrode current due to the addition of the anaerobic mitochondria.

ADP is now added (25 μl of a 40 mM solution, i.e. 1 μmole) and the trace of the electrode current falls rapidly. As the mitochondrial preparation is tightly coupled and there is an excess of substrate present, exhaustion of the added ADP causes respiration to cease. If more ADP were added at this stage then respiration would resume and continue until either the ADP, substrate or oxygen in the medium was exhausted. However, as shown in Fig. 10.7a, in this experiment the uncoupling agent 2,4-dinitrophenol is added and this permits oxidation of α-oxoglutarate to continue until the oxygen in the medium is exhausted, i.e. anaerobic conditions prevail. This latter value enables the total oxygen concentration of the air-saturated medium to be computed since it is proportional to electrode current L.

The respiration rates of the mitochondria are obtained from the slopes of the trace and are measured in scale units per minute. Then, assuming the oxygen concentration of the air-saturated medium to be 237 μM,

$$\text{Respiration rate} = \frac{\text{Slope of trace}}{L} \times \frac{237}{10^3} \times v \; \mu\text{moles O}_2/\text{min}$$

where v is the volume of the total reaction mixture in ml. If x mg of protein were added in the mitochondrial preparation then the rate could be expressed as μmoles O$_2$/min/x mg protein. If desired this could be given as a Q$_{0_2}$ value, i.e. μl O$_2$/h/mg protein.

The amount of oxygen consumed is obtained from the height l of the electrode

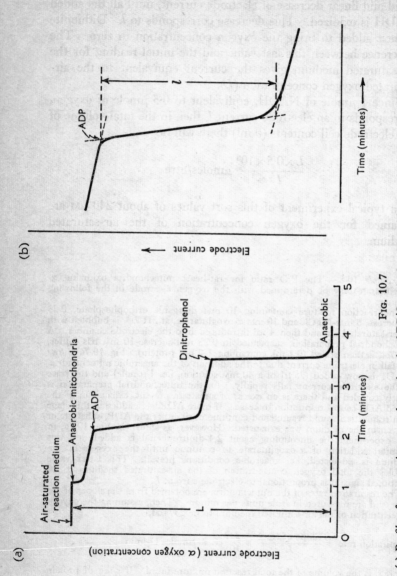

FIG. 10.7

(a) Profile of an experiment to determine the ADP/O ratio for the oxidation of α-oxoglutaric acid by a mitochondria preparation using the oxygen electrode.

(b) Enlargement of trace to show method of determining the total oxygen consumed following the addition of a known amount of ADP. The decrease in electrode current l is proportional to the amount of oxygen utilized.

current curve following the addition of ADP; this is shown in enlarged form in Fig. 10.7b. Thus the amount of oxygen utilized

$$= \frac{l}{L} \times \frac{237}{10^3} \times v \times 2 \,\mu\text{atoms.}$$

Since the amount of ADP added is known, the ADP/O ratio may be calculated.

Fermentation Balances

The anaerobic fermentations of sugars and related compounds carried out by various micro-organisms result in the formation of a diversity of products. However, an essential feature of all these anaerobic processes is the necessity for the oxidation-reduction balance to be maintained, so that if some oxidized products are formed they must be accompanied by correspondingly reduced ones.

Careful analysis of the fermentation products enables the stoicheiometry of the fermentation to be ascertained and, unless assimilation of substrate carbon by the cells has occurred, the total carbon content of the products should equal that of the added substrate carbon. The ratio of recovered carbon to that added is referred to as the *carbon balance*, and is a measure of the carbon recovery in the fermentation. The carbon balance and the oxidation-reduction balance are always determined when accurate quantitative analyses of fermentations are carried out.

Carbon Balance.—It is customary to express the results of an analysis as millimoles of product per 100 millimoles of substrate fermented. These values in millimoles are then multiplied by the number of carbon atoms in the respective molecules to obtain millimoles of substrate or product carbon i.e. expressed on the basis of C_1 units. The resulting values for products are then totalled and compared with the substrate value, which they should equal unless assimilation by the cells has occurred. In this event, determination of the amount of carbon assimilated should complete the balance.

Example 10.9.—Baker's yeast, *Saccharomyces cerevisiae*, was allowed to ferment 25 m-moles of glucose to completion. The ethanol and carbon dioxide yields were 49·2 and 49·6 m-moles respectively. Determine the carbon recovery.

100 m-moles of glucose would yield 196·8 m-moles of ethanol and 198·4 m-moles of carbon dioxide.

Product	Yield of product	
	m-moles per 100 m-moles of glucose	m-moles of carbon
Ethanol	196·8	393·6
CO_2	198·5	198·4
Total	—	592·0

As 100 m-moles of glucose are equivalent to 600 m-moles of carbon, the carbon balance is 592/600 and the carbon recovery is therefore $\frac{592}{600} \times 100$

$$= 98\cdot7 \text{ per cent.}$$

Oxidation-Reduction Balance.—The oxidation-reduction (O/R) balance is the ratio of the number of equivalents of oxidized products to those of reduced products which have been formed during the fermentation. Under anaerobic conditions it is clear that this ratio should have a value of unity.

The O/R balance is obtained in the following manner. The amount of each product in millimoles is multiplied by a factor which relates its degree of oxidation or reduction to that of glucose, taken as zero. If the compound is more oxidized than glucose the value is conventionally regarded as positive, and if more reduced, as negative. The sum of the positive values is divided by the sum of the negative values to obtain the O/R index for the fermentation.

The appropriate factors are derived by reference to the molecular formula and by comparing the ratio of hydrogen to oxygen in the compound with the corresponding ratio for glucose. In glucose the ratio of hydrogen to oxygen is 2, as in water, and the factor for this ratio is taken as zero. Consequently any compound possessing this same ratio of hydrogen to oxygen, e.g. acetic acid, $C_2H_4O_2$, and lactic acid, $C_3H_6O_3$, does not contribute to the O/R balance since the factor of zero is involved.

In the case of ethanol, C_2H_6O, there is an excess of 4H atoms over the 2H:1O ratio and hence the compound is more reduced than glucose. The appropriate factor is therefore -4. Carbon dioxide, CO_2, with a deficiency of 4H, is more oxidized than glucose and the factor is accordingly $+4$. Formic acid, CH_2O_2, requires two H atoms to achieve the appropriate ratio and the factor is therefore $+2$.

If fermentations are allowed to proceed under aerobic conditions O/R balances cannot be calculated unless the amount of oxygen utilized is known. This can be determined by carrying out the fermentation in a Warburg manometer and measuring the gas exchange. Each mole of oxygen utilized is equivalent to two moles of hydrogen produced, so the factor is -4. The O/R balance value is usually slightly higher than unity because the cells, which are more reduced than glucose, are not accounted for in the calculation.

Example 10.10.—The following data were obtained in an experiment to determine a fermentation balance for *Escherichia coli* in the anaerobic dissimilation of glucose at pH 6·2. Cells were grown on a medium containing 5 per cent. glucose and the pH was automatically controlled.

(Data from BLACKWOOD, NEISH & LEDINGHAM (1956), *J. Bact.* 72, 497.)

Product	Yield m-moles per 100 m-moles glucose fermented	Product carbon m-moles	O/R factor	Milli-equivalents of	
				Oxidized (+)	Reduced (−)
Acetic acid	36·5	73·0	0	—	—
Acetoin	0·059	0·236	−4	—	0·236
2,3-Butanediol	0·30	1·2	−6	—	1·8
Carbon dioxide	88·0	88·0	+4	352·0	—
Ethanol	49·8	99·6	−4	—	199·2
Formic acid	2·43	2·43	+2	4·86	—
Glycerol	1·42	4·26	−2	—	2·84
Hydrogen	75·0	—	−2	—	150·0
Lactic acid	79·5	238·5	0	—	—
Succinic acid	10·7	42·8	+2	21·4	—
Total		550·026		378·26	354·076

$$\text{Carbon balance} = \frac{550}{600} \text{ and carbon recovery} = \underline{91 \cdot 7\%}$$

$$\text{Oxidation-reduction balance} = \frac{378 \cdot 3}{354 \cdot 1} = \underline{1 \cdot 07}$$

REFERENCES AND SUGGESTED READING

CLARK, W. M. (1960). *Oxidation-Reduction Potentials of Organic Systems.* Baltimore: Williams & Wilkins.

DIXON, M. (1949). *Multi-Enzyme Systems.* London: Cambridge University Press.

FRIEDHEIM, E. A. H. & MICHAELIS, L. (1931). Potentiometric study of pyocyanine. *J. biol. Chem.,* **91**, 355.

GRIFFITHS, D. E. (1965). *Essays in Biochemistry*, **1**, 91. London : Academic Press.

HEWITT, L. F. (1950). *Oxidation-Reduction Potentials in Bacteriology and Biochemistry*, 6th ed. Edinburgh : Livingstone.

MICHAELIS, L. (1951). *The Enzymes*, II, Part 1, 1, ed. J. B. Summer & K. Myrbäck. New York : Academic Press.

MITCHELL, P. (1966). Chemiosmotic coupling in oxidative and photosynthetic phosphorylation. *Biol. Rev.*, **41**, 445.

NEISH, A. C. (1952). *Analytical Methods for Bacterial Fermentations.* Saskatoon : National Research Council of Canada, Report No. 46-8-3.

PURDY, W. C. (1965). *Electroanalytical Methods in Biochemistry.* New York : McGraw-Hill.

WURMSER, R. & BANERJEE, R. (1964). Oxidation-Reduction Potentials. In *Comprehensive Biochemistry*, **12**, 62, ed. M. Florkin & E. H. Stotz. Amsterdam : Elsevier.

YAMAZAKA, I. (1970). One-electron and two-electron transfer processes in enzymic oxidation and reduction. *Advances in Biophysics* 2. Baltimore : University Park Press.

PROBLEMS

10.1. Determine the percentage of the reduced form of cresyl violet at pH 7·0 when the oxidation-reduction potential is (*a*) $-0·128$ volt and (*b*) $-0·179$ volt. E_0' for cresyl violet at pH 7·0 is $-0·166$ volt and n is 2. Assume a temperature of 30° throughout.

10.2. What is the ratio of reduced to oxidized forms of the pigment pyocyanine if the measured oxidation-reduction potential of the system is $-0·025$ volt ? E_0' for the reaction is $-0·034$ volt at the same temperature (25°) and pH. ($n = 2$ for pyocyanine.)

10.3. The lactate dehydrogenase system, catalysing the oxidation of lactate to pyruvate, has an E_0' value of $-0·180$ volt at 35° and pH 7·01. What will be the potential of this system when the oxidation has gone to 95 per cent. completion ?

10.4. The dye 2,6-dichlorophenolindophenol is used for the quantitative estimation of ascorbic acid. The respective E_0' values at pH 7·0 and 30° are 0·217 and 0·060 volt. What is the potential of the dye when it has been 99·8 per cent. reduced by ascorbic acid ?

If equal concentrations of ascorbic and dehydroascorbic acids are reacted with the same concentration of oxidized and reduced forms of the dye, what will be the final oxidation-reduction potential of the system ?

10.5. Oxidation-reduction potentials and related measurements can afford valuable information for the biochemist interested in intermediary metabolism. Discuss the validity of this statement.

Measurements with the formate hydrogenlyase system of *Escherichia coli* revealed the standard electrode potential to be $-0·420$ volt at pH 7·0 and 30° C. What is the oxidation-reduction potential of this system at 30° C under conditions where the equilibrium constant for the reaction is 2·25 moles/litre ?

$$(2·303 \ RT/F = 0·060 \text{ at } 30° \text{ C})$$

(Hull Honours Course Finals, Part II, 1967.)

10.6. A Thunberg experiment is carried out to follow the activity of a lactate dehydrogenase preparation. In the tube is placed 3 ml of a solution containing lactate and pyruvate (both at 0·002 M concentration) and the enzyme, while the side arm contains 1 ml of a mixture of methylene blue and leucomethylene blue in equal concentration at 0·006 M. Determine the percentage oxidation of lactate after mixing the solutions and allowing to attain equilibrium. The experiment is conducted at 35° and pH 7·0, under which conditions E'_o for the lactate-pyruvate system is −0·180 volt and E'_o for the methylene blue-leucomethylene blue system is 0·011 volt.

10.7. The following E'_o values were obtained for ascorbic acid at different pH values and at 30° :

pH	1·05	2·16	3·04	4·00	5·19	6·32	7·24	8·57
E'_o (volt)	+0·326	+0·260	+0·209	+0·154	+0·115	+0·078	+0·051	+0·012

From these figures deduce graphically the effect of pH on the course of the oxidation of ascorbic acid. Obtain a value for the standard electrode potential, $E°$, of ascorbic acid.

(After BALL (1937), *J. biol. Chem.*, 118, 219.)

10.8. The following standard electrode potentials have been determined at 25°.

System :	Fe − Fe²⁺	Fe²⁺ − Fe³⁺	Fe(CN)₆⁴′ − Fe(CN)₆³′
$E°$ (volt) :	−0·44	+0·77	+0·36

Determine the rH values and the standard free energy changes of these systems.

10.9. Determine the percentage oxidation of thionine if the electrode potential measured at 30° and pH 7·0 is 0·10 volt. E'_o for thionine is 0·063 volt under the same conditions. What is the r'H value for thionine at pH 7·0 and 30° ?

10.10. The enzymes lactate dehydrogenase and alcohol dehydrogenase are coenzyme-linked by nicotinamide-adenine dinucleotide (NAD). At pH 7·5 and 35°

$$lactate^- \rightleftharpoons pyruvate^- \quad r'H = 8·0$$
$$ethanol \rightleftharpoons acetaldehyde \quad r'H = 6·8$$

If equal concentrations of all reactants are mixed in the presence of NAD and the two enzymes, determine the free energy change of the reaction.

What percentage of lactate will remain at equilibrium ?

10.11. β-Hydroxybutyric acid, which is formed during ketogenesis, is oxidized by the kidneys and muscles. The first step in its oxidation is acetoacetic acid. The hydrogen is transferred by means of NAD and flavoprotein to cytochrome *a*, which may be assumed to be maintained in the half-reduced state by the cell. If equal concentrations of β-hydroxybutyrate and acetoacetate are present initially, determine the free energy liberated by the oxidation of β-hydroxybutyrate and reduction of cytochrome *a*. E'_o for the β-hydroxybutyrate-acetoacetate system is −0·293 volt and for cytochrome *a* is 0·290 volt at pH 7·5 and 37°. Reduced cytochrome *a* reacts with atmospheric oxygen under the influence of the enzyme cytochrome oxidoreductase. Assuming the potential E'_o of atmospheric oxygen to be 0·80 volt, determine the energy liberated by the final reaction with oxygen.

10.12. The following data were obtained in an experiment in which 0·002 M ascorbic acid was titrated with potassium ferricyanide (0·04 M) at 30°. The acid was in 0·1 M-acetate buffer of pH 4·581 and a small quantity of thionine was added to act as a mediator for the reaction of ascorbic acid with the electrode. (This is necessary because otherwise ascorbic acid reacts very sluggishly with the electrode.)

Per cent. oxidation	Observed E_h volts
25·37	+0·1224
35·43	+0·1284
45·50	+0·1338
50·55	+0·1364
60·46	+0·1417
70·79	+0·1478
80·75	+0·1552
90·79	+0·1670
95·80	+0·1776

Determine a value for E'_o at pH 4·581 and also the number of electrons involved in the oxidation process at this pH value.

(After BALL (1937), *J. biol. Chem.*, 118, 219.)

10.13. The overall reaction of biological oxidations may be represented as

$$H_2 + \tfrac{1}{2}O_2 \rightleftharpoons H_2O$$

and the standard free energy of this reaction is $-56,560$ calories. Use this information to deduce the oxidation-reduction potential of the oxygen-activating system catalysing this final reaction at pH 7·0 and 30°. Assume the pressure of atmospheric oxygen to be 0·2 atmosphere.

Note.—In aqueous solutions the concentration of H_2O may be taken as unity.

10.14. The oxidation-reduction potentials of the haemoglobin-methaemo-globin system have been investigated at various pH values at 30°. The following E'_o values were obtained in such a study :

pH	5·08	5·46	5·85	6·04	6·06	6·12
E'_o volt	0·1673	0·1657	0·1679	0·1681	0·1671	0·1661

pH	6·22	6·36	6·76	7·48	7·56	7·63
E'_o volt	0·1654	0·1615	0·1507	0·1150	0·1126	0·1009

pH	8·50	8·64	8·72	8·75	9·18
E'_o volt	0·0554	0·0517	0·0444	0·0417	0·0202

From these data determine the pK value of the system and hence the apparent acidic dissociation constant. What can you deduce about the oxidation of haemoglobin from the E'_o-pH dependence curve ?

(After TAYLOR & HASTINGS (1939), *J. biol. Chem.*, 131, 649.)

10.15. Cytochrome *a* provides a one electron oxidation-reduction system with a standard potential, E'_o, at pH 7 of about +0·29 volt.

$E°$ for the oxygen reaction :

$$O_2(g) + 4H^+ + 4e \rightarrow 2H_2O$$

is +1·23 volts at 25° C.

Compute the electrode potential for this reaction at pH 7. Compute the theoretical pressure of oxygen required to maintain the cytochrome a system 99·9 per cent. in the oxidized form at 25° C.

(After BALL (1939), *Symposium Quant. Biol.*, **7**, 100. Harvard Medical Sciences 201 ab.)

10.16. Discuss the concept of rH and compare critically the merits and demerits of this scale with those of the oxidation-reduction potential scale, as applied to biological systems.

What is the electrode potential of the hydrogen electrode at 30° in equilibrium with a partial pressure of 10^{-6} atmosphere of hydrogen at pH 6·0 ?

($R = 8·314$ joules per mole per degree ; $F = 96,494$ coulombs.)

(Glasgow Honours Course Finals, 1956.)

10.17. Outline briefly the principles involved in the measurement of pH using the quinhydrone electrode.

At 18° C. the e.m.f. of a quinhydrone electrode dipping into a solution of 0·01 M-potassium tetroxalate, which has a pH of 2·15, is +0·330 v. with respect to a saturated calomel electrode. The e.m.f. becomes +0·212 v. when this solution is replaced by one containing 0·106 mole per litre of p-chloro-benzoic acid together with 0·0632 mole per litre of sodium p-chloro-benzoate.

Calculate the pH of the latter solution and hence the dissociation constant of p-chloro-benzoic acid. Mention any errors which may affect these results.

$$(2·303 \times RT/F = 0·0578 \text{ volt at } 18° \text{ C.})$$

(Special Degree of B.Sc., Biological Chemistry, University of Bristol, 1958.)

10.18. Explain the meaning of rH.

The systems lactate-pyruvate and ethanol-acetaldehyde are both enzyme catalysed and require NAD. The r′H values at the same temperature and pH are 8·5 and 7·0, respectively. If equimolar amounts of lactate, pyruvate, ethanol and acetaldehyde are mixed in the presence of their enzymes and NAD, calculate the percentage change in the concentration of the pyruvate.

(Leeds Honours Course Finals, 1961.)

10.19. The following data were obtained in an analysis of the products of the anaerobic dissimilation of glucose by *Serratia marcescens*. All values are given as m-moles of product formed per 100 m-moles of substrate glucose dissimilated.

Acetic acid	. . .	0·0	Acetoin . . .	0·81
2,3-Butanediol	. .	51·45	Carbon dioxide . .	106·1
Ethanol	. . .	42·24	Formic acid . . .	39·80
Glycerol	. . .	4·54	Hydrogen . . .	0·52
Lactic acid	. . .	33·09	Succinic acid . . .	3·41

Determine the carbon recovery and the oxidation-reduction balance for the fermentation. Suggest the metabolic pathways by which the products are derived from glucose.

(Data from NEISH (1952), *Analytical Methods for Bacterial Fermentations*. Report No. 46-8-3, National Research Council of Canada.)

10.20. The following analytical data were obtained for glucose fermentation by three different species of lactic acid bacteria. All values are expressed as millimoles of product per 100 millimoles of glucose fermented.

N

Product	Leuconostoc mesenteroides	Lactobacillus pentoaceticus	Lactobacillus plantarum
Lactic acid	102·0	90·6	197·2
Ethanol	112·0	61·2	0·0
Acetic acid	0·0	35·4	0·0
Carbon dioxide	96·0	86·1	0·0

Determine the carbon recoveries and the oxidation-reduction balance for each organism. Suggest the enzymic basis for the observed differences in fermentation patterns.

10.21. What is meant by (a) the carbon balance and (b) the oxidation-reduction balance of a fermentation ? What is the importance of such values to the biochemist ?

The following data were obtained in an experiment to determine a fermentation balance for *Serratia marcescens* in the anaerobic dissimilation of glucose.

Fermentation products (m-moles per 100 m-moles of glucose fermented)

Acetic acid	0·0	Formic acid	39·80
Acetoin	0·81	Glycerol	4·54
2,3-Butanediol	51·45	Hydrogen	0·52
Carbon dioxide	106·10	Lactic acid	33·09
Ethanol	42·24	Succinic acid	3·41

Calculate (a) the carbon balance (recovery) and
(b) the oxidation-reduction balance for the fermentation.

(Hull Special Biochemistry I Course, 1967.)

10.22. Distinguish between (a) substrate-level phosphorylation and (b) oxidative phosphorylation, and give examples of each.

What will be the ATP yield per mole when the following substrates are oxidized completely to CO_2 and water ? (Show clearly how you arrive at your conclusions.)

(a) Glucose
(b) Undecylic acid, $CH_3(CH_2)_9CO_2H$.

Compare the biologically useful energy yields per gram from these substrates, assuming that the free energy change for the reaction

$$ATP + H_2O \rightarrow ADP + H_3PO_4$$

is −8 kcal/mole.

(Hull Biochemistry Special I Course, 1967.)

CHAPTER XI

ISOTOPES IN BIOCHEMISTRY

SOME of the most spectacular advances in biochemistry have stemmed from the application of isotopes to the study of metabolic processes, a technique which gained impetus after the Second World War due to the rapidly increasing commercial availability of suitable isotopes. Compounds have been 'labelled' or 'tagged' by the incorporation of either stable or radioactive isotopes into their molecules and then the subsequent fate of the molecules studied by tracing the movements of the labelled atoms.

The major use of labelled compounds has been to follow metabolic pathways. The appearance of the labelling in some isolated product of a metabolic sequence and the determination of its position in the molecule by suitable degradative procedures has afforded evidence for the existence of certain pathways. The feeding of labelled compounds and their incorporation in the animal body has demonstrated that bodily components are not necessarily inert simply because their concentration remains constant. The classical researches of Schoenheimer and Rittenberg paved the way to the concept of dynamic equilibria in the body tissues. Isotopes have also been used to determine the rates of reactions in living organisms where, despite the constancy of concentration of the bodily components, continual metabolic activity, termed 'turnover', occurs and can be measured by certain isotopic techniques. They are particularly valuable too in the investigation of the kinetics of transport of molecules or ions across membranes.

Selected aspects of the theoretical background to isotopic tracer work will now be considered.

Stable Isotopes

The atomic number of an atom, i.e. the number of protons in the nucleus, determines its chemical properties. Almost all known elements exist as mixtures of atoms having two or more different atomic weights. Such atoms, having identical atomic number but different atomic mass, are termed isotopes of the element. Any sample of the element consists of a mixture of isotopes and the observed chemical atomic weight is therefore the average weight of the mixture of different atoms and depends upon the abundance of each isotope in nature. Thus chlorine has isotopes of mass 35 and 37 in such proportion that the determined atomic weight is almost 35·5, and carbon consists of isotopes of mass 12 and 13 in the proportion of 99·3 to 0·7.

Isotopes have almost identical chemical properties yet may often be separated by difference in physical properties such as rates of diffusion and evaporation. In the case of deuterium, the isotope of hydrogen with mass 2, the situation is very favourable because it has a mass 100 per cent. greater than hydrogen, and heavy water, D_2O, may actually be estimated by the difference in its density compared with ordinary water. If by a suitable enrichment process the proportion of one of the less abundant isotopes of an element can be increased, then a 'labelled' element is produced and compounds containing the element will similarly be labelled. Alteration of the natural abundance ratio is achieved by the same physical techniques that are used for the separation of isotopes, and of these fractional distillation and electrolysis are the main methods used.

The determination of the amounts of stable isotopes present in an element necessitates use of the mass spectrometer for all cases excepting deuterium. The mass spectrometer produces a beam of rays of positively charged gaseous ions of homogeneous energy. By the application of electrostatic and magnetic fields these ions are separated according to their mass, which determines the trajectory. An electrometer guarded by a slit constitutes the measuring device. Application of a potential causes the mass spectra produced to traverse the slit and each ion beam in turn enters the electrometer, where the ionization current is measured. Information is thereby obtained as to the amount of the element of a particular mass present. It is on this type of determination that the use of stable isotopes in biochemistry depends. The abundance of the labelling isotope as compared with the normal abundance, in the labelled starting material and in the product of the reaction, is measured.

The degree of labelling of a compound is always indicated by the *atom per cent. excess*. If an element normally contains y atoms of a particular, say heavy, isotope per 100 atoms of the element, then there will be $100 - y$ atoms of the other isotope or isotopes and the normal abundance is y atom per cent. Suppose that this abundance is increased to z atom per cent. in the process of labelling. The difference between the enriched and normal abundance, $z - y$, is termed the atom per cent. excess of the heavy isotope. Suppose that during the course of an experiment the labelled material is diluted n times with the element possessing the normal abundance ratio. For every 100 atoms of labelled element there will be a total of $100n$ atoms of the element. Of these, z atoms of the heavy isotope are derived from the enriched material and $y(n - 1)$ atoms of heavy isotope are derived from the diluting material. The resultant element therefore contains $z + y(n - 1)$ atoms of heavy isotope in a total of $100n$ atoms, i.e. $(z + y(n - 1))/n$ atom per cent., and the atom per cent. excess of the isotope is

$$\frac{z + y(n - 1)}{n} - y = \frac{z - y}{n}.$$

Since $z - y$ was the atom per cent. excess of the starting material, comparison of the atom per cent. excess of starting and final material gives a measure of the dilution factor n.

The use of stable isotopes has largely been superseded by radioactive ones, wherever there is a choice between the two, on account of the easier methods of measurement and their greater sensitivity. In the case of nitrogen and oxygen, no suitable radioactive isotopes exist and the stable isotopes ^{15}N and ^{18}O are used. Stable isotopes possess the advantage that they are permanent and do not disintegrate with time and furthermore produce no radiation effects on the tissues. Combination of both stable and radioactive isotopes permits the double labelling of a molecule, so that, for example, two different carbon atoms of a molecule may be labelled with stable ^{13}C and radioactive ^{14}C respectively.

Radioactive Isotopes

Radioactive isotopes for tracer studies may be prepared artificially from non-radioactive elements by means of either the cyclotron or the atomic pile. In this manner considerable quantities of radioactive isotopes have been made available for biochemical and medical research. Radioactive isotopes disintegrate spontaneously, giving rise to new elements and emitting radiation or particles or both.

The disintegration may occur initially in one of the following ways :

1. Alpha (α) particle emission : α particles are helium nuclei (^4_2He or He^{++}) which are ejected from the nucleus of the atom. They are emitted only by radionuclides of high mass number and, in consequence, are of relatively little biological interest.

2. Beta (β^-) particle emission : these particles are emitted as a result of the nuclear transformation of a neutron into a proton

$$^1_1\text{n} \rightarrow {}^1_0\text{p} + {}_{-1}^{\ 0}\text{e}$$

3. Positron (β^+) emission : these accompany the conversion of a proton to a neutron

$$^1_1\text{p} \rightarrow {}^1_0\text{n} + {}^0_1\text{e}$$

Positron emission is always associated with γ-ray photons which arise from the annihilation reaction

$$^0_1\text{e} + {}_{-1}^{\ 0}\text{e} \rightarrow 2h\nu$$

For example $^{11}_{6}\text{C}$, with a half-life of 20.5 min, is a positron emitter.

4. Electron capture (K-capture) : an orbital electron is captured by the nucleus resulting in a proton to neutron conversion. This transition is characterized by the emission of X-rays, e.g. ^{55}Fe has a very weak K-capture X-ray of 0·0065 MeV.

5. Isomeric transition : an unstable nucleus isomerizes without alteration in the proton/neutron complement of the nucleus. The emission of γ-rays accompanies the change.

In cases 1-4 there may be additionally the production of γ-rays where the energy of the emitted particle or X-ray does not account for the energy difference between the initial and final states. There is also the possibility that the emitted γ-rays may interact with the orbital electrons, resulting in the expulsion of homo-energetic electrons (e^-), a process known as *internal conversion*.

Apart from their greater energy, and hence shorter wavelength,

γ-rays do not differ fundamentally from X-rays. Both form part of the electromagnetic spectrum (p. 259) and possess strong penetrative powers ; their absorption by matter results in the secondary production of β^--particles and so enables their detection by gas-ionization devices primarily designed for the detection of β-particles.

β^--Particles carry a charge of one electron unit (4.803×10^{-10} absolute electrostatic units) and their energies are recorded in electron-volt (eV) units. An eV unit is defined as the kinetic

TABLE 11.1

Beta Emitters used in Biological Research

Type of β-emission	Element		E_{max} MeV	Other radiation emitted	Half-life
'Soft'	Iron	$^{55}_{26}$Fe	nil	K*	2·94 years
	Tritium	$^{3}_{1}$H	0·018		12·4 years
	Carbon	$^{14}_{6}$C	0·155		5568 years
	Sulphur	$^{35}_{16}$S	0·167		87·1 days
'Hard'	Iron	$^{59}_{26}$Fe	0·46, 0·27	γ	45·1 days
	Iodine	$^{131}_{53}$I	0·61	γ	8·1 days
	Potassium	$^{40}_{19}$K	1·36	γ	$1·3 \times 10^9$ years
	Sodium	$^{24}_{11}$Na	1·39	γ	15 hours
	Phosphorus	$^{32}_{15}$P	1·71		14·3 days
	Potassium	$^{42}_{19}$K	3·6, 2·0	γ	12·5 hours

* K-capture X-ray

energy gained by an electron when under the influence of a potential gradient of 1 volt. In practice, the MeV (10^6 eV) and KeV (10^3 eV) are usually employed. The β-particles emitted by radionuclides are not generally homo-energetic but occur with a continuous distribution of energies from zero to an upper maximum, E_{max}, characteristic of each isotope. The energy of the β-particle determines its penetrating power and so influences the choice of assay system. β-Particles are classified as 'soft' or 'hard' emission depending on their energies. Table 11.1 records isotopes which have been widely used as biological tracers, their E_{max} for β-emission, half-life and whether or not they additionally emit γ-radiation.

In the decay process the elements undergo transmutation and the new element product of the process may be radioactive or non-radioactive ; in the former case the product will undergo further decay until eventually a non-radioactive product is formed. For example, ^{14}C is converted to ^{14}N, ^{32}P to ^{32}S and ^{40}K to a mixture of ^{40}Ca and ^{40}A. The conversion has a negligible effect on reactions being followed by the tracers, because the radioactive isotopes form but a minute proportion of the number of non-radioactive atoms of the same element present in the labelled compound.

A radioactive isotope disintegrates at a rate which is a function only of the constitution of its nucleus and which cannot be altered in any way by chemical or physical means. As already seen in Chapter V, the radioactive disintegration process is a first order reaction and the decay constant (rate constant) k is given by the expression

$$k = \frac{2 \cdot 303}{t} \log \frac{n_0}{n} \qquad . \qquad . \qquad . \qquad (5.5)$$

where n_0 is the number of atoms of an element at time $t = 0$ and n the number after time t. The half-life period, $t_{\frac{1}{2}}$, or time required for the concentration of the decomposing element to reach half its original concentration, is expressed by equation 5.6

$$t_{\frac{1}{2}} = \frac{0 \cdot 693}{k} \qquad . \qquad . \qquad . \qquad (5.6)$$

Decay constants can be conveniently obtained by the use of semi-logarithmic graph paper, i.e. paper with one axis marked out proportionally to the logarithms of numbers and the other marked out linearly. A straight line through points $2n, t_1$ and n, t_2, where $2n$ and n are the number of atoms (log axis) at times t_1 and t_2, gives the half-life period $(t_2 - t_1)$ directly.

Units and Definitions

The *curie* (Ci) is the amount of radioactive isotope necessary to produce the same number of nuclear disintegrations as 1 gram of radium, namely $3 \cdot 7 \times 10^{10}$ disintegrations per second. This amount is far too great for normal biological use, and microcurie (μCi) and millicurie (mCi) quantities, 10^{-6} and 10^{-3} curies respectively, find application as tracers.

SPECIFIC ACTIVITY is the ratio of the radioactive atoms of an element to the total atoms of the same element present in the mixture, e.g. $^{35}S/(^{32}S + ^{35}S)$, and is usually expressed as curies, millicuries or microcuries per gram or milligram of element. Similar notation is used for compounds of the element, e.g. mCi of radioactive isotope per mg of compound, although a better notation is mCi per millimole, which facilitates comparison with compounds of different molecular weight.

Radioactivity, when measured by Geiger-Müller counter, is usually expressed as counts per minute (c.p.m.) for a given weight of material. These counts are purely arbitrary units, and in biological work usually no effort is made to convert the count rates into millicuries. Commercially produced radioactive isotopes are supplied with the specific activity given in terms of μCi or mCi per ml of solution or per mg or g of solid. Consequently the term specific activity is applied to two different types of measurement. The count rate on material emitting soft β-particles if made at infinite thickness (see Fig. 11.1) is itself a measure of specific activity.

RELATIVE SPECIFIC ACTIVITY.—Specific activities are frequently expressed relative to the specific activity of a reference compound, e.g. organic phosphates relative to inorganic phosphate of the tissue or of the blood; they are then termed relative specific activities.

TURNOVER.—This term is used to denote the renewal of a given substance by synthesis or by exchange or by entering of a labelled molecule into a tissue. Net synthesis can, of course, be studied by methods other than isotopic, but the tracer technique is particularly suited for demonstrating the incorporation of new molecules which is being balanced by removal of molecules already present, i.e. where there is no net increase in concentration and classical analytical methods are of no avail.

TURNOVER TIME.—The turnover time is the time interval required for the amount of a substance transferred into or out of a compartment or 'pool' in the steady state to be numerically equal to the amount present in the compartment.

TURNOVER RATE.—The turnover rate is defined as the absolute

amount of a substance that is turned over in unit time, i.e. the rate of appearance (or disappearance) of substance A in the steady state, and has the dimensions of mass time^{-1}. This may be denoted by v, the velocity of the reaction. Turnover rates are therefore expressed as grams or gram-moles per unit time.

STEADY STATE.—This applies to the situation where the rates of removal of the substances being studied from compartments are equalled by their rates of replacement, so that the concentrations and amounts of the substances are constant during the period of observation.

In the steady state, and with constant transfer rates for the unlabelled substance, the behaviour of a tracer is independent of the reaction order, and equations describing the transfer all have the form of equations characteristic of first order reactions. The simplest demonstration of this is as follows :

For a catenary system in which the tracer is not recycled, consider a one compartment open system

$$\longrightarrow A \xrightarrow{k} $$

into which an isotope is instantaneously introduced. The rate of disappearance of the isotope at any instant is proportional to the concentration of label, (A*), at that instant and if first order kinetics are assumed

$$\frac{d(A^*)}{dt} = -k(A^*)$$

Hence $\qquad (A^*) = (A_0^*)e^{-kt}$

i.e. $\qquad a_t = a_0 e^{-kt}$ . . . . (11.1)

where a_0 and a_t are the specific activities of A at times 0 and t (i.e. $(A_0^*)/(A_0)$ and $(A^*)/(A)$ respectively)

and $\qquad\qquad v = k(A)$

If no assumption is made about the order of the reaction and v is the rate of disappearance of A, then

$$\frac{d(A^*)}{dt} = v a_t$$

i.e. $\qquad \frac{da_t}{dt} = \left(\frac{v}{(A)}\right) a_t$

Hence
$$a_t = a_0 e^{\left(\frac{v}{(A)}\right)t} \qquad . \qquad . \qquad . \qquad (11.2)$$

$v/(A)$ can be computed from graphical procedures and then v obtained by multiplying this value by (A). Comparison of the equations 11.1 and 11.2 derived in the two cases reveals that $k \equiv v/(A)$.

Thus, for mathematical convenience, it is possible to treat steady state processes as though they are first order reactions. Consequently a graph plotting the logarithm of the specific activity of A against time is linear with a negative gradient equal to the rate constant of A, i.e. the turnover rate.

The description of turnover rate in these conditions is analogous to the mathematical treatment of radioactive decay processes previously discussed and has led in some cases (notably the plasma proteins) to the expression of the turnover rate as the 'half-life' $(t_{\frac{1}{2}})$. The half-life of A is the time taken for the specific activity of A to fall to half of its original value (see pp. 154 and 378). It should be appreciated, however, that the term half-life $(t_{\frac{1}{2}} = 0.693/k)$ only has significance if the order of reaction is first.

Measurements and Corrections to Observed Counts

Radioactivity is measured with suitable apparatus which records a counting rate for the isotope for that particular equipment. The observed disintegrations or counts per minute are related to the *absolute disintegration rate* by a factor which determines the *overall efficiency* of the counting equipment. Thus some disintegrations may not be recorded on account of poor geometry, the backscattering effect, self-absorption by the source and the intrinsic efficiency of the detector itself. Techniques are available for determining the overall efficiency so that, if required, an observed count rate can be converted into the absolute disintegration rate of the isotope. For most biochemical work, however, no attempt is made to convert count rates to the absolute rates of disintegration.

The disintegration rate of a radioactive isotope may be measured experimentally by means of a Geiger-Müller counter-tube. In its simplest form a Geiger-Müller tube is either a metal or glass cylinder with a coaxial wire of platinum or tungsten which forms the anode. The cylinder of metal tubes forms the cathode, while glass tubes either contain a cylinder of metal foil or are coated with colloidal graphite or evaporated metal films. A high potential difference is produced between anode and cathode. The tube is filled with an inert gas plus ethanol or ethyl formate vapour. When the radiation enters the counter via the end window of mica or duralumin it produces ionization and the discharge of the negative ions on the highly charged anode wire causes a pulse. The pulses are amplified and operate a register which records the number of charged particles entering the tube. Tubes for liquid specimens comprise a jacketed, borosilicate-walled Geiger-Müller tube, the sample being placed in the annular space. These tubes are suitable only for fairly high-energy β-emitters, e.g. ^{32}P, on account of absorption by the relatively thick walls.

By such measurements the radioactive isotopes used as tracers are detected and quantitatively assayed. End window counters are sensitive to β-radiation but relatively insensitive to any other radiation. With thinner windows of mica β-particles of low energy (0·1 MeV) may be detected. Increased sensitivity is obtained with windowless flow type or proportional counters and also when the isotope is measured as a gas, e.g. carbon may be assayed as carbon dioxide, tritium as tritiomethane or tritiobutane. The gas ionization technique suffers, however, from serious limitations both in ease of detection and ease of sample preparation of those isotopes which display only weak energetic radiations, e.g. ^{3}H and ^{55}Fe. For the high-efficiency assay of these and other isotopes, in particular ^{14}C and ^{35}S, the liquid scintillation technique is now the method of choice. It also permits the assay of α- and γ-emitters.

In liquid scintillation methods the sample for counting consists of three components, the radioisotope, an organic solvent or solvent mixture, and one or more organic phosphors. The particle or radiation is absorbed in and transfers its energy to the solvent (xylene, toluene and 1,4-dioxane are widely used), which in turn transfers it to the phosphor. In response the phosphor fluoresces or scintillates, i.e. emits light photons which are collected by the photomultiplier and converted into an amplified pulse, which is recorded as a count corresponding to the particle or radiation emitted. The efficiency may approach 100 per cent. since the isotope is in intimate contact or in solution with the phosphor, thus eliminating the problems of self-absorption encountered with solid samples and as discussed below. The rapid expansion of the use of tritium as a tracer is largely attributable to the development of liquid scintillation techniques.

Although liquid scintillation is a highly efficient technique it is nonetheless subject to a variety of errors which must be fully appreciated.

Some or all of the following corrections to the observed counts may be necessary in order to secure a reliable value for the amount of radioactive isotope present in the sample assayed.

1. RESOLVING TIME.—Geiger-Müller tubes become 'dead' for a short period after each ionizing event takes place and, during this time, any other particle entering the tube will not be detected. Following the 'dead time' there is a recovery phase and for the initial portion of this phase radiation will likewise not be recorded. The resolving time of the tube is the dead time plus the insensitive period of the recovery phase, and its effect is to lower the observed count. In practice, ignoring the resolving time correction (which is usually referred to as the *coincidence correction*) at count rates of 3,000 counts/min. will introduce an error of about 1-2 per cent. This correction is not necessary with detectors other than Geiger counters unless very high count rates, e.g. 50,000 counts/min. are being measured. In some circumstances it may also be necessary to correct for the resolving time of the electronic circuits.

2. BACKGROUND COUNT.—The background count is the count recorded when the tube is operated without any known radioactive source in position. Such counts are caused mainly by cosmic rays which produce ionizations in the counter tube. The background count must be subtracted from the observed count. Liquid scintillation assemblies are subject to very high background counts which arise from three main sources, high energy pulses from cosmic radiation, low energy pulses from chemiluminescence of the sample bottle glass and low energy pulses generated by the random thermal emission of electrons from the photocathode of the photomultiplier tube. Operation at low temperature significantly reduces the third contributory factor but introduces additional problems such as the reduced solubility of the sample in the liquid scintillant. However, a very important advance has been made with the introduction of coincidence circuitry which eliminates thermal noise without the need for refrigeration.

3. SELF-ABSORPTION.—Soft radiation, e.g. β-particle emission,

is absorbed within the radioactive source itself. Thus while particles emitted from the surface have only to traverse the gas phase to reach the detector, those arising from lower levels have to pass through varying thicknesses of the source itself. It is found that particles emitted below a certain minimum thickness of material are completely absorbed and never reach the detector. The source is then said to be 'infinitely thick' and it follows that all infinitely thick samples of the same material give the same count

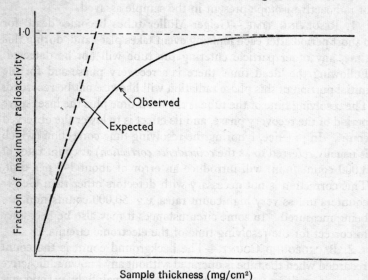

Sample thickness (mg/cm²)

FIG. 11.1

Typical self-absorption curve showing effect of sample thickness on the observed count rate.

rate—addition of more material has no further effect on the rate. Self-absorption curves for emitters such as ^{14}C and ^{35}S are prepared by spreading on identical sample holders increasing amounts of a uniform source of activity, such as $Ba^{14}CO_3$ or $Ba^{35}SO_4$, and then determining the count rate. The observed count rate is plotted against the weight or thickness of the sample as shown in Fig. 11.1. If no self-absorption occurred, a linear relationship would exist between the count rate and the sample weight; in the region where this proportionality exists the sample is said to be 'infinitely thin'. To correct for self-absorption it is necessary to prepare a curve expressing the count rate at various thicknesses of the source as a percentage of the count rate at zero thickness. These curves also allow for certain other errors such as the nature of the mounting of the source, the degree of back-scattering it produces and the geometrical arrangement of the source relative to the detector.

When an infinitely thick source is used the count rate observed is proportional to the *concentration* of the radioactive material in the sample and not to the *absolute amount* in the sample counted, and consequently a correction for self-absorption is not required. However, one consideration requires attention when using the infinite thickness technique. Since the count rate is proportional to concentration, dilution of the isotope with inert material, or conversion to a different chemical form in which the percentage of the element under investigation differs, will alter the observed count rate. Consequently the specific activities of different samples can be compared directly only if all contain the same percentage of the element; otherwise the specific activity of the element in the material must be calculated by dividing the count rate by the percentage of the isotopic element present in the sample.

4. QUENCHING.—Liquid scintillation counters are subject to 'quenching' errors, i.e. to interference with the production of light in the liquid scintillant and its transmission to the photo-multiplier tube. Two types of quenching may occur, chemical and colour. With chemical quenching, compounds in the liquid scintillant interfere with the transfer of energy from the particle or emitted radiation to the organic phosphor and the energy is lost to processes which do not emit light. Colour quenching occurs when coloured compounds in the liquid scintillant absorb light emitted by the organic phosphor and thus prevent its detection by the photomultiplier tube. Many compounds, including dissolved oxygen, produce quenching so that it is difficult to predict the efficiency of counting of any sample. Consequently it is always essential to determine the counting efficiency of samples and this is now usually achieved by one of three methods, viz. internal standard, channels ratio or external standard, details of which are given by Peng (1966).

5. RADIOACTIVE DECAY.—If the half-life of the radioactive element is sufficiently short for appreciable decay to have occurred during the period of the experiment when count rates are measured,

a correction must be applied. This is conveniently made from a suitable decay curve or decay table expressing the percentage of the original isotope remaining as a function of the time.

In addition to making the foregoing corrections it is also necessary to assess the reliability or accuracy of the observed count rate, and this demands an appreciation of the statistical considerations involved. These are treated in the following section.

The various corrections of the observed count rate must be applied in the following sequence :

1. The 95/100 error is determined from the measured count, as discussed in the next section.
2. The quench correction is applied in the case of scintillation counting.
3. The 'dead time' (coincidence) correction is made to the observed count.
4. The background correction is applied to the corrected count from 3.
5. The self-absorption correction is applied to the count from 4.
6. If there are corrections for counter efficiency or for sample geometry these are applied to the count from 5.

Statistical Aspects of Radioactive Decay

It is impossible to make physical measurements without error and consequently it is essential that whenever a quantitative measure is reported an indication be given of the probability of the correct value being within stated limits of the observed figure. This consideration applies with particular force to radiochemical assays where, on account of the randomness of the occurrence of nuclear disintegrations, successive estimates scatter widely about a mean value. It is this mean value (the average of an infinite number of observations) of which an estimate is required. In practice it is important to know how good an estimate of this mean is given by any one particular observation.

In any series of measurements of a variable, x, the frequency of occurrence of the mean and the infinite other possible values can be described by a probability function the nature of which will depend, amongst other factors, on whether x is a continuous or a discontinuous variable. For random nuclear events where x is a discontinuous variable confined to integer values only, and where the probability of occurrence of a disintegration is very small and constant, the *Poisson Distribution* equation :

$$P_x = \frac{m^x e^{-m}}{x!} \qquad . \qquad . \qquad . \qquad (11.3)$$

gives the probability, P_x, of finding a value of x counts when the mean value is m counts.

If after a number n of successive determinations on the one sample the frequency of occurrence of each observation is plotted against count magnitude a histogram of the type shown in Fig. 11.2a may be obtained. Where both m and n are large ($m \geqslant 100$;

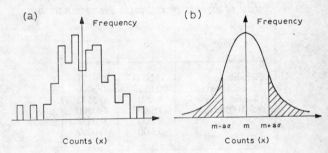

FIG. 11.2

(*a*) Histogram of frequency of occurrence of each observation on a single sample plotted against magnitude (x) of the count.
(*b*) Normal distribution curve for a large number of observations made on a single sample.
The mean value is in counts.

$n \to \infty$) the resultant histogram approximates closely to a smooth symmetrical curve, a normal distribution curve (Fig. 11.2b). The units in which the frequencies are measured are such that the sum of the frequencies for all possible values of x (i.e. the total area under the curve) equals unity.

Distributions are characterized by a number of factors including the *Standard Deviation* (σ) which is a quantity defined to give information about the breadth of scatter of the observations about the mean. The greater the scatter the less likely will any single observation be a good estimate of the mean value. The standard deviation is defined as follows :

$$\sigma = \sqrt{\sum_{i=1}^{i=n} (x_i - m)^2 . P_x} \qquad . \qquad . \qquad (11.4)$$

For a Poisson Distribution

$$\sigma = \sqrt{\sum_{i=1}^{i=n} (x_i - m)^2 \cdot m^{x_i} e^m / x_i!} \qquad . \qquad (11.5)$$

i.e.

$$\sigma = \sqrt{m}$$

For values of $m \geqslant 100$, Table 11.2 gives the proportions of observations (A) lying outwith the limits $m + a\sigma$ (i.e. $m + a\sqrt{m}$) for varying values of a. The values of A are obtained by determining the

TABLE 11.2

Probability of Errors of a Single Determination

Error	a	A
Reliable error	0·5000	0·6171
Probable error	0·6745	0·5000
Standard error	1·000	0·3173
'95/100' error	1·9600	0·0500
	2·000	0·0455
	3·000	0·0027

fraction of the area under the curve lying outwith the stated limits (see the shaded areas in Fig. 11.2b). For a normal distribution the area outwith the limits $m + a\sigma$ is given by

$$2 \int_{-\infty}^{m-a} \frac{1}{\sigma\sqrt{2\pi}} \cdot e^{\frac{-(x-m)^2}{2\sigma^2}} \cdot dx$$

It will be seen that approximately 32 per cent. of the measures lie outwith the limits $m + \sigma$. An equivalent statement is that there is an approximately 68 per cent. chance in one trial of finding a value of x within the limits $m + \sigma$ (i.e. $m + \sqrt{m}$). Similarly there is a 95 per cent. chance in one trial of finding a value of x within the limits $m \pm 1·96\sqrt{m}$. In practice m is not known and in the case of a single observation x_1 it is customary to state that there is a 95 per cent. chance of the mean (i.e. correct) value lying within the limits $x_1 \pm 1·96\sqrt{x_1}$ ($\simeq x_1 + 2\sqrt{x_1}$). The expression $1·96\sqrt{x_1}$, or more generally $a\sqrt{x_1}$, can be referred to as the error of the count, and in practice one tries to ensure that $a\sqrt{x_1}$ is small in relation to

x_1. Various errors can be defined depending on the magnitude of a (see Table 11.2).

A proportional error can be defined as $\dfrac{a\sqrt{x_1}}{x_1}$, or $\dfrac{100a\sqrt{x_1}}{x_1}$ if expressed as a percentage. Thus an observed count of 1000 counts can be stated to have a '95/100' error of $1.96\sqrt{1000} \simeq 64$ counts. The '95/100' proportional error will be 6·4 per cent. Table 11.3 records '95/100' proportional errors as a function of the magnitude (x_1) of the count.

TABLE 11.3

Counts (x_1)	'95/100' Proportional error per cent.
100	19·6
1,000	6·4
1,600	5·0
10,000	1·96

It is recommended that at least 1600 counts be accumulated in any single determination, thus ensuring a '95/100' proportional error of $\not> 5$ per cent. (i.e. to ensure that the correct value has a 95 per cent. chance of lying within ±5 per cent. of the observed value).

It is usually the case that the information desired is a count corrected for background count. Consequently it is important to consider the ways in which errors are propagated. If x_1 is an observed count and b the background count determined in the same time interval, then the '95/100' error of the net count $(x_1 - b)$ is $1.96\sqrt{(x_1 + b)}$. In general for a series of counts c, d, e etc. the '95/100' error of the algebraic sum $c + d + e +$ etc. is

$$1.96\sqrt{(c + d + e + \text{etc.}).}$$

From the foregoing relationships it will be obvious that the time required to secure the desired percentage error for a known counting rate can be calculated

$$E\% = \frac{100a\sqrt{x_1}}{x_1} = \frac{100a}{\sqrt{x_1}} \qquad . \qquad . \quad (11.6)$$

Since $x_1 = Rt$ where R is the rate and t the time of measurement

$$E\% = \frac{100a}{\sqrt{(Rt)}}$$

and
$$t = \frac{10^4 a^2}{R(E\%)^2} \qquad . \qquad . \qquad (11.7)$$

If the sample and the background are counted for the same length of time, then the following formulae hold

$$E = \sqrt{(E_b^2 + E_s^2)} = a\sqrt{(b + x_1)} \qquad . \qquad (11.8)$$

where E, E_b and E_s are the errors (in counts) for net counts, background and sample respectively and b and x_1 are the background and sample counts. The percentage error of net activity is

$$E(\%) = \frac{100a\sqrt{(b + x_1)}}{x_1 - b} \qquad . \qquad . \qquad (11.9)$$

It is possible to count the background for long periods and thus reduce its contribution to the error. Where background and sample are counted for different periods the relationship holding is

$$E_R = a\sqrt{\left(\frac{R_s}{t_s} + \frac{R_b}{t_b}\right)} \qquad . \qquad . \qquad (11.10)$$

or
$$E_R\% = \frac{100a\sqrt{\left(\frac{R_s}{t_s} + \frac{R_b}{t_b}\right)}}{R_s - R_b} \qquad . \qquad . \qquad (11.11)$$

where all the R values refer to count rates, e.g. counts per minute.

Example 11.1. A sample had an activity of 200 counts in 10 minutes, with a background of 100 counts in the same period. Calculate the error in the net count from these data and also from the information that in a background count carried out over 1,000 minutes the rate was 10 counts per minute.

Firstly, it is essential to decide the probability level at which to work. Thus, if we wish to use the 95 per cent. reliable error level then we are deciding that the calculated errors in percentages will not be exceeded in 95 out of 100 cases. From Table 11.2, at this level $a = 1 \cdot 9600$ and

$$E\% = \frac{100 \times 1 \cdot 96 \sqrt{(100 + 200)}}{200 - 100}$$

$$= 34\%$$

With the long background period

$$E_R\% = \frac{100 \times 1 \cdot 96 \sqrt{\left(\frac{20}{10} + \frac{10}{1000}\right)}}{20 - 10}$$

$$= 28\%$$

The large error of these counts indicates that the sample should be counted for a longer period. Suppose 2,000 counts are collected in 100 minutes, with $t_s = t_b$

$$E\% = \frac{100 \times 1 \cdot 96 \sqrt{(1000 + 2000)}}{2000 - 1000} = 10 \cdot 8\%$$

With the long background count of 1000 minutes

$$E_R\% = \frac{100 \times 1 \cdot 96 \sqrt{\left(\frac{20}{100} + \frac{10}{1000}\right)}}{20 - 10} = 9 \cdot 0\%$$

These values emphasize the relationship between the error, the magnitude of the count and period of observation. For assessing the error at any other probability level the appropriate value of a is selected from Table 11.2.

Isotope Dilution

A very elegant and valuable method of analysis has been placed at the disposal of the biochemist with the advent of tracer techniques. Many compounds are present in mixtures of biological substances (amino acids in protein hydrolysates, for example) in such small amount that quantitative isolation of the pure substance is extremely difficult or even impossible. However, if a pure labelled sample of the same compound is added to the biological material, it mixes with and becomes indistinguishable from the compound already present. If now a sample of the substance is isolated in pure form, regardless of yield, the specific radioactivity may be measured, and knowing the specific activity and weight of the added material, it is possible to calculate the amount of unlabelled compound originally present in the mixture.

Suppose that a grams of labelled compound displaying n_1 counts per minute, and therefore having a specific activity of n_1/a, are added to a mixture containing b grams of the unlabelled compound. A pure sample of the compound is now isolated, say c grams, and found to give n_2 counts per minute. The specific activity is therefore n_2/c. Measurement of the specific activity before and after dilution permits b to be determined since

$$\frac{n_2}{c} = \frac{n_1}{(a + b)} \qquad . \qquad . \qquad . \quad (11.12)$$

whence
$$b = \frac{n_1 c - n_2 a}{n_2} \qquad . \qquad . \qquad . \quad (11.13)$$

and if the specific activities of added and isolated material are denoted by S_1 and S_2 respectively, then equation 11.13 becomes

$$b = a\left(\frac{S_1}{S_2} - 1\right) \qquad . \qquad . \qquad . \quad (11.14)$$

Although the weight of the isolated material is required for the determination of its specific activity, this does not necessarily involve weighing the sample. Microgram quantities of the isolated material may be determined, for example, by spectro-photometry or other suitable techniques.

With radioactive isotopes the difference in atomic weight between the radioactive and stable isotope, and hence between the molecular weight of labelled and unlabelled compound, becomes vanishingly small and may be neglected. This is not so with stable heavy isotopes, where the atom per cent. excess of labelling element used may be sufficient to justify taking it into considera-tion. In these cases equation 11.14 becomes

$$b = a\left(\frac{e_1}{e_2} - 1\right)\frac{M_2}{M_1} \qquad . \qquad . \qquad . \quad (11.15)$$

where e_1 and e_2 are the atom per cent. excess of added and isolated material and M_1 and M_2 their molecular weights. When the difference in molecular weights is insignificant, M_2/M_1 in equation 11.15 becomes unity. Note that with stable isotopes it is not even necessary to know the weight of the isolated pure material because the atom per cent. excess determined by mass spectrometry is all that is required, whereas the weight is necessary for the deter-mination of specific radioactivity. It is desirable that the added material should have an atom per cent. excess greater than 5 for isotope dilution analysis.

Example 11.2.—5·1 mg of ^{14}C-labelled arginine giving 2623 counts per minute were added to a protein hydrolysate and then a sample of arginine was isolated from the mixture and purified to yield 12·5 mg. The resultant material gave 1047 counts per minute when assayed with a Geiger-Müller tube. If the counter had a background correction of 15 counts per minute, determine the amount of arginine present in the hydrolysate.

Applying the background correction, the true counts per minute are 2608 and 1032 for added and isolated materials respectively, and the specific activities are 2608/5·1 and 1032/12·5. Let x mg be the amount of arginine in the hydrolysate then, using equation 11.9

$$x = 5·1 \left(\frac{2608 \times 12·5}{5·1 \times 1032} - 1 \right) = 31·6 - 5·1 = \underline{26·5 \text{ mg.}}$$

There are circumstances in which direct dilution analysis may prove difficult or unsuitable. For example, if the recovery of the material falls to a fraction of a microgram it is difficult to determine its specific activity with accuracy. Further, when it is necessary to analyse a number of related complex substances it may be impossible to obtain pure radioactive tracers for each of them. Alternative procedures have therefore been devised to deal with these situations.

DERIVATIVE DILUTION.—The compound to be determined is treated with a radioactive reagent of known specific activity and the product of the reaction is isolated and purified. The total radioactivity of the pure product is a measure of the amount of radioactive reagent it contains and knowledge of the stoicheiometry of the reaction enables the amount of compound present to be calculated. The technique clearly demands that the reagent and component to be determined react quantitatively and in high yield, and that the reaction product be completely separated from the other radioactive components of the mixture and especially from excess of the reagent.

If a radioactive reagent of specific activity S (μCi per μmole) is added to a mixture containing the component to be determined, and with which it reacts, the product of the reaction will also have a specific activity of S μCi per μmole, or will bear a simple relationship to it. The radioactive product is now separated from all traces of the excess reagent and other radioactive products and its radioactivity measured, say A. The amount W, in μmoles, of the original product can then be calculated from the expression

$$W = \frac{A}{S} \quad . \quad . \quad . \quad . \quad (11.16)$$

The most powerful of the derivative dilution techniques is that involving the double label procedure, which is capable of extraordinary sensitivity. For example, it has been successfully applied to the determination of testosterone in peripheral plasma by using tritium-labelled testosterone and [35S]thiosemicarbazide, and one part of testosterone in 10^{10} parts of sample have been measured. The method is as follows :

1. A small amount of tritium-labelled testosterone of high specific activity is added to a sample of peripheral plasma. The radioactive and inactive species are then equilibrated thoroughly.
2. The steroids are separated from the plasma chemically.
3. The steroids are then treated with [^{35}S]thiosemicarbazide which converts the keto steroids to their corresponding thiosemicarbazones.
4. The testosterone thiosemicarbazone is purified by repeated chromatography, conversion to a derivative and further chromatography.
5. The tritium content is then measured.
6. The ^{35}S content is measured.

The salient features of the method are, first, that since the [^{3}H] testosterone is added to the sample and equilibrated before any separations are carried out, losses of material will be reflected by losses of tritium activity. Thus the very low recoveries attendant on the exceptionally rigorous purification of the testosterone semicarbazone which is necessary, can be readily corrected for.

Second, the original testosterone content can be determined in the normal way by measuring the ^{35}S activity in the purified product and using equation 11.16.

SATURATION OR SUB-STOICHEIOMETRIC DILUTION ANALYSIS.— The principle of this analytical method is as follows. A solution of the sample containing an unknown weight (W micrograms) of the component to be analysed is thoroughly mixed with a known weight (W_r micrograms) of the radioactive form of the same component and then treated with a compound which reacts specifically with it. The quantity of the specific reagent used must be just less than equivalent to W_r micrograms of the component, i.e. there must be a slight excess of the component being analysed. The product of the reaction is then isolated by a suitable procedure, e.g. solvent extraction, precipitation etc., and its radioactivity counted, say x counts/min.

The foregoing procedure is now repeated, but using *only* the radioactive form of the component being analysed ; the activity is measured as before, say y counts/min.

Now the weight of reaction product isolated in the two cases will be identical, since a sub-stoicheiometric quantity of reagent was used in each procedure. Consequently, since the specific activity is given by the ratio of activity to weight, the ratio of the specific activities of the pure (S) to the diluted (S_d) component will be simply the ratio of the observed counts/min, i.e.

$$\frac{S}{S_d} = \frac{v}{x} \text{ and also since } \frac{S}{S_d} = \frac{W + W_r}{W_r}$$

then
$$W = W_r\left(\frac{y}{x} - 1\right) \qquad . \qquad . \qquad . \qquad (11.17)$$

and the weight W of the component can be determined.

Turnover of Plasma Proteins

The turnover of plasma proteins is amenable to isotopic investigation and it might be expected that any one of a number of isotopes could be used for this purpose, e.g. isotopes of hydrogen (deuterium and tritium), carbon (^{13}C and ^{14}C), nitrogen (^{15}N) and sulphur (^{35}S). However, the stable isotopes have not been used on account of the greater difficulty of assay, and deuterium and tritium are open to objection because their double and triple mass relative to hydrogen might affect reaction rates. Furthermore, since almost all amino acids enter into transamination reactions, ^{15}N is of little use as a label for plasma protein turnover studies. This leaves ^{14}C and ^{35}S, which have been employed although their use is open to one serious objection, namely the possibility of recycling of the label. To illustrate this aspect consider ^{14}C or ^{35}S administered to an animal, either as labelled amino acids or in a form which is incorporated into amino acids before the latter compounds are incorporated into the animal proteins. In the course of metabolism the protein is broken down and free amino acids are released, but a significant amount of these is once more reincorporated into new protein. Even if the initial amino acid is catabolized, unless it is broken down to CO_2 there is the possibility that the isotope could be reincorporated into another amino acid and so recycled.

These objections can be largely overcome by the use of radioactive iodine ^{131}I, a β and γ emitter with a half-life of 8·04 days. Although iodine is not a normal constituent of plasma proteins such as albumin, under appropriate conditions it can be induced to combine with the aromatic nucleus of the amino acid tyrosine present in protein and, in the correct proportions and under mild conditions, does so without modifying the properties of the plasma proteins (McFarlane, 1957). The experimental procedure adopted is to remove plasma from the animal, fractionate the protein without denaturation, iodinate and then return the labelled protein to the animal by intravenous injection. When the protein is subsequently broken down, the monoiodotyrosine liberated cannot be reincorporated into new protein and consequently the label cannot be recycled. Monoiodotyrosine is degraded and inorganic iodide appears as the major excretion product.

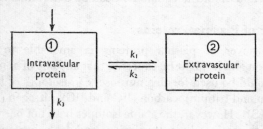

FIG. 11.3

Plasma protein turnover: example of an open two-compartment system showing the appropriate rate constants k_1, k_2 and k_3. The fractional turnover rate is given by k_3.

The site and stages of breakdown of plasma proteins are at present not precisely known (although the liver may be involved) and therefore studies have required that a model system be postulated and the results appraised by 'goodness of fit' to the model. By successive approximations the best model system can be selected ; various systems have been postulated at different times by workers in this field of research (McFarlane, 1964).

The theory of tracer experiments using ^{131}I-labelled plasma proteins was presented by Matthews in 1957. The simplest model is an open two-compartment system of the type shown in Fig. 11.3. Compartment 1 represents the intravascular protein and there is evidence that protein breakdown occurs in this compartment only. Compartment 2 represents protein in the intravascular space. (More complicated systems take into account the existence of n extravascular compartments.) It is assumed that the system is in dynamic equilibrium so that the total protein in each compartment is constant, and that the rate of breakdown (equal to rate of synthesis) and rates of transfer from one compartment to the other are constant. It is further assumed that all newly synthesized protein enters the intravascular compartment and that an equal amount of protein passes in each direction between intra- and extravascular compartments. Radioactivity is lost from the intravascular compartment and appears in the urine and faeces in breakdown products. For mathematical convenience, the model is treated on the basis of first order kinetics although, as we have already seen (p. 380), the same turnover rates are computed irrespective of any assumptions about the order of the reaction.

For the model

$$A \underset{k_2}{\overset{k_1}{\rightleftharpoons}} B$$

$$\downarrow k_3$$

let (A^*) be the concentration of labelled atoms and a the specific activity of A. Then

$$\frac{d(A^*)}{dt} = -k_1(A^*) + k_2(B^*) - k_3(A^*)$$

$$= -(k_1 + k_3)(A^*) + k_2(B^*)$$

$$\frac{d(B^*)}{dt} = k_1(A^*) - k_2(B^*)$$

Hence $\quad \dfrac{d^2(A^*)}{dt^2} + (k_1 + k_2 + k_3)\dfrac{d(A^*)}{dt} + k_2 k_3(A^*) = 0$

and $\quad\quad\quad (A^*) = c_3 e^{-pt} + c_4 e^{-qt}$

$$\frac{(A^*)}{(A)} = c_1 e^{-pt} + c_2 e^{-qt}$$

i.e. $\quad\quad\quad\quad a = c_1 e^{-pt} + c_2 e^{-qt} \quad . \quad\quad . \quad\quad . \quad (11.18)$

where $\quad\quad\quad p + q = k_1 + k_2 + k_3 \quad . \quad\quad . \quad\quad . \quad (11.19)$

and $$pq = k_2k_3 \qquad . \qquad . \qquad . \quad (11.20)$$

Now if the specific activity a of the labelled plasma proteins is set equal to unity when $t = 0$, then

$$c_1 + c_2 = a(= 1) \qquad . \qquad . \qquad . \quad (11.21)$$

and $$c_1p + c_2q = k_1 + k_3 \qquad . \qquad . \qquad . \quad (11.22)$$

Solution of equations 11.13 to 11.16 gives

$$k_1 = \frac{c_1c_2(q-p)^2}{pc_2 + qc_1} \qquad . \qquad . \qquad . \quad (11.23)$$

$$k_2 = pc_2 + qc_1 \qquad . \qquad . \qquad . \quad (11.24)$$

$$k_3 = \frac{pq}{pc_2 + qc_1} \qquad . \qquad . \qquad . \quad (11.25)$$

From the foregoing treatment it will be apparent that a semi-logarithmic plot of intravascular plasma protein activity, expressed as a fraction of the zero time activity, as a function of time yields a curve of the type shown in Fig. 11.4. It is found experimentally that, after a given time, the curve becomes linear indicating that intravascular activity is now a single exponential function of time, assumed to be equal to c_1e^{-pt}. Thus $-p$ is the slope of the line. If this line is extrapolated to zero time it cuts the ordinate at a value representing c_1. The extrapolated line is then subtracted from the original curve at suitable time intervals to give a second curve, which is linear. This new exponential function is assumed to be c_2e^{-qt}, with slope $-q$ and, when extrapolated to zero time, cuts the ordinate at c_2.

Where there are more than two compartments subtraction of the first extrapolated line yields a curve which, after a time, becomes linear. Extrapolation of this line to zero time enables a further subtraction to be carried out and another line plotted. The process is repeated until subtraction gives a single exponential. The number of exponentials obtained in this way is assumed to be equal to the number of extravascular compartments plus one intravascular compartment. The number of extravascular compartments is therefore taken to be the smallest number that will give a curve corresponding to the experimental results, which is clearly a simplification.

The method thus involves the graphical determination of values for p, q, c_1 and c_2 from the semi-logarithmic plot of the experimental results and then the use of these values in equation 11.25 to find k_3, the fractional turnover rate.

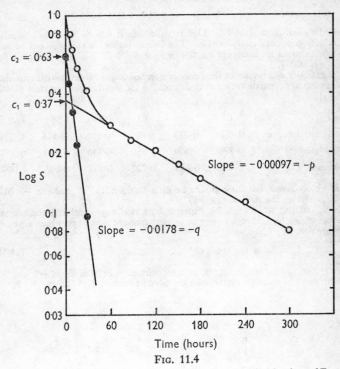

Fig. 11.4

Plot of logarithm of plasma albumin activity against time for the data of Example 11.3.   The curve obtained is characteristic of an open two-compartment system and is of the form $a = c_1 e^{-pt} + c_2 e^{-qt}$

Example 11.3.   In a study of plasma protein turnover in rats ^{131}I-labelled serum albumin was injected and the plasma radioactivity measured at intervals over a period of 300 hours.   The following values are the plasma radioactivities expressed as fractions of the plasma radioactivity at zero time.

Time (h)	6	12	18	30	60	90	120	150	180	240	300
Fraction of activity in plasma (S)	0·65	0·47	0·36	0·26	0·19	0·17	0·16	0·15	0·14	0·13	0·12

Determine the turnover rate of the plasma albumin.

The experimental data are plotted in the form of log S against time, where S is the serum albumin activity.   To avoid negative values in the plots the values of S are multiplied by 10 before taking logarithms.

Time (h)	6	12	18	30	60	90
S	0·65	0·47	0·36	0·26	0·19	0·17
log S	0·813	0·672	0·556	0·415	0·279	0·230

Time (h)	120	150	180	240	300
S	0·16	0·15	0·14	0·13	0·12
log S	0·204	0·176	0·146	0·114	0·079

These data are plotted in Fig. 11.4 from which it can be seen that linearity occurs after approximately 60 hr. The linear region is extrapolated to zero time which gives an intercept on the ordinate of 0·37. The gradient of this line is −0·00097.

The extrapolated region of the curve is now subtracted from the experimental curve at the appropriate time intervals and the values obtained are then plotted in Fig. 11.4.

Time (h)	0	6	12	18	30	60
Experimental curve	1·000	0·813	0·672	0·556	0·415	0·279
Extrapolated curve	0·370	0·360	0·350	0·338	0·320	0·270
Difference	0·630	0·453	0·322	0·218	0·095	0·009

It will be observed that these points lie on a straight line of gradient −0·0178 and intercept on the ordinate of 0·63.

These results, which can be expressed as two exponential functions, are therefore characteristic of an open two-compartment system to which equation 11.18 applies, i.e.

$$a = c_1 e^{-pt} + c_2 e^{-qt} \qquad . \qquad . \qquad . \qquad . \qquad . \qquad . \quad (11.18)$$

where c_1 and c_2 are the intercepts on the ordinate of lines of slope $-p$ and $-q$ respectively. For such a system the fractional turnover rate is given by k_3

$$\begin{array}{c} \downarrow k_1 \\ A \rightleftharpoons B \\ \downarrow k_2 \\ \downarrow k_3 \end{array}$$

where $k_3 = \dfrac{pq}{pc_2 + qc_1}$ $\qquad . \qquad . \qquad . \qquad . \qquad . \qquad . \qquad . \quad (11.25)$

Values for p, q, c_1 and c_2 have been obtained graphically from Fig. 11.4 and are as follows:

$$p = 0.00097 \qquad\qquad q = 0.0178$$
$$c_1 = 0.37 \qquad\qquad c_2 = 0.63$$

The factor of 2·303 must be introduced since $\log_{10}$ plots were made.

$$\text{Hence } k_3 = \frac{0.00097 \times 0.0178 \times 2.303}{0.00097 \times 0.63 + 0.0178 \times 0.37}$$

$$= \underline{0.0055}$$

expressed as a fraction of the total intravascular albumin per hour, or

$$0.0055 \times 24 = \underline{0.132} \text{ per day.}$$

Determination of Metabolic Pathways

The use of isotopes for the qualitative and quantitative determination of metabolic pathways has led to some of the outstanding discoveries of the last decade. The pathway of carbon in photosynthesis was mapped by Calvin and his collaborators and, later, Kornberg, by use of similar methods, obtained results which led to the discovery of the glyoxylate cycle in micro-organisms. The use of uniquely labelled glucose has permitted the recognition not only of pathways of carbohydrate metabolism other than glycolysis, but also the estimation of their quantitative significance. Space does not permit a full survey of these applications but certain aspects of the methods will be discussed.

In its simplest form the isotopic tracer technique affords a method for the study of the conversion of a compound A into compound B, or for the utilization of part of A in the synthesis of B. Labelled substance A is administered to the intact animal or bacterial suspension, or to tissue slices, homogenates or cell-free extracts, followed by analysis to discover whether labelling occurs in compound B. Such experiments are usually of qualitative nature since labelled A will be diluted with any non-labelled A present in the system or formed from other sources during the course of the experiment.

Example 11.4. *Torulopsis utilis* was allowed to grow for a short period in a basal medium supplemented with [2-^{14}C]acetate. Aspartic acid and threonine were then isolated from total cell proteins.

Aspartic acid was degraded by treatment with ninhydrin to give a quantitative yield of acetaldehyde and CO_2(A). A portion of the acetaldehyde was isolated as a nitrophenylhydrazone (B) and the remainder treated with I_2 and KOH to give CHI_3(C).

Threonine was degraded (*a*) by treatment with ninhydrin to give CO_2(D) and (*b*) by periodate oxidation to yield acetaldehyde. As before, part of the acetaldehyde was isolated as a nitrophenylhydrazone (E) and part degraded to yield CHI_3(F).

All compounds assayed for activity were oxidized to CO_2 and counted as infinitely thick layers of barium carbonate.

Compound oxidized for assay	Counts/min/ planchet of $BaCO_3$
Threonine	184
CO_2(A)	137
Acetaldehyde-nitrophenylhydrazone (B)	92
CHI_3(C)	360
CO_2(D)	90
Acetaldehyde-nitrophenylhydrazone (E)	46
CHI_3(F)	104

Derive as much information as possible about the distribution of activity in the carbon chains of the two amino acids and discuss what information this experiment yields concerning the pathway of threonine biosynthesis.

As all the samples are counted at infinite thickness the specific activities are proportional to the counts observed, and the total counts of each combusted molecule will be proportional to the observed counts/min. in the CO_2 multiplied by n, where n is the total number of carbon atoms in the combusted molecule.

Aspartic acid

1 COOH   1 CO_2 (A)

|

2 CHNH$_2$ Ninhydrin 2 CHO ⟶ Nitrophenylhydrazone(C-2 + C-3)(B)

| ⟶

3 CH$_2$ 3 CH$_3$

| ↘ 3

4 COOH   4 CO_2 (A)   CHI$_3$(C)

CO_2(A) is derived from C-1 and C-4 of aspartate and for both atoms the activity is $137 \times 2 = 274$ counts/min.

The acetaldehyde-nitrophenylhydrazone contains C-2 and C-3 of aspartate diluted with six carbon atoms from the nitrophenylhydrazine, i.e. 8 carbon atoms in all. Hence the true value for (C-2 + C-3) = $92 \times 8 = 736$ counts/min.

But CHI$_3$(C) is derived from C-3 and yields 360 counts/min. Therefore C-2 yields $736 - 360 = 376$ counts/min.

As aspartate is derived via the symmetrical compounds succinate and fumarate it is reasonable to assume that the labelling in (C-1 + C-4) is equally distributed between these two carbon atoms, i.e. 137 counts/min in each. In this event the aspartate molecule would contain activity distributed as follows :

$$\begin{array}{cccc} 1 & 2 & 3 & 4 \\ C - & C - & C - & C \\ 137 & 376 & 360 & 137 \end{array}$$

Threonine

1 COOH Ninhydrin 1
        ⟶ CO_2(D) = 90 counts/min.

|

2 CHNH$_2$

|

3 CHOH   3 CHO ↗ Nitrophenylhydrazone(C-3 + C-4)(E)

| Periodate

4 CH$_3$ ⟶   4 CH$_3$ ↘ 4
                              CHI$_3$(F) = 104 counts/min.

Threonine itself contained $184 \times 4 = 736$ counts/min (C-3 + C-4) are diluted by the nitrophenylhydrazine carbon as before and the true value is $46 \times 8 = 368$ counts/min. Of these, 104 counts/min are attributable to C-4 and therefore

$$\text{C-3} = 368 - 104 = 264 \text{ counts/min.}$$

The C-2 activity can now be obtained by difference, thus

$$736 - (90 + 368) = 736 - 458 = 278 \text{ counts/min.}$$

Hence the distribution of activity in threonine is as follows

$$\begin{array}{cccc} 1 & 2 & 3 & 4 \\ C - & C - & C - & C \\ 90 & 278 & 264 & 104 \end{array}$$

In each compound, therefore, the activities of C-1 and C-4 are about equal, as are C-2 and C-3, although the latter carbon atoms have a greater activity than the former. If aspartate is the precursor of threonine then one would expect to find that the ratios of the activities of C-2/C-1 and C-3/C-4 are similar for each compound. C-2/C-1 for aspartate is 2·7 and for threonine 3·1. C-3/C-4 for aspartate is 2·6 and for threonine 2·5. This agreement is quite reasonable and supports the concept that aspartate is a precursor of threonine. The finding that threonine is not labelled to the same extent as aspartate is a reflexion of the fact that the cells were grown for only a short period in the presence of $[2\text{-}^{14}C]$ acetate and the threonine has been diluted by unlabelled material to a greater extent than aspartate. The heavier labelling of the C-2 and C-3 positions is explicable in terms of the behaviour of the C-2 atom of acetate in the tricarboxylic acid cycle.

The route of threonine synthesis, as at present envisaged, may be outlined as :

Acetate → oxaloacetate → aspartate → aspartyl
semi-aldehyde → homoserine → threonine

(Note that each arrow does not in every case represent a single reaction.)

Identification of Precursors.

—A mathematical approach to the precursor problem has been made by Zilversmit, Entenman and Fishler (1943) by considering the specific activities of precursor and product as a function of time. Certain assumptions are made. It is necessary to assume that the compound studied is in the steady state over the interval of time considered, i.e. its rate of appearance is equal to its rate of disappearance. Furthermore, there must be random appearance and disappearance of all molecules, i.e. the organism does not distinguish between 'old' and 'newly formed' molecules.

If compound A is the immediate precursor of B, and the labelled material administered is some compound other than A, the ideal relationship between them is shown in Fig. 11.5. It will be noticed that both A and B have zero specific activity at zero time and that when the specific activity of B has reached its maximum it is equal to that of the precursor A. Prior to this time the specific activity of A has reached its maximum, and after the intersection of the curves the specific activity of A is less than that of B.

The mathematical expression for this situation is

$$\frac{d(B^*)}{dt} = va_t - vb_t = v(a_t - b_t) \quad . \quad (11.26)$$

where (B^*_t) = concentration of radioactive B at time t

a_t and b_t are the specific activities of A and B at time t, i.e. $(A^*_t)/(A)$ and $(B^*_t)/(B)$ respectively.

v = constant velocity of reaction

o

Integration between the limits of t_1 and t_2 gives

$$(B_{t_2}^*) - (B_{t_1}^*) = v \left[\int_{t_1}^{t_2} a_t dt - \int_{t_1}^{t_2} b_t dt \right]$$

$$= v \text{ [shaded area of Fig. 11.5]}$$

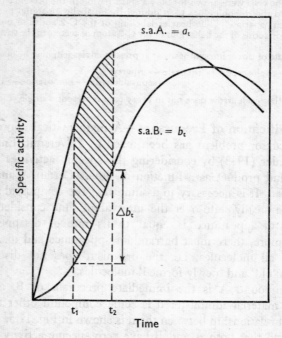

FIG. 11.5

Relationship between specific activity and time for compounds A and B, where A is the precursor of B and the labelled material administered is a compound other than A. (ZILVERSMIT, ENTENMAN & FISHLER, 1943.)

i.e.

$$b_{t_2} - b_{t_1} = \Delta b_t = \frac{v}{(B)} \text{ [shaded area]} \quad . \quad (11.27)$$

Hence we obtain $\dfrac{(B)}{v} = \text{turnover time} = \dfrac{\text{shaded area}}{\Delta b_t} \quad . \quad (11.28)$

where Δb_t is the difference in specific activity of B at times t_2 and t_1 ($\Delta b_t = b_{t_2} - b_{t_1}$).

Unfortunately, it is not always possible to obtain all the numerical values required by this approach on a single animal. Where it can be done, the turnover time of B is, as shown, given by the ratio of the area between the two specific activities of B, over any time period, i.e. it may be in either the rising or falling portions of the curve. Knowledge of the amount of B present in the tissue is not necessary for the calculation of the turnover time but permits determination of the turnover rate, which may be expressed as μg or mg of B turned over per gram of tissue per hour. The turnover time is equal to $(B)/v$.

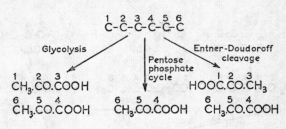

FIG. 11.6

Distribution of the carbon atoms of glucose in pyruvate produced via glycolysis, pentose phosphate cycle and Entner-Doudoroff cleavage.

In the majority of experiments it does not prove possible to obtain all the required data from a single animal and it becomes necessary to compare relative specific activities for the animals employed. The obvious limitations of this method curtail its value.

Pathways of Glucose Metabolism.—At the present time, three major pathways of glucose metabolism are generally recognized. These are glycolysis and the pentose phosphate cycle, which occur in animals, plants and micro-organisms, and the 2-oxo-3-deoxy-6-phosphogluconate cleavage pathway discovered by Entner and Doudoroff which, to date, has been found only in micro-organisms. Both glycolysis and the Entner-Doudoroff pathway yield two moles of pyruvate per mole of glucose catabolized, but they differ in the manner in which the carbon atoms of pyruvate are derived from the hexose (Fig. 11.6). In glycolysis the carboxyl groups of the pyruvate are derived from carbon atoms 3 and 4 whereas they arise from carbon atoms 1 and 4 in the Entner-Doudoroff sequence. In the pentose phosphate cycle, a version of which is shown in Fig. 11.7, three molecules of glucose 6-phosphate are oxidatively decarboxylated (via 6-phosphogluconate) to pentose phosphate, and the three pentoses then undergo carbon transformations which result in the formation of two moles of fructose 6-phosphate and one of glyceraldehyde 3-phosphate. As the fructose 6-phosphate molecules can be converted to glucose 6-phosphate the net result of one turn of the cycle may be written as

Glucose 6-phosphate→Glyceraldehyde 3-phosphate + $3CO_2$.

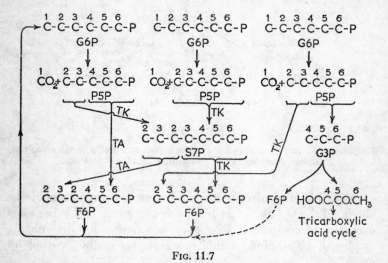

FIG. 11.7

Fate of the carbon atoms of glucose in the pentose phosphate cycle.

KEY.—*Substrates*: G6P, glucose 6-phosphate: F6P, fructose 6-phosphate; P5P, pentose 5-phosphate; S7P, sedoheptulose 7-phosphate; G3P, glyceraldehyde 3-phosphate.
 Enzymes: TK, transketolase; TA, transaldolase.

The fate of the individual carbon atoms of the original hexose phosphate is mapped in Fig. 11.7. It will be seen that carbon dioxide is derived exclusively from the C-1 position of the three participating hexose molecules and the glyceraldehyde 3-phosphate from carbon atoms 4, 5 and 6. The glyceraldehyde 3-phosphate may then either be converted to pyruvate via the reactions common to glycolysis or two molecules can re-form glucose 6-phosphate, as shown in Fig. 11.8. The latter re-formation would demand that one of the triose phosphate molecules be isomerized to dihydroxyacetone phosphate. The two isomers then undergo an aldol condensation to yield fructose 1, 6-diphosphate, under the influence of aldolase. A phosphatase specific for fructose 1, 6-diphosphate forms fructose 6-phosphate which is then converted to glucose 6-phosphate. Any pyruvate formed via the pentose phosphate cycle must therefore be derived from carbon atoms 4, 5 and 6 of the hexose and the carboxyl group always arises from C-4.

It is evident from the redistribution of carbon atoms which the pentose phosphate cycle effects (Fig. 11.7) that the re-formed glucose 6-phosphate now contains carbon atom 2 of the original

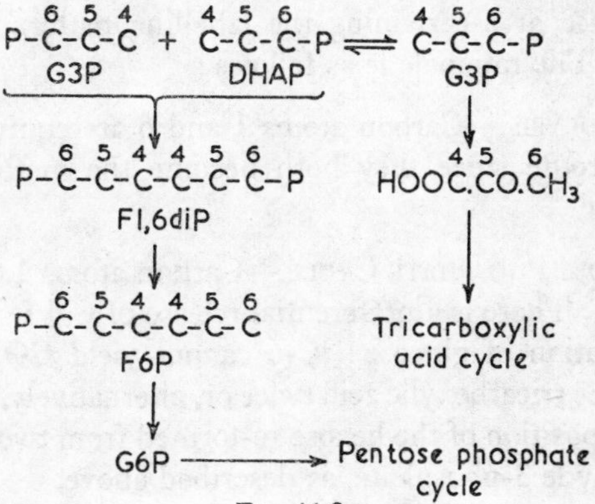

$$
\overset{6\ \ 5\ \ 4}{\text{P-C-C-C}} + \overset{4\ \ 5\ \ 6}{\text{C-C-C-P}} \rightleftharpoons \overset{4\ \ 5\ \ 6}{\text{C-C-C-P}}
$$

G3P DHAP G3P

$$
\overset{6\ \ 5\ \ 4\ \ 4\ \ 5\ \ 6}{\text{P-C-C-C-C-C-C-P}} \qquad \overset{4\ \ 5\ \ 6}{\text{HOOC.CO.CH}_3}
$$

Fl,6diP

$$
\overset{6\ \ 5\ \ 4\ \ 4\ \ 5\ \ 6}{\text{P-C-C-C-C-C-C}} \qquad \text{Tricarboxylic}
$$

F6P acid cycle

G6P ⟶ Pentose phosphate cycle

FIG. 11.8

Fate of triose phosphate produced in the pentose phosphate cycle.

KEY.—G3P, glyceraldehyde 3-phosphate ; DHAP, dihydroxyacetone phosphate ; Fl,6diP, ructose 1,6-diphosphate ; F6P, fructose 6-phosphate ; G6P, glucose 6-phosphate.

hexose in the C-1 position, and, on re-entering the cycle, this will be released as CO_2. As a result of recycling, carbon atom 3 of the original hexose will also appear in the C-1 position and be converted to CO_2 ; this continues until all the original C-2 and C-3 have been oxidized to CO_2. Now consider the hexose re-formed by aldol condensation of triose phosphates. In this case C-6 appears in the C-1 position (Fig. 11.8) and thus undergoes oxidation. As a result of recycling, original carbon atoms 4 and 5 (now occupying the C-3 and C-2 positions respectively) eventually appear in the C-1 position and so are oxidized. The student should trace for himself the fate of individual carbon atoms for successive recycles, using Fig. 11.7 as a model.

Under conditions of aerobic metabolism terminal oxidation

usually occurs via the tricarboxylic acid cycle, and the pyruvate formed by the different routes of glucose metabolism is subjected to identical catabolism, e.g. oxidation to acetyl-coenzyme A which gains entry to the cycle by condensing with oxaloacetate to yield citrate. The initial stage is therefore the loss of the carboxyl group of the pyruvate, followed by the carbonyl carbon and, finally, the methyl carbon atom.

The evaluation of the quantitative significance of pathways of glucose metabolism is carried out with glucose specifically labelled in particular carbon atoms and then either determining the rate of $^{14}CO_2$ release or ascertaining the labelling pattern of metabolic products. The rationale is as follows :

1. GLYCOLYSIS.—Carbon atoms 1 and 6 are equivalent by this metabolic route since they both become the methyl groups of pyruvic acid.

2. PENTOSE PHOSPHATE CYCLE.—Carbon atoms 1 and 6 are not equivalent. There is a preferential release of C-1 in the oxidative decarboxylation of glucose ; C-6 cannot yield CO_2 until it has traversed the tricarboxylic acid twice or, alternatively, has appeared in the C-1 position of the hexose re-formed from two molecules of glyceraldehyde 3-phosphate, as described above.

3. ENTNER-DOUDOROFF SCHEME.—Carbon atoms 1 and 6 are not equivalent and C-1 is preferentially released since it appears in the carboxyl group of pyruvate whereas C-6 appears in the methyl group.

Glucose 6-phosphate ⟶ 6-Phosphogluconate
↓

    1 2   3   4   5   6

$HOOC.CO.CH_2.CHOH.CHOH.CH_2OPO_3H_2$

2-Oxo-3-deoxy-6-phosphogluconate
↓

   1 2   3   4 5   6

$HOOC.CO.CH_3$ + $OHC.CHOH.CH_2OPO_3H_2$
Pyruvate   Glyceraldehyde 3-phosphate
↓

     4 5 6

$HOOC.CO.CH_3$
Pyruvate

Methods based on the release of $^{14}CO_2$ from differently labelled substrates usually demand that the specific activity of the CO_2 be measured. This is not always simple because of the small amounts of CO_2 involved and the presence of adventitious CO_2. Where products of triose phosphate metabolism accumulate, as for example with fermentations, these compounds can be isolated and their specific activities determined. Additional information can be obtained by subjecting these products to chemical or enzymic degradation to ascertain the distribution of the isotope.

A possible source of error in experiments where the specific activities of metabolic products are determined is dilution with non-labelled compounds, or with exogenous carbon dioxide in fixation reactions. It is important, therefore, to obtain information concerning possible endogenous dilution of labelled products. This may be achieved by the use of uniformly labelled [U-^{14}C] glucose. If there is no endogenous dilution the specific activity per carbon atom of all metabolic products should be identical with that of the substrate carbon. The extent to which the value is lowered is a measure of the dilution which has occurred.

Example 11.5. Washed suspensions of (a) *Zymomonas mobilis* and (b) *Saccharomyces cerevisiae* were the subject of experiments to determine the pathways of glucose fermentation in operation. Warburg manometers were set up containing the washed suspensions in buffer at pH 5·5 under an atmosphere of nitrogen. After temperature equilibration (30°) 10 μmoles of appropriately labelled glucose were added and the fermentations allowed to proceed until gas evolution ceased. The total CO_2 outputs were measured and then the gas was absorbed in alkali and precipitated as $BaCO_3$ for counting.

Ethanol was distilled from the reaction mixture and estimated by dichromate oxidation. Further portions of distillate, after suitable additions of carrier ethanol, were oxidized to acetic acid and the acetate was then degraded to obtain each carbon atom as CO_2. This was collected in NaOH and precipitated as $BaCO_3$ for counting. The total radiochemical yield was calculated from the results. The data obtained are recorded below:

Organism	Substrate glucose	CO_2 evolution (μl)		Ethanol (μmoles)	Radiochemical yield (nCi)	
		Control	Reaction		CO_2	Ethanol
Z. mobilis	[1-^{14}C] 20 nCi	4	405	17·7	19·8	0·0
	[3,4-^{14}C$_2$] 15 nCi	5	410	17·5	7·0	7·2
S. cerevisiae	[1-^{14}C] 20 nCi	24	443	18·3	0·0	19·5
	[3,4-^{14}C$_2$] 15 nCi	17	431	18·1	14·5	0·01

Degradation of ethanol

Organism	Substrate glucose	CH_3- nCi	$-CH_2OH$ nCi
Z. mobilis	$[3,4-^{14}C_2]$	7·0	0·05
S. cerevisiae	$[1-^{14}C]$	19·3	0·2

Determine the pathways of glucose metabolism operating in these organisms and predict the labelling you would expect to find in the very small amount of lactate (about 0·4 μmole) produced by *Z. mobilis* in this fermentation.

(a) *Z. mobilis*. The CO_2 evolutions from $[1-^{14}C]$ and $[3,4-^{14}C_2]$ glucose are respectively $\dfrac{(405 - 4)}{22·4} = 17·9$ μmoles and $\dfrac{(410 - 5)}{22·4} = 18·1$ μmoles.

The fermentation may thus be written as

$$1 \text{ Glucose} \rightarrow 1·76 \text{ Ethanol} + 1·80 \text{ } CO_2$$

This shows that ethanol and CO_2 are produced in equimolar amounts and are the major products of the fermentation. When $[1-^{14}C]$ glucose is fermented all the radioactivity is located in the CO_2, but the specific activity of the CO_2 is $19·8/17·9 = 1·1$ nCi per μg-atom of carbon as compared with $20/10 = 2$ nCi per μg-atom of C of the substrate, i.e. it has been diluted with an equal amount of unlabelled CO_2. With $[3,4-^{14}C_2]$ glucose the radioactivity is distributed almost equally between the CO_2 and ethanol, and degradation of the ethanol reveals that all the activity is located in the methyl group. The specific activity of the methyl carbon is $7·2/17·5 = 0·41$ as compared with $15/10 \times 2 = 0·75$ for the specific activity of either C-3 or C-4 of glucose (assuming equal activity in C-3 and C-4). Again this indicates dilution with approximately an equal amount of unlabelled ethanol; the CO_2, with a specific activity of $7·0/18·1 = 0·39$, is similarly diluted.

The information derived from the experimental data may therefore be summarized as

```
        1 2 3 4 5 6              1 2 3 4 5 6
        C-C-C-C-C-C              C-C-C-C-C-C
        *                            *|*
        │                             │
        ↓                             ↓
    *                            *   *
    CO₂   CH₃CH₂OH           CO₂   CH₃CH₂OH
    CO₂   CH₃CH₂OH           CO₂   CH₃CH₂OH .
```

The only known pathway of anaerobic glucose metabolism which can satisfy these requirements is Entner-Doudoroff cleavage in which C-1 becomes the carboxyl group of one of the two molecules of pyruvate produced. This could then be decarboxylated to yield CO_2. On this scheme C-3 becomes the methyl group of one pyruvate molecule and C-4 the carboxyl of the other. Thus

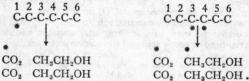

Any lactate formed would be expected to arise by reduction of pyruvate and hence it could be predicted that [1-^{14}C] glucose would yield lactate labelled in the carboxyl group whereas [3,4-^{14}C$_2$] glucose would yield lactate labelled in both carboxyl and methyl groups.

(b) *S. cerevisiae*. The CO$_2$ evolutions from [1-^{14}C] and [3,4-^{14}C$_2$] glucose are respectively $(443 - 24)/22\cdot4 = 18\cdot7$ μmoles and $(431 - 17)/22\cdot4 = 18\cdot5$ μmoles. Here again ethanol and CO$_2$ are clearly produced in equimolar amounts and are the major products

$$1 \text{ Glucose} \longrightarrow 1\cdot82 \text{ Ethanol} + 1\cdot86 \text{ CO}_2.$$

Fermentation of [1-^{14}C] glucose yields radioactivity exclusively in the ethanol, and degradation shows this to be virtually all in the methyl carbon. The specific activity of the methyl carbon is $19\cdot3/18\cdot3 = 1\cdot05$ nCi per μg-atom as compared with $20/10 = 2$ nCi for C-1 of the labelled substrate. Hence dilution with an equal amount of unlabelled ethanol has occurred.

With [3,4-^{14}C$_2$] glucose the radioactivity is exclusively in the CO$_2$ and its specific activity is $14\cdot5/18\cdot5 = 0\cdot78$ nCi per μg-atom as compared with $15/10 \times 2 = 0\cdot75$ for C-3 or C-4 of the substrate. Hence the CO$_2$ has been produced without dilution and, on account of the high yield, from both C-3 and C-4 of glucose. These findings are compatible with Embden-Meyerhof glycolysis, e.g.

1 C*	*CH$_3$	*CH$_3$	*CH$_3$
2 C	CO $\longrightarrow$	CHO $\longrightarrow$	CH$_2$OH
3 C•	•COOH	•CO$_2$	
4 C•	•COOH	•CO$_2$	
5 C	CO $\longrightarrow$	CHO $\longrightarrow$	CH$_2$OH
6 C	CH$_3$	CH$_3$	CH$_3$

The methods available for the quantitative estimation of pathways of glucose metabolism have been discussed in detail in a series of papers by Wood, Katz and their associates. It is not feasible to consider all their derivations here and the reader is referred to their review articles (see p. 425) for further details. We shall content ourselves therefore with an example of their approach, using steady state conditions, to a case where metabolism occurs exclusively by the pentose phosphate cycle and glycolysis. The situation is that depicted in Fig. 119. where the rates of the individual reactions are designated by V_1, V_2, V_3 and V_4. It is assumed that V_1 and V_2 are great compared with V_3 and V_4, i.e. the rate at which glucose 6-phosphate and fructose 6-phosphate are brought into isotopic equilibrium by the hexosephosphate isomerase is rapid compared with the rate of utilization of hexose, whether by oxidation of glucose 6-phosphate to pentose 5-phosphate and CO$_2$ (V_3) or by phosphorylation of fructose 6-phosphate to fructose 1,6-diphosphate (V_4).

Under steady state conditions the inflow of glucose equals the outflow and the outflow is via fructose 6-phosphate (V_4) in glycolysis, and by CO_2 and glyceraldehyde 3-phosphate in the pentose phosphate cycle $(\frac{1}{3}V_3)$. From the operation of the cycle shown in Fig. 11.7 it will be appreciated that $\frac{2}{3}V_3$ of fructose 6-phosphate is re-formed while $\frac{1}{3}V_3$ is converted to glyceraldehyde 3-phosphate. Hence

$$\text{glucose input} = \text{glucose output}$$

$$= V_4 + \frac{1}{3}V_3$$

By the definition of Wood & Katz, glucose which is converted to fructose 1,6-diphosphate is metabolized via the Embden-Meyerhof

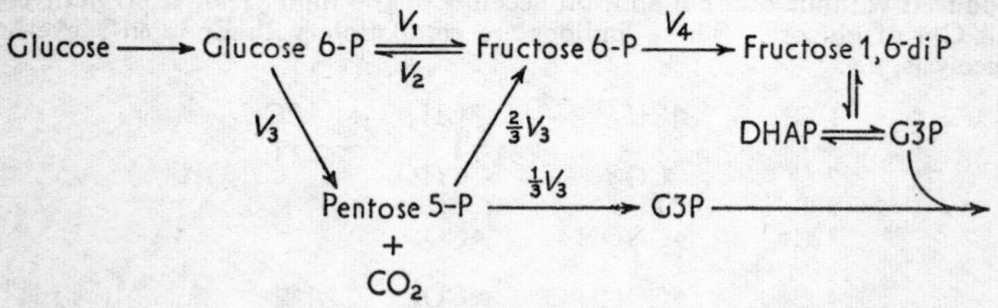

FIG. 11.9

Relationships of the rates of reactions (V) of glycolysis and the pentose phosphate cycle in glucose metabolism. The abbreviations are those used in Fig. 11.7. (WOOD, KATZ & LANDAU, 1963.)

(EM) glycolytic pathway, and the pentose which is converted to glyceraldehyde 3-phosphate is metabolized via the pentose phosphate cycle (PC),

i.e. $V_4 = \text{EM and } \frac{1}{3}V_3 = \text{PC}$

Thus $\text{EM} + \text{PC} = V_4 + \frac{1}{3}V_3$

Now when [1-^{14}C] glucose is metabolized via the pentose phosphate cycle, ^{14}C is lost in CO_2 and the fructose 6-phosphate regenerated by the cycle does not contain isotope; consequently dilution of ^{14}C occurs in the hexosemonophosphate pool. In order to be able to estimate pathways the extent of this dilution must be known. An equation to secure this information will now be derived (Wood, Katz & Landau, 1963).

Consider glucose to be metabolized exclusively by glycolysis and the pentose phosphate cycle and in the proportions $\text{EM} = 0·6$ and $\text{PC} = 0·4$. If the inflow of glucose is taken as 1 μmole per unit time and the glucose has a specific activity of A, the inflow would be equal to $1A$. The specific activity of the equilibrated glucose 6-phosphate and fructose 6-phosphate in the pool at the steady state is designated as a_1 for [1-^{14}C] glucose as the substrate and as α_6 for [6-^{14}C] glucose.

The outflow of ^{14}C when [1-^{14}C] glucose is metabolized would be equal to

$$0.6 \times a_1 + 3.0 \times 0.4 \times a_1$$

Note that the factor 3.0 enters the expression because 3 moles of $^{14}CO_2$ are produced from C-1 per net hexose phosphate utilized via the pentose phosphate cycle. With [6-^{14}C] glucose as substrate the corresponding expression will be

$$0.6 \times a_6 + 0.4 \times a_6$$

Algebraically, $1 - PC$ (equivalent to EM) can be substituted for 0.6 and PC for 0.4. Then, with the two specifically labelled substrates, we have for [1-^{14}C] glucose

$$1A = 0.6a_1 + 3 \times 0.4a_1$$
$$= (1 - PC)a_1 + (3PC)a_1$$
$$= (1 + 2PC)a_1$$

and for [6-^{14}C] glucose

$$1A = 0.6a_6 + 0.4a_6$$
$$= (1 - PC)a_6 + (PC)a_6$$
$$= a_6$$

Since equivalent amounts of labelled substrate are utilized, the outflows are equal and

$$(1 + 2PC)a_1 = a_6$$

Thus the ratio of the specific activity of hexose 6-phosphate in an experiment with [1-^{14}C] glucose to that in an experiment with [6-^{14}C] glucose will be

$$\frac{a_1}{a_6} = \frac{1}{(1 + 2PC)} = Q \qquad . \qquad . \quad (11.29)$$

This ratio is designated Q, or the *dilution factor* for [1-^{14}C] glucose, by Katz & Wood. It will be apparent that the greater the contribution the pentose phosphate cycle makes to glucose metabolism the greater will be the dilution effect ; this is illustrated by the data of Table 11.4.

TABLE 11.4

The relative dilution of ^{14}C in the hexose 6-phosphate pool which occurs with [1-^{14}C] glucose as compared with [6-^{14}C] glucose expressed as a function of the proportion of metabolism occurring by the pentose phosphate cycle.

Values are calculated from equation 11.29, i.e.

$$\frac{\alpha_1}{\alpha_6} = \frac{1}{1 + 2PC} = Q$$

Percentage contribution of pentose phosphate cycle	0	10	20	30	40	50	60	70	80	90	100
Dilution factor Q	1·0	0·83	0·72	0·625	0·56	0·50	0·45	0·42	0·39	0·36	0·33

Application of the dilution factor to the calculation of pathways will now be considered for the situation illustrated in Fig. 11.10. Suppose the glucose has a specific activity of 5000 counts/min/μmole, that 1 μmole of glucose is utilized per minute and 0·6 is

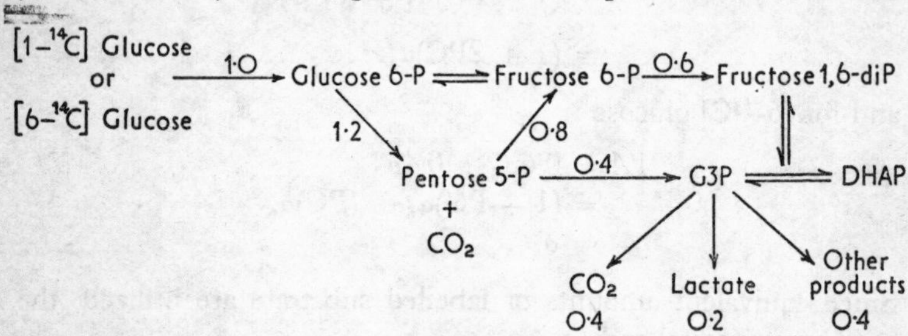

FIG. 11.10

Metabolism of glucose exclusively by glycolysis and the pentose phosphate cycle, showing the fate of glyceraldehyde 3-phosphate. The values shown are those discussed in the example on page 412.

metabolized by glycolysis and 0·4 by the pentose phosphate cycle, then the appropriate dilution factor Q will be 0·56 (Table 11.4). Assume that 0·4 part of the glyceraldehyde 3-phosphate produced is oxidized to CO_2, 0·2 part is recovered as lactate and 0·4 part is converted to other products such as amino acids and fatty acids. As previously mentioned, it is possible to base the calculations on

the ^{14}C yields in either CO_2 or a product of triose phosphate metabolism (lactate in this instance), and the two derivations will now
be discussed.

Method 1. *Calculation based on triose phosphate derivatives
from* [1-^{14}C] *and* [6-^{14}C] *glucose.*

[1-^{14}C] Glucose yields [^{14}C]glyceraldehyde 3-phosphate only
by glycolysis, since this compound is formed only from carbon
atoms 4, 5 and 6 of glucose by the pentose phosphate cycle (Fig.
11.7). Consequently 0·6 μmole of [^{14}C] glyceraldehyde 3-phos-
phate would be formed per minute. But the pentose phosphate
cycle would cause dilution of the ^{14}C in the hexose monophosphate
pool and the specific activity would be 5000 × 0·56.

As only 0·2 part of glyceraldehyde 3-phosphate is converted to
lactate, the total ^{14}C yield in lactate would be

$$5000 \times 0\cdot56 \times 0\cdot6 \times 0\cdot2 = 336 \text{ counts/min.}$$

[6-^{14}C] Glucose would be converted to glyceraldehyde 3-phos-
phate by both pathways and as no dilution factor is involved in
this case the yield in lactate would be

$$5000 (0\cdot6 + 0\cdot4) 0\cdot2 = 1000 \text{ counts/min.}$$

It is assumed that the experiments with the two differently labelled
substrates are comparable with respect to both glucose utilized
and lactate formed so that when the yields of ^{14}C in lactate are expressed as a ratio the 5000 and 0·2 cancel leaving as the ratio (expressed algebraically as before)

$$\frac{[1\text{-}^{14}C] \text{ Glucose, triose-P deriv.}}{[6\text{-}^{14}C] \text{ Glucose, triose-P deriv.}}$$

$$= \frac{0\cdot56 \times 0\cdot6}{(0\cdot6 + 0\cdot4)} = \frac{Q \times \text{EM}}{(\text{EM} + \text{PC})} \quad . \quad (11.30)$$

If metabolism is solely by EM and PC, i.e. EM + PC = 1, then
since $Q = 1/(1 + 2\text{PC})$, the above ratio may be expressed as

$$\frac{[1\text{-}^{14}C] \text{ Glucose, triose-P deriv.}}{[6\text{-}^{14}C] \text{ Glucose, triose-P deriv.}} = \frac{1 - \text{PC}}{1 + 2\text{PC}} \quad . \quad (11.31)$$

For the example considered this ratio would have the value
0·336. Thus if the yields of ^{14}C in lactate are determined, the

proportion of metabolism occurring by the pentose phosphate cycle can be calculated by means of equation 11.31.

Method 2. *Calculation based on* $^{14}CO_2$ *yields from* [1-^{14}C] *and* [6-^{14}C] *glucose.*

It is not possible to carry out a calculation of the percentage participation of the two pathways simply on the ratio of the yields of ^{14}C in carbon dioxide. The reason is that $^{14}CO_2$ arises from two sources, namely the pentose phosphate cycle and the tricarboxylic acid cycle. In contrast, the sole source of lactate is glyceraldehyde 3-phosphate so that in Method 1 the factor (assumed to be 0·2 in the example) which is the fraction of the glyceraldehyde 3-phosphate converted to lactate is common to both labelled sugars and cancels in the ratio. This is not true of the corresponding factor for CO_2, i.e. the extent of oxidation of C-3 of glyceraldehyde 3-phosphate to CO_2, so that this factor (termed N by Wood *et al*, 1963) remains in the ratio. Consequently the expression for the ratio contains two unknowns, PC and N, and the value of N, which varies considerably according to physiological conditions, is frequently the predominating factor.

Radiorespirometry

Because of the inherent difficulties of CO_2 release methods based on specific radioactivity measurements, Wang and his collaborators introduced what they have termed the *radiorespirometric method* for the estimation of concurrent catabolic pathways of glucose metabolism (for a good survey, see Wang, 1971). It consists of two major methods based respectively on the $^{14}CO_2$ yield and the catabolic rate. The former technique measures the yields of $^{14}CO_2$ from various metabolites, making use of a modified Warburg apparatus to permit the collection of CO_2 at desired intervals throughout the experiment. Plots of the interval yields of $^{14}CO_2$ from different substrates reflect the kinetics of the process. As it is the total yield of $^{14}CO_2$ which is measured the method effectively eliminates the vagaries of metabolism which cause dilution of individual carbon atoms and so affect the specific activities obtained ; these effects can thus be ignored.

The radiorespirometric method considers glucose catabolism in micro-organisms primarily in the light of the immediate catabolic fate of the glucose and recognizes three sequences, namely glycolysis, pentose cycle and the Entner-Doudoroff pathway. They are defined as follows :

Glycolytic (EM) pathway

Glucose 6-phosphate $\rightarrow$ fructose 1,6-diphosphate $\rightarrow$
2 triosephosphate $\rightarrow$ x pyruvate

where x is the actual yield of pyruvate and has an upper limit of 2.

Pentose phosphate (PP) pathway

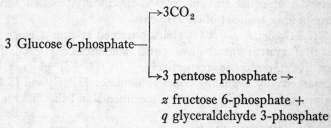

3 Glucose 6-phosphate —

$\rightarrow 3CO_2$

$\rightarrow 3$ pentose phosphate $\rightarrow$

z fructose 6-phosphate +
q glyceraldehyde 3-phosphate

where z and q represent the actual yields of fructose 6-phosphate and glyceraldehyde 3-phosphate respectively. The upper limits are $z = 2$ and $q = 1$. In the unlikely event of all the pentose phosphate being utilized for biosynthesis then $z = 0$ and $q = 0$.

Entner-Doudoroff (ED) pathway

Glucose 6-phosphate $\rightarrow$ 1 glyceraldehyde 3-phosphate +
1 pyruvate

A set of equations has been derived on the basis of certain assumptions, the principal of which are as follows :

1. Glucose is metabolized either by the operation of concurrent EM and PP pathways or by concurrent ED and PP pathways, for respiratory and biosynthetic functions.
2. The preferential conversion of C-1 of glucose to CO_2 via the PP is a virtually irreversible process.
3. The C_3 units formed by EM glycolysis are equivalent to each other with respect to further metabolic reactions.
4. The formation of hexose by recombination of trioses, and the randomization of the hexose skeleton via transketolase or transaldolase exchange reactions, do not occur to any significant extent.

5. Gluconate presented as a substrate is utilized by the biological system in a manner identical with 6-phosphogluconate derived from glucose 6-phosphate *in vivo*.

The following terminology is used by Wang, where all yields are expressed as fractions of unity.

G_T = total amount of each labelled glucose administered, expressed on a percentage basis as unity, i.e. $G_T = 1.00$. $G_{T'}$ = fraction of labelled glucose administered that is engaged in anabolic processes, expressed as a fraction of unity. It is understood that equal amounts of ^{14}C-specifically labelled substrates are used in each set of experiments.

G_1, G_6 or G_4 = $^{14}CO_2$ yields observed at time t when a biological system metabolizes equal amounts of [1-^{14}C], [6-^{14}C] or [4-^{14}C] glucose respectively. As [4-^{14}C] glucose is not commercially available, equal amounts of [3-^{14}C] and [3,4-$^{14}C_2$] glucose are used in separate experiments and the difference obtained.

A_6 = $^{14}CO_2$ yield observed at time t when the system metabolizes a given amount of [6-^{14}C] gluconate. G_e, G_p, G_{ed} = fraction of administered glucose catabolized respectively via EM PP and ED pathways.

With this technique it is only practicable to base the calculations on the cumulative $^{14}CO_2$ yield data obtained at the time of exhaustion of the substrate glucose, to which the time interval t refers (this period is termed 1 relative time unit or RTU), or thereafter.

Consider now the concurrent operation of EM and PP pathways. If it were assumed that the pentose phosphate derived from glucose via the PP route is not further catabolized, then the participation of the PP would be given by

$$G_p = \frac{G_1 - G_6}{G_T - G_{T'}} \qquad . \qquad . \qquad . \quad (11.32a)$$

When $G_{T'}$ is small, i.e. little diversion of the substrate for anabolic purposes occurs, then $G_T - G_{T'} = 1$ and equation 11.32a becomes

$$G_p = G_1 - G_6 \qquad . \qquad . \qquad . \quad (11.32b)$$

However, when the catabolism of pentose phosphate is taken into consideration the possibility arises that a portion of the C-6 of glucose is converted to CO_2 via the PP route, in addition to that derived from C-6 via the EM sequence and the tricarboxylic acid cycle. The amount of CO_2 derived from C-6 via the PP pathway can be estimated as the product of the amount of glucose that has traversed the PP pathway and the CO_2 yield from C-6 of glucose, i.e. $G_p.G_6$. This follows since theoretically C-6 of glucose is converted to C-3 of glyceraldehyde 3-phosphate via either EM or PP routes in the same yield (Figs. 11.6 and 11.7). In practice, however, since the biosynthetic functions of the EM and PP pathways differ, the yield of C-6 of glucose in the triose phosphate via these routes may differ, depending on the magnitude of the factors x and q in the foregoing equations for the EM and PP pathways. To estimate the extent of such diversion in the PP pathway, the yield of CO_2 derived from C-6 of glucose may be equated to $G_p.A_6$, i.e. the product of the amount of substrate glucose that traversed the hexose monophosphate oxidation pathway to pentose phosphate and the CO_2 yield from C-6 of gluconate.

Equation 11.32a can thus be modified to read

$$G_p = \frac{G_1 - (G_6 - G_p.G_6)}{G_T - G_{T'}} \qquad . \qquad . \quad (11.33a)$$

and again, when $G_{T'}$ is small, we have

$$G_p = G_1 - G_6 + G_p.A_6 \qquad . \qquad . \quad (11.33b)$$

which becomes

$$G_p = \frac{G_1 - G_6}{1 - A_6} \qquad . \qquad . \qquad . \quad (11.34)$$

If the assumption is made that the triose phosphates derived from glucose via the PP route and from administered gluconate via the PP route are catabolized in an identical manner, then equation 11.34 becomes

$$G_p = \frac{G_1 - G_6}{1 - G_6} \qquad . \qquad . \qquad . \quad (11.35)$$

This expression is particularly useful when systems which cannot utilize gluconate as an exogenous carbon source are being investigated in radiorespirometric experiments.

Where EM and PP are the only pathways in operation then the extent of the EM contribution can be obtained by difference, thus

$$G_e = 1 - G_p \qquad . \qquad . \qquad . \quad (11.36)$$

We may now consider this method in relation to the example previously quoted where the dilution factor Q for [1-^{14}C] glucose is 0·56. $3CO_2$ are formed from C-1 per μmole of glucose utilized via the pentose phosphate cycle and 0·4 part of the C-3 of glyceraldehyde 3-phosphate is oxidized to CO_2 (N). The yield of ^{14}C from [1-^{14}C] glucose will be

$$= 5000 \times 0.56 \, (0.6 \times 0.4 + 3 \times 0.4)$$
$$= 5000 \times 0.56 \times 1.44$$

and the yield from [6-^{14}C] glucose

$$= 5000 \times 0.4 \, (0.6 + 0.4)$$
$$= 5000 \times 0.4$$

Thus
$$G_p = \frac{G_1 - G_6}{G_T} = \frac{5000 \, (0.56 \times 1.44 - 0.4)}{5000}$$
$$= 0.80 - 0.40$$
$$= 0.40$$

and
$$G_e = 1 - 0.40 = 0.60$$

The equations for the specific yield of $^{14}CO_2$ can, of course, be expressed algebraically in the terms previously employed. The yield from [1-^{14}C] glucose is equivalent to

$$A \times Q \, (EM \times N + 3PC)$$

and that from [6-^{14}C] glucose is

$$A \times N \, (EM + PC) = A \times N$$

The specific yields, denoted by $G1_{CO_2}$ and $G6_{CO_2}$ respectively in the terminology of Wood et al, are therefore

$$G1_{CO_2} = Q(EM \times N + 3PC) \qquad . \qquad . \quad (11.37)$$

$$G6_{CO_2} = N \qquad . \qquad . \qquad . \qquad . \quad (11.38)$$

As before, $1/(1 + 2PC)$ may be substituted for Q, and $G6_{CO_2}$ for N in the equation 11.25. On rearranging

$$\frac{G1_{CO_2} - G6_{CO_2}}{1 - G6_{CO_2}} = \frac{3PC}{1 + 2PC} \qquad . \qquad . \quad (11.39)$$

i.e.
$$G_p = \frac{3PC}{1 + 2PC}$$

and so, by determining the specific yields of $^{14}CO_2$ from [1-^{14}C] and [6-^{14}C] glucose the percentage participation of the pentose phosphate cycle may be determined. In the above example $G1_{CO_2} = 0.80$ and $G6_{CO_2} = 0.4$, therefore

$$\frac{0.80 - 0.4}{1 - 0.4} = \frac{3PC}{1 + 2PC}, \text{ whence PC} = 0.4,$$

It is fortunate that the Entner-Doudoroff pathway does not appear to function simultaneously with glycolysis. The $^{14}CO_2$ yield method must, however, allow for the simultaneous operation of the Entner-Doudoroff and pentose phosphate pathways. In this case the difference between the yield of $^{14}CO_2$ from C-1 and C-4 of glucose is measured, since exclusive Entner-Doudoroff activity would result in equal yields from these carbon atoms

$$G_p = \frac{G_1 - G_4}{G_T}$$

This equals $G_1 - G_4$ when $G_{T'} = 0$. G_{ed}(Entner-Doudoroff) is the remainder ($G_{ed} = 1 - G_p$).

The application of the $^{14}CO_2$ yield technique necessitates the assumption that the bulk of the assimilated glucose has been catabolized, i.e. $G_{T'} = 0$, an assumption which is not always justified. For example, many micro-organisms, and especially fungi, synthesize significant quantities of endogenous reserves of carbon. In these cases it is necessary to evaluate $G_{T'}$, a difficult task which, in fact, constitutes the major disadvantage of the method.

THE CATABOLIC RATE METHOD.—Developments in techniques for the high-resolution measurement of the rate of production of respiratory $^{14}CO_2$ have enabled Wang and his colleagues to devise a radiorespirometric method that permits the determination of the amount of substrate glucose engaged in anabolic processes. A second advantage is the independent and direct measurement of the relative participation of the pathways instead of measuring only G_p and obtaining G_e by difference.

The basis for the technique is as follows. It is assumed that in the glycolytic-tricarboxylic acid cycle (EM-TCA) sequence C-3 and C-4 of glucose are rapidly converted to $^{14}CO_2$ via pyruvate oxidation to acetyl CoA, the key intermediate for entry to the TCA cycle. If it is further assumed that triose phosphate and pyruvate derived via the EM pathway do not significantly engage as intact

units in anabolic processes then, for each mole of glucose catabolized via the EM-TCA pathway, C-3 and C-4 will each yield 1
mole of CO_2. The rate of production of $^{14}CO_2$ from C-3 and C-4
of glucose should thus reflect directly the catabolic rate of the EM
pathway. Clearly the method depends on the validity of the
assumptions made and which Wang believes to be justified.

While the net rate of formation of respiratory CO_2 from C-1 of
glucose should, on similar premises, represent directly the catabolic
rate of the PP pathway, there is the complication that the EM-TCA
sequence can additionally produce CO_2 from the C-1 position.
However, whereas the conversion of C-1 to CO_2 via PP is rapid,
involving but three enzymic steps, its conversion via EM-TCA
involves traversing the whole of the EM sequence and cycling
more than once in the TCA cycle. Hence the appearance of C-1
as CO_2 via the EM-TCA route is relatively slow compared with
its appearance in the PP pathway. Consequently if the rate of
formation of $^{14}CO_2$ from [1-^{14}C] glucose is extrapolated to the
time of administration of the labelled substrate, it may be taken
as the rate of operation of the PP pathway.

These experiments are carried out with proliferating cells and
samples of the culture are withdrawn at intervals to monitor growth
and also to determine the radioactivity of the cells and the medium.
From this information the corrected rates of $^{14}CO_2$ production
are calculated and the rate of glucose assimilation ascertained.
These data enable the relative participation of the pathways of
glucose metabolism to be calculated.

Example 11.6.—Jacobsen and Wang have carried out experiments to evaluate
the concurrent operation of the Embden-Meyerhof and pentose phosphate
pathways for glucose metabolism in baker's yeast, *Saccharomyces cerevisiae*,
employing the catabolic rate radiorespirometry technique.

Three substrates were used, 747 μmoles of each being administered.

[1-^{14}C]glucose 5·04 μCi. Specific activity = 5·04/747
= 0·00675 μCi/μmole.
[3,4-$^{14}C_2$]glucose 3·04 μCi. Specific activity = 3·04/747
= 0·00407 μCi/μmole.
[6-^{14}C]glucose 4·85 μCi. Specific activity = 4·85/747
= 0·00649 μCi/μmole.

By assaying the glucose content of the medium during the mid-exponential phase
following isotopic substrate addition, the assimilation rate of glucose (G_{ar}) was
found to be linear and equal to 0·540 μmole/min.

The rate of $^{14}CO_2$ production from each labelled glucose was recorded, corrected for the increase in cell numbers and extrapolated to zero time, i.e. the
time of administration of the labelled substrate. The values were then expressed

as a percentage of the total substrate radioactivity administered. They are recorded as G_{r-1}, G_{r-6}, $G_{r-3,4}$ etc. Thus

G_{r-1} = 0·020 per cent of the total substrate activity/min.
     i.e. $0 \cdot 00020 \times 5 \cdot 04$ μCi = $0 \cdot 00101$ μCi/min.

G_{r-6} = 0
$G_{r-3,4}$ = 0·022 per cent of the total substrate activity/min.
     i.e. $0 \cdot 00022 \times 3 \cdot 04$ μCi = $0 \cdot 00067$ μCi/min.

Jacobsen and Wang use the following terminology :

G_{pr} = catabolic rate of the pentose phosphate pathway
G_{er} = catabolic rate of the Embden-Meyerhof-pyruvate decarboxylation pathway
G_{cat} = rate of glucose catabolism
G_{ana} = rate of glucose anabolism

The relative participation of the glucose pathways can now be determined from the experimental data.

The overall rate of glucose metabolism is given by

$$G_{ar} = G_{cat} + G_{ana}$$

Assuming that only the EM and PP pathways operate in this organism we have

$$G_{cat} = G_{er} + G_{pr}$$

and the catabolic rate of the individual catabolic reactions can be calculated as follows :

$$G_{er} = \frac{G_{r-3,4}}{\text{sp. act. of } [3,4\text{-}^{14}C_2] \text{ glucose}}$$

$$= \frac{0 \cdot 00067}{0 \cdot 00407} = 0 \cdot 165 \ \mu\text{mole/min.}$$

$$G_{pr} = \frac{G_{r-1}}{\text{sp. act. of } [1\text{-}^{14}C] \text{ glucose}}$$

$$= \frac{0 \cdot 00101}{0 \cdot 00675} = 0 \cdot 150 \ \mu\text{mole/min.}$$

Therefore $G_{cat} = 0 \cdot 165 + 0 \cdot 150 = 0 \cdot 315 \ \mu\text{mole/min.}$

Since $G_{ana} = G_{ar} - G_{cat}$

$$G_{ana} = 0 \cdot 540 - 0 \cdot 315 = 0 \cdot 225 \ \mu\text{mole/min.}$$

From these rate values the relative participation of the glucose pathways can be expressed on the basis of the rate of glucose assimilation, G_{ar}.

Thus :

Catabolic pathways $= \dfrac{0 \cdot 315 \times 100}{0 \cdot 540} = 58$ per cent.

Anabolic pathways $= \dfrac{0 \cdot 225 \times 100}{0 \cdot 540} = 42$ per cent.

Embden-Meyerhof pathway $= \dfrac{0 \cdot 165 \times 100}{0 \cdot 540} = 30$ per cent.

Pentose phosphate pathway $= \dfrac{0 \cdot 150 \times 100}{0 \cdot 540} = 28$ per cent.

The relative participation of these pathways of glucose catabolism expressed on the basis of the net amount of substrate glucose which is routed into catabolic mechanisms is consequently

$$\text{Embden-Meyerhof} = \frac{30 \times 100}{58} = 52 \text{ per cent.}$$

$$\text{Pentose phosphate} = \frac{28 \times 100}{58} = 48 \text{ per cent.}$$

The very substantial proportion of the substrate glucose utilized for anabolic processes is noteworthy and it emphasizes the fallacy of one of the major assumptions underlying the $^{14}CO_2$ yield method.

(Data from JACOBSEN & WANG (1967), *Federation Proc.*, 26 (2).

Limitations and Precautions

REACTION RATES OF ISOTOPES.—Although isotopes are almost identical in chemical properties there are some slight differences which must be borne in mind when using isotopes as tracers. For instance, there are slight differences in reaction rates: although isotopes undergo the same reactions, a heavy isotope reacts more slowly than a lighter one. The most pronounced examples are deuterium and tritium. Deuterium has a mass double that of protium (hydrogen, mass 1) and may, under the most unfavourable conditions, react at one-eighteenth of the rate of protium. Tritium (T or ^3H) has treble the mass of protium and the discrepancy in reaction rate is even further accentuated. For example, tritium may react at one-sixtieth the rate of protium. These effects are particularly pronounced in reactions involving labilization of the hydrogen atom, e.g. dehydrogenation reactions, and Thorn has shown that tetra-deuterium substituted succinic acid is oxidized by a succinate oxidoreductase preparation at only 40 per cent. of the rate of normal succinate. The activation energy was computed to be 1450 ± 450 cal. higher than the activation energy of the reaction with the normal substrate. Where the labelling atom is not directly involved in the reaction, as for example when it maintains its position in the molecule throughout the reactions followed, there is no effect on the reaction rate. For a thorough treatment of the effects of isotopes on reaction rates the reader is referred to Bigeleisen and Wolfsberg (1958).

EXCHANGE REACTIONS.—Care must be taken to ensure that the labelled atom occupies a position in the molecule where it will not undergo spontaneous exchange with unlabelled atoms present in the medium. This consideration excludes the use of deuterium in carboxyl, hydroxyl and amino groups because rapid exchange occurs between the deuterium and the hydrogen of the aqueous medium. Many other types of exchange reaction are known. To quote an example, deuterium-labelled valine may be prepared with the isotope in the β and γ positions by means of an exchange reaction with D_2SO_4.

REFERENCES AND SUGGESTED READING

ARONOFF, S. (1956). *Techniques of Radiobiochemistry.* Ames : Iowa State College Press.

BIGELEISEN, J. & WOLFSBERG, M. (1958). Theoretical and experimental aspects of isotope effects in chemical kinetics. *Adv. chem. Phys.*, 1, 15. New York : Interscience Publishers.

CHELDELIN, V. H., WANG, C. H. & KING, T. E. (1962). Saccharides : alternate routes of metabolism. In *Comparative Biochemistry*, 3, 427, ed. M. Florkin & H. S. Mason. New York : Academic Press.

CROSBIE, G. W. (1971). Ionization methods of counting radio-isotopes. In *Methods in Microbiology*, 6B, Ch. 4, ed. J. R. Norris & D. W. Ribbons. New York : Academic Press.

FRANCIS, G. E., MULLIGAN, W. & WORMALL, A. (1959). *Isotopic Tracers*, 2nd ed. London : Athlone Press.

HASH, J. H. (1971). Liquid scintillation counting in microbiology. In *Methods in Microbiology.*, 6B, Ch. 5, ed. J. R. Norris & D. W. Ribbons. New York : Academic Press.

KAMEN, M. D. (1957). *Isotopic Tracers in Biology*, 3rd ed. New York : Academic Press.

KATZ, J. (1961). The use of glucose-^{14}C in the study of the pathways of glucose metabolism in mammalian tissues. In *Radioactive Isotopes in Physiology Diagnostics and Therapy*, 705, ed. H. Schwiegk & F. Turba. Berlin : Springer Verlag.

MCFARLANE, A. S. (1957). The behaviour of ^{131}I-labelled plasma proteins *in vivo*. *Ann. N.Y. Acad. Sci.*, 70, 19.

MCFARLANE, A. S. (1964). Metabolism of plasma proteins. In *Mammalian Protein Metabolism*, vol. I, 297, ed. H. N. Munro & J. B. Allison. New York : Academic Press.

MATTHEWS, C. E. M. (1957). The theory of tracer experiments with ^{131}I-labelled plasma proteins. *Physics Med. Biol.*, 2, 36.

NEUBERGER, A. & RICHARDS, F. F. (1964). Protein biosynthesis in mammalian tissues. II Studies on turnover in the whole animal. *Mammalian Protein Metabolism*, vol. I, 243, ed. H. N. Munro & J. B. Allison. New York : Academic Press.

PENG, C. T. (1966). *Advances in Tracer Methodology*, 3, 81. New York : Plenum Press.

QUAYLE, J. R. (1971). The use of isotopes in tracing metabolic pathways. In *Methods in Microbiology*, 6B, Ch.6, ed. J. R. Norris & D. W. Ribbons. New York : Academic Press.

ROBERTSON, J. S. (1957). Theory and use of tracers in determining transfer rates in biological systems. *Physiol. Rev.*, 37, 133.

WANG, C. H. (1971). Radiorespirometric methods. In *Methods in Microbiology*, 6B, Ch. 7, ed. J. R. Norris & D. W. Ribbons. New York : Academic Press.

WOOD, H. G., KATZ, J. & LANDAU, B. R. (1963). Estimation of pathways of carbohydrate metabolism. *Biochem. Z.*, 338, 809.

ZILVERSMIT, D. B., ENTENMAN, C. & FISHLER, M. C. (1943). On calculation of 'turnover time' and 'turnover rate' from experiments involving use of labelling agents. *J. gen. Physiol.*, 26, 325.

PROBLEMS

11.1. ^{32}P has a half-life of 14·3 days. Calculate the decay constant and determine what percentage of the initial radioactivity remains after 10, 20, 30 and 50 days.

11.2. In an experiment with ^{32}P (half-life = 14·3 days) a sample containing 50 μg phosphorus was assayed in a Geiger-Müller counter with a background count of 14 counts per minute. The recorded count was 2640 in 10 minutes. If the sample was assayed 14·3 days after commencement of the experiment what was the specific activity of the phosphorus (in counts/min./100 μg total phosphorus) at the start of the experiment ?

(Glasgow Double Science Course, 1952.)

11.3. Radioactive sulphur ^{35}S was assayed over a period of 15 days and the following counts recorded in a Geiger-Müller counter. Determine the half-life period and the decay constant for this isotope.

Time (days)	0	1	2	3	4	5	10	15
Counts/min	4280	4245	4212	4179	4146	4113	3952	3798

11.4. Shemin and Rittenberg demonstrated that glycine is a specific precursor of the haem present in red blood cells. Glycine labelled with ^{15}N was ingested over a period of 3 days by the subject (Shemin) and the isotope content of haemoglobin isolated from red blood cells determined during the course of several months. The following data were obtained :

Time (days)	0	4	18	77	86	99
^{15}N atom per cent. excess	0·000	0·134	0·422	0·466	0·462	0·448
Time (days)	127	134	154	170	192	231
^{15}N atom per cent. excess	0·342	0·200	0·164	0·112	0·096	0·062

Express these data graphically. What explanation can you offer for the observed behaviour of the isotope content of the haem ?

(After SHEMIN & RITTENBERG (1946), *J. biol. Chem.*, 166, 627.)

11.5. 3·0 grams of dry haemoglobin were hydrolysed and racemized by treatment with 33 per cent. sulphuric acid in a sealed ampoule at 170°. The hydrolysate was mixed with 0·392 g DL-leucine containing deuterium. The sulphuric acid was then removed and the amino acid mixture dried and powdered. The residue was extracted to remove unwanted material, and the leucine isolated as the copper salt and purified by recrystallization from water.

The excess density of water formed by combustion of the heavy leucine was 518 parts per million and that of the leucine isolated from the hydrolysate 161 parts per million.

Calculate the weight of leucine present in the hydrolysate and hence the leucine content of haemoglobin as a percentage.

(After USSING (1939), *Nature*, 144, 977.)

11.6. An experiment was carried out in order to study the biosynthesis of phospholipids in the laying hen. Radioactive inorganic phosphate was administered intravenously at frequent intervals so as to maintain a constant blood level of inorganic ^{32}P over a period of 72 hours. At the end of this time the animal was sacrificed and the phospholipids of the liver, blood and yolks were isolated for analysis. The following results were obtained :

Substance	P in sample counted (μg)	Counts/minute
Blood lecithin .	23·2	474
Blood cephalin .	14·4	495
Liver lecithin .	17·4	360
Liver cephalin .	12·0	246
Yolk lecithin .	24·0	1440
Yolk cephalin .	19·2	379

What conclusions can you draw as to the biosynthesis of liver, blood and yolk phospholipids ?

(Glasgow Double Science Course, June 1954.)

11.7.   The palmitic acid content of a mixture of fatty acids derived from rat fat has been determined by the isotope dilution method.   Deuteriopalmitic acid was added to the mixture and then a sample of pure palmitic acid isolated and its isotope content determined by mass spectrometer.   Results from two experiments are recorded below.

Total fatty acids in mixture, g	14·641	14·135
Labelled palmitic acid added, g	0·2163	0·1757
Deuterium content of added palmitic acid, per cent. .	21·5	21·5
Deuterium content of isolated palmitic acid, per cent. .	1·28	1·18

Calculate the amount of palmitic acid present from the figures obtained in each experiment.   The difference in molecular weight between palmitic and deuteriopalmitic acids may be neglected.

(After RITTENBERG & FOSTER (1940), *J. biol. Chem.*, **133**, 737.)

11.8.   Amino acid analyses of horse haemoglobin have been carried out by Foster using the isotope dilution technique.   A weighed sample of the protein was hydrolysed for 18-20 hours with 6 N-HCl, followed by addition of a weighed amount of isotopic DL-amino acid of known ^{15}N excess to the hydrolysate.   Isolation of the natural (L) isomer of the amino acid in a high state of purity was achieved and the atom per cent. ^{15}N excess determined by the mass spectrograph.   The following data were obtained :

Compound	Amount of L-amino acid g	^{15}N excess atom per cent.	Mean ^{15}N excess in compound isolated atom per cent.	Weight of protein hydrolysed g
Tyrosine . .	0·0740	6·85	0·699	21·65
Phenylalanine .	0·1296	6·79	0·472	25·60
Arginine . .	0·0850	8·72	0·838	21·60
Lysine . .	0·1372	9·09	0·558	24·58
Glutamic acid .	0·2030	4·52	0·383	25·60
Aspartic acid .	0·1642	6·75	0·450	22·13
Glycine . .	1·1869	1·15	0·570	22·13
Leucine . .	0·0767	6·73	0·638	4·85

Calculate the percentage composition of haemoglobin with respect to the analysed amino acids and, assuming the molecular weight of the protein to be 66,700, determine the number of residues of each amino acid per mole of haemoglobin.

(After FOSTER (1945), *J. biol. Chem.*, **159**, 431.)

11.9. Give an account of the mechanism of urea formation in the mammalian organism.

α-^{15}N-Ornithine or δ-^{15}N-ornithine was fed over several days to a group of mice and the amino acids of the tissue proteins then assayed for ^{15}N. Some of the results obtained are given in the following table :

Contributions of α- and δ-nitrogen atoms of ornithine to some amino acids

Amino acid	N derived from δ-N of ornithine per cent.	N derived from α-N of ornithine per cent.
Arginine		
Amidine N	0·56	0·42
α-Amino N	0·02	6·48
δ-Amino N	6·24	0·00
Glutamic acid		0·75
Aspartic acid		0·56
Proline		1·41
Hydroxyproline		0·46

Discuss the significance of these results.

(After STETTEN (1951), *J. biol. Chem.*, **189**, 499 ; Glasgow Double Science Course, March 1954.)

11.10. Amino acid analyses of β-lactoglobulin have been carried out by Foster using the same experimental techniques as for haemoglobin (Problem 11.8).

Data obtained in isotope dilution experiments are recorded below. From this information express the amino acid contents as percentages and also determine the number of amino acid residues per mole of β-lactoglobulin, the molecular weight of which may be assumed to be 42,000.

Compound	Amino acid added		Mean ^{15}N excess in compound isolated atom per cent.	Weight of protein hydrolysed g
	Amount of L-amino acid g	^{15}N excess atom per cent.		
Glutamic acid	0·1484	4·62	0·448	7·05
Aspartic acid	0·0681	6·75	0·534	7·05
Lysine	0·0660	9·09	0·686	7·05
Leucine	0·0688	6·73	0·396	7·05
Glycine	0·3147	1·99	1·485	7·05

(After FOSTER (1945), *J. biol. Chem.*, **159**, 431.)

11.11. Experiments to discover the immediate phosphorus precursor of phospholipins (lecithin and cephalin) have been carried out by Popjak and Muir. Rats in groups of four to six were injected subcutaneously with Na_2HPO_4 labelled with ^{32}P. The members of the group were killed at different intervals after the injection, the liver excised, frozen and ground to a powder. The powdered liver was extracted and the extract fractionated for acid-soluble phosphates. Five fractions were obtained corresponding to inorganic phosphate, phosphoglyceric acid, α- and β-glycerophosphates, adenylic acid and phospholipin, and each was assayed in a Geiger-Müller counter. Results obtained are recorded below.

Use these data to deduce the possible immediate precursor of liver phospholipins and also the turnover time of the phospholipins.

Time after injection of ^{32}P (min)	Fraction A Inorganic P	Fraction B ?Residual inorganic P ?Phosphoglyceric acid	Fraction C α- and β-glycerophosphate	Fraction D Adenylic acid	Fraction E Phospholipin P
50	4·4	2·80	2·65	1·20	0·15
60	5·2				
100	5·1	3·80	2·45	1·85	0·30
182	3·55		2·45		
190	3·38		2·20		
240					0·70
250	3·02		2·10	1·95	
255		2·85			
440	2·22		1·70	1·90	1·35
530	2·05	1·90	1·60	1·75	1·30
685			1·48		1·48
875	1·40		1·20	1·40	1·35
1325	1·10	0·90	0·95	1·10	1·03

(After POPJAK & MUIR (1950), *Biochem. J.*, 46, 103.)

11.12 *Torulopsis utilis* is grown on [1-^{14}C] glucose as sole carbon source. After harvesting, phenylalanine is isolated from a protein fraction and degraded (a) by treatment with ninhydrin to yield CO_2 and (b) by permanganate oxidation to yield benzoic acid. The benzoic acid is decarboxylated to give benzene.

All compounds studied are combusted to CO_2 prior to counting and all specific activities in the following table are given as counts per min per μmole of CO_2 of combustion.

Compound studied	Specific activity of CO_2 of combustion
[1-^{14}C] glucose in medium	340
phenylalanine	220
CO_2 (ninhydrin)	10
benzoic acid	280
benzene	170

Calculate the specific activities of each of the side chain carbon atoms and suggest a possible origin of this three carbon fragment.

(Glasgow Double Science Course, June 1955.)

11.13. The amino acid composition of crystalline human serum albumin has been determined by the isotope dilution technique. After hydrolysis of a weighed amount of the protein, ^{15}N-labelled amino acids were added to the hydrolysate and, after mixing, either metal or other derivatives of the amino acids isolated and purified. The data obtained are tabulated below.

| | Amino acid added | | Mean ^{15}N excess in compound isolated | Weight of protein hydrolysed |
Compound	Amount of L-amino acid mg	^{15}N excess atom per cent.	atom per cent.	g
Glutamic acid .	89·37	18·69	1·187	7·80
Aspartic acid .	91·05	11·82	1·269	7·80
Tyrosine . .	77·65	6·85	1·159	7·98

Calculate the percentage of each amino acid present and also determine the number of residues of each amino acid per molecule of human serum albumin of molecular weight 70,000.

(After SHEMIN (1945), *J. biol. Chem.*, 159, 439.)

11.14. A radioactive sample was counted for 10 minutes and 2000 counts were registered. The background for the same period gave 100 counts. The background count was also extended to 1000 minutes and gave a count rate of 10 counts/min.

Calculate the percentage error for both short and long background counts at (*a*) the 95 per cent. and (*b*) 99 per cent. reliable error levels.

(*a* (95/100) = 1·9600 and *a* (99/100) = 2·5758.)

11.15. A sample of adenylic acid has been labelled with radioactive phosphorus (^{32}P). If its initial activity was one millicurie, what would be its activity after six days ? Derive any expression used. What is the meaning of the following terms used in radiotracer work : geometry, backscattering, self-absorption, back-ground, and paralysis time ?

(Half-life of ^{32}P is 14·3 days ; $\log_e 10 = 2·303$.)

(Leeds Honours Course Finals, 1957.)

11.16. Explain how the characteristics of the isotope ^{14}C determine the methods necessary for its manipulation and assay.

In a counting apparatus which registered one-eighth of the emitted radiation, a sample containing 1 microgram of iodine registered 1840 counts after 20 minutes and 2896 counts after 40 minutes. Calculate the strength of the sample in microcuries, and the abundance of the radioactive isotope at the time when counting was begun.

(I = 127, 1 Curie = $3·7 \times 10^{10}$ disintegrations per second, 1 g atom contains $6·02 \times 10^{23}$ atoms.)

(Special Degree of B.Sc., Biological Chemistry, University of Bristol, 1958.)

11.17. In an experiment to determine the half-life of serum albumin in a rabbit 2 ml of a 2 per cent. (w/v) solution of albumin labelled with ^{131}I to an activity of 0·5 $\mu Ci/mg$ protein was injected intravenously into a 2 kg rabbit. Samples of serum were obtained at intervals after injection and the radioactivity of the albumin assayed. The following data were obtained :

Days after injection	Albumin activity $10^{-3} \times Counts/min$
2	3588
4	2548
8	1271
13	608·7
18	257·2
22	136·2

The efficiency of counting was 30 per cent. and the half-life of ^{131}I is 8·04 days.

Calculate (a) the observed counts/min for the sample injected and (b) use this figure to determine the half-life of serum albumin in the rabbit.

(After Reeve & Roberts (1959), J. gen. Physiol., 43, 415.)

11.18. What is meant by the term specific activity in relation to the use of radioactive isotopes in biochemistry ?

A rat received on 6th May an intramuscular injection of $200 \mu Ci$ ^{32}P as inorganic orthophosphate and was killed two hours later, a sample of blood being collected and the liver excised at the time of killing. Inorganic phosphate was isolated from the blood, and phospholipid and ribonucleic acid from the liver. A sample of the blood inorganic phosphate containing 12·6 μg P was assayed for radioactivity on 8th May and gave 12,460 counts in a 10 minute period. Similarly, a sample of the liver phospholipid containing 6·3 μg P recorded 2680 counts in 10 minutes on the same day, while the RNA sample, which contained 5·8 μg RNA-P, gave 3040 counts in 20 minutes on 11th May. All the samples were counted with the same counter which recorded a background of 10·6 counts per minute. Calculate the relative specific activities of the liver phospholipid and RNA, using the blood inorganic phosphate as the standard of reference.

Decay factors for ^{32}P :

Days	Factor
0	1·000
1	1·049
2	1·101
3	1·156
4	1·214
5	1·274

(Glasgow Double Science Course, 1961.)

11.19. E. coli was grown in a basal medium containing $H^{14}COOH$. Total cell proteins were isolated and hydrolysed to the constituent amino acids. After a chromatographic separation, the amino acids were degraded with ninhydrin to give CO_2. Aspartic acid was treated with β-aspartic acid decarboxylase to give CO_2 and arginine was treated with arginase to give urea. All compounds assayed for activity were oxidized to CO_2 and counted as such.

Compound oxidized for assay	Counts/min/μmole of CO_2 of combustion
Serine	0
Alanine	0
Aspartic acid	187
CO_2 (ninhydrin)	375
CO_2 (β-aspartic acid decarboxylase)	530
Glutamic acid	149
CO_2 (ninhydrin)	752
Arginine	256
CO_2 (ninhydrin)	739
Urea (arginase)	800
Proline	152
CO_2 (ninhydrin)	750

Are the results obtained consistent with the suggestion that the $H^{14}COOH$ is utilized only after conversion to $^{14}CO_2$? Indicate any possible metabolic interrelationships revealed by the results and discuss briefly the information which the data give concerning the mechanism of aconitase action.

(Glasgow Double Science Course, 1956.)

11.20. If to the steady state system

a tracer dose of labelled A is added at time $t = 0$ it can be shown that the turnover rate (i.e. the quantity of A transferred to B per unit time) is given by the expression :

$$\text{turnover rate} = \frac{2 \cdot 303ab}{t(a + b)} \log_{10} \frac{b}{(s_t/s_0)(a + b) - a}$$

where a and b are the pool sizes of A and B, respectively, and where s_0 and s_t are the specific activities of A at zero time and time t, respectively.

Krebs et al. have sought to use this relationship in a study of ATP turnover in a liver homogenate oxidizing α-oxoglutarate (0·02M) in the presence of added steady state concentrations of ^{32}P-labelled inorganic phosphate (18·8 μmoles per flask) and ATP (13·7 μmoles per flask). Samples were removed at intervals for chromatographic separation of ^{32}P-labelled compounds. The following results were obtained :

Time (min)	2	3	4	8	10
Per cent. total counts in inorganic phosphate	88·2	83·5	79·5	71·6	67·7

Comment on the applicability of the expression for turnover rate to the results obtained. Discuss in detail the nature of those experimental complications which would tend to vitiate the application to the system studied of kinetic analysis based on the above model and indicate the probable detailed design of Krebs' experiment.

During the experiment oxygen consumption was linear with time (4·9 μl/min) and some 80 per cent. of the α-oxoglutarate disappearing was accounted for by the accumulation of fumarate and malate. Calculate values for the P/O ratios at the various time intervals and comment on the significance of the figures obtained.

(Glasgow Honours Course Finals, 1961. After KREBS, RUFFO, JOHNSON, EGGLESTON & HEMS (1953), *Biochem. J.*, **54**, 107.)

11.21. In an experiment to establish the route by which tartaric acid is metabolized by a pseudomonad, cells in the exponential phase of growth with *dl*-tartaric acid as carbon source were incubated with *dl*-$[1,4-^{14}C_2]$ tartaric acid. Samples of culture were discharged into hot ethanol at short time intervals and the ethanol-soluble fraction analysed by two-dimensional chromatography and radioautography. Data obtained are summarized in the following table.

Reaction time (sec)		0	5	10	15	20	30	60	105
Counts/min as	Malate	0	9	16	18	22	8	32	17
	Glycerate	0	23	30	18	10	16	46	102
	Alanine	0	7	19	37	39	93	339	566
	Lactate	0	0	0	0	5	11	56	59
Counts/min	Total	0	39	65	73	76	128	473	744

Use these data to express the distribution of ^{14}C in each compound as a percentage of the total activity in the ethanol-soluble fraction. Present the results in the form of a graph. How do you interpret the findings ?

(Unpublished data of DAGLEY & TRUDGILL.)

11.22. Discuss very briefly the usefulness of specifically labelled [^{14}C]glucose in the elucidation of the pathway(s) of glucose metabolism.

A strain of *Acetobacter xylinum* accomplishes the oxidation of glucose carbon atoms to CO_2 at the following relative rates:

$$C-1 > C-2 > C-6 = C-3 > C-4$$

When allowed to metabolize specifically labelled glucose in the presence of arsenite, pyruvate accumulated. Degradation of the pyruvate yielded the following results :—

	Specific activity of pyruvate carbon atoms as percentage of pyruvate activity		
Labelled substrate	C-1	C-2	C-3
[1-^{14}C] Glucose	97	1	2
[2-^{14}C] Glucose	12	85	3
[6-^{14}C] Glucose	64	7	29

What light do these results throw on the pathway(s) of glucose metabolism in this micro-organism ?

(Hull Special Biochemistry I Course, 1964.)

11.23. What is meant by the expression 'Isotope Dilution Analysis' ? 13·7 mg of crystalline pancreatic ribonuclease were hydrolysed to amino acids, 1 μmole of ^{35}S-methionine of specific activity 1 μCi per μmole was added to the hydrolysate and a sample of methionine was isolated chromatographically. The specific activity of this methionine was determined 40 days after addition of ^{35}S-methionine to the hydrolysate in a counting system with an efficiency of 12 per cent. The value obtained was $37·3 \times 10^3$ counts/min/μmole. Calculate the number of methionine residues per mole ribonuclease. The half-life of ^{35}S is 87 days. The molecular weight of ribonuclease is 13,700. 1 curie $\equiv 3\cdot7$ $\times 10^{10}$ disintegrations per second.

(Glasgow Honours Course Finals, 1964.)

11.24. A micro-organism metabolizes glucose with the accumulation of acetate in the medium. Investigation of the enzymic constitution failed to reveal enzymes specific for the Entner-Doudoroff pathway of metabolism but enzymes characteristic of glycolysis and the pentose phosphate cycle were present. Experiments were carried out with [1-^{14}C] and [6-^{14}C]glucose in an effort to estimate the relative contribution of these two metabolic pathways to glucose metabolism.

When 10 μmoles of each labelled sugar were used in duplicate experiments the total yields of ^{14}C in the acetate formed were 5470 counts/min from [1-^{14}C] glucose and 11,200 counts/min from [6-^{14}C] glucose.

The specific activities of the substrates were : [1-^{14}C] glucose, 6250 counts/ min/μmole and [6-^{14}C] glucose, 5600 counts/min/μmole.

Calculate the percentage participation of the pentose phosphate pathway.

11.25. A pure sample of ^{32}P labelled ATP in dilute acid had an extinction of 0·840 at 257 nm, and 5 ml of the solution counted in a Geiger counter of 10 per cent efficiency gave rise to 9640 counts per min. Following hydrolysis of 15 ml of the solution with dilute acid at 100° for 10 min and isolation of the inorganic phosphate in a total volume of 50 ml, a sample of 4 ml of this gave rise to 1540 counts per min when counted as above. Calculate the specific activities of the ATP, the two terminal phosphates of the ATP and the ester phosphate in counts per min per μmole. What conclusions can you draw from the results ? Assuming that 1 curie gives rise to $3\cdot7 \times 10^{10}$ disintegrations per second, calculate the total amount of activity in μCi present in 100 ml of the ATP solution. The molar extinction coefficient of ATP at pH 2 is $14\cdot7 \times 10^{3}$.

(Glasgow Double Science Course, 1963.)

11.26. In the ruminant acetate turnover is very rapid. The metabolism of acetate can be studied, however, by the technique of continuous infusion of the labelled substrate. The following data were obtained in an experiment on a 37 kg sheep.

[1-^{14}C]Acetate was infused at a rate of $10\cdot08 \times 10^{6}$ counts/min for a total infusion time of 721 minutes, after which the fall in specific activity of the blood acetate was determined with the following results:

Time after infusion ceased (sec)	25	50	75	100	125
Percentage of initial specific activity . . .	80	63	49	38	30

During the period of continuous infusion the average blood acetate concentration was 0·73 m-equiv/l and the mean specific activity of the blood acetate was 2220 counts/min/μequiv.

Determine (a) the half-life of acetate, (b) the turnover rate of acetate in m-equiv/hr/kg body weight, (c) the turnover time for acetate, (d) the pool size of acetate, and (e) the acetate space (i.e. the volume of fluid in which the acetate pool is dissolved).

Note. For the purposes of this calculation it may be assumed that equilibration of intravascular and extravascular acetate is instantaneous.

(After SABINE & JOHNSON (1964), *J. biol. Chem.*, **239**, 89.)

11.27. A sample of [U-^{14}C; ^{32}P] adenosine 5′-phosphate (5′-AMP) was assayed for activity 6·5 days after the start of an experiment and yielded, after correction, 2458 counts per minute. After a further 7·7 days the same sample gave 2000 counts per minute (corrected). Calculate the ^{14}C activity and the

[32]P activity of the sample at zero time. The half-lives of ^{32}P and ^{14}C are 14·2 days and 5500 years, respectively.

Fibroblast cells were incubated in a bicarbonate-phosphate buffer for 2 hours in the presence of 5′-AMP double labelled as above. The cells were then harvested, washed thoroughly and disrupted. 5′-AMP was isolated from the free nucleotide pool by ion-exchange chromatography. On assay it gave an activity ratio ^{14}C/^{32}P of 5000 as compared with the corresponding ratio of 25 for the added extracellular nucleotide. Suggest possible explanations for this observation.

(Hull Honours Course Finals, Part I, 1965.)

11.28. In a series of experiments with *Escherichia coli* grown with uniformly-labelled [^{14}C] glucose in the absence and presence of unlabelled amino acids, the following relative specific activities were found in the amino acids isolated from the bacterial cells.

	Aspartic	Lysine	Methionine	Threonine	Isoleucine
[U-^{14}C] Glucose	100	100	100	100	100
[U-^{14}C] Glucose +aspartate	20	15	12	10	8
[U-^{14}C] Glucose +homoserine	100	100	20	15	15
[U-^{14}C] Glucose +threonine	100	100	100	12	8
[U-^{14}C] Glucose +methionine	100	100	7	100	100

What light does this information throw upon the biosynthetic pathways leading to the formation of the amino acids determined, and what principles are involved in experiments of this type?

(Hull Pass Degree Finals, 1966.)

11.29. To a growing culture of yeast was added 1 mCi each of L-[Me[14]C] methionine and L-(^{35}S) methionine. After incubation for 48 h the culture was killed and radiochemically pure thiamine was isolated. An aliquot of thiazole obtained by degradation of the isolated thiamine was plated on a planchet and counted for approximately 10 min at intervals over a period of 3 months, with the following results:

Days after addition of labelled methionine	30	35	48	65	109
Radioactivity (disintegrations per min)	9130	8960	8590	8175	7095

Discuss the implications of this result and suggest how a sequence of intermediates postulated to lie between methionine and thiamine (i.e. methionine ⟶ X ⟶ Y ⟶ thiamine) may be established or disproved.

[Half lives: [35]S, 87 days; ^{14}C, 5,760 years]

(University of Wales, Cardiff, Honours Course Finals, 1966.)

11.30. Give examples to illustrate the value of auto-radiography in the solution of biochemical problems.

P

Bacteriophage were grown utilizing phosphate from a medium containing 1·0 millicurie ^{32}P/100 mg ^{32}P. A sample of phage gave, after washing, 10^9 plaques and 500 counts/min. How many atoms of phosphorus are contained in each phage particle?

(Counting efficiency $= 25$ per cent. Avogadro's number $= 6 \times 10^{23}$. 1 curie $= 3·7 \times 10^{10}$ d/sec)

(Leeds Honours Course Finals, 1966.)

11.31. A washed cell suspension of a homolactic fermentative bacterium grown on glucose as an energy source did not ferment gluconate. Cells grown in the presence of gluconate fermented glucose, gluconate and 2-keto-gluconate. The fermentation balance for gluconate could be written as

$$0·1 \text{ Gluconate} \rightarrow 0·5 \ CO_2 + 1·8 \text{ Lactate}$$

Traces of formate, acetate and ethanol were also found. Gluconate fermentation was sensitive to fluoride but not to arsenite.

Experiments were carried out with uniquely-labelled gluconate with the results indicated in the Table.

Substrate	[^{14}C] *in Fermentation Products* (per cent. added radioactivity)			
	CO_2	*Lactic acid*		
		CH_3^-	$=CHOH$	$-COOH$
[1-^{14}C] Gluconate	53	0	0	36
[6-^{14}C] Gluconate	0	98	0·9	0·5
[2-^{14}C] Gluconate	0·6	8·5	44	17·8
[2-^{14}C] Gluconate + 10^{-3} M arsenite	0	24	62	17·8

From these data deduce the probable mechanism(s) involved in the induced fermentation of gluconate by this homolactic organism, giving in full the reasoning that you employ.

(Hull Special Biochemistry I Course, 1968.)

11.32. Discuss briefly the rationale of the scintillation counting of a sample doubly-labelled with ^{3}H(β^-, $E_{max} = 0·018$ MeV) and ^{35}S(β^-, $E_{max} = 0·167$ MeV).

A method for plasma testosterone assay involves the addition of a negligible amount of [^{3}H]testosterone (t_1 counts/min) to a plasma sample. The steroids are then separated from the plasma and allowed to react with excess [^{35}S] thiosemicarbazide (s_1 counts/mole) to give a quantitative yield of steroid thiosemicarbazones. Testosterone thiosemicarbazone is then isolated without regard to yield and assayed for ^{3}H(t_2 counts/min) and ^{35}S(s_2 counts/min).

Derive an expression giving x, the amount of testosterone in the initial plasma sample, as a function of t_1, t_2, s_1 and s_2.

What factors limit the sensitivity and accuracy of such an analytical procedure ?

(Hull Honours Course Finals, Part II, 1970.)

11.33. A sample containing 0·1 μg was withdrawn from a specimen of ^{35}S-labelled methionine having specific activity 50 μCi/mg and the activity was counted with 10 per cent. efficiency.

(1 curie $= 22 \cdot 2 \times 10^{11}$ disintegrations/min). Calculate the number of counts per minute given by the sample.

A further sample of 1 mg from the same specimen of methionine was added to a protein hydrolysate and a pure sample of methionine was isolated from the mixture. The activity of a sample of 0·1 μg of this methionine was measured under the same conditions in the same counter as above, seven days after the previous measurement of activity and 65 counts/min were obtained. The half-life of ^{35}S is 87·1 days.

(a) Calculate the decay constant (k) for the isotope where

$$ k = \frac{2 \cdot 303 \log \dfrac{n_0}{n}}{t} $$

($n_0 =$ counts/min at time $t = 0$; $n =$ counts/min at time t in days.)

(b) Calculate the weight of methionine in the protein hydrolysate.

(University of Manchester, Biological Chemistry Honours Course Finals, 1967.)

11.34. The radioactivity of a sample of valine uniformly labelled with ^{14}C was measured using a Geiger counter. 1 μg of the material gave 20,000 counts/min. The efficiency of counting was 10 per cent. Calculate the specific activity of the valine in μCi/mg.

[1 Curie $= 22 \cdot 2 \times 10^{11}$ disintegrations/min]

30 μg of the valine was injected into each of a batch of mice. The mice were killed and pure ribonuclease was isolated from the pancreas glands. The ribonuclease was hydrolysed and valine isolated in pure form. 10 μg of the isolated valine gave 15,000 counts/min measured in the same counter as above. Calculate the relative specific activity of the isolated valine.

The above experiment was repeated except that the pancreas was homogenized and separated into fractions containing respectively microsomes and zymogen granules. Ribonuclease was prepared in pure form from each fraction and the valine isolated as before. The specific activities of the valine from such samples prepared after varying intervals of time following injection of the valine were as shown in the Table. Comment on these results.

Time (min)	Specific activity (μCi/mg)	
	Microsomes	Zymogen granules
5	0·3	0·05
10	0·55	0·1
15	0·75	0·15
30	0·5	0·35
40	0·3	0·35

(University of Manchester, Biological Chemistry, Second B.Sc. Examination, 1968.)

11.35. Indicating clearly the units which may be used, define specific radioactivity.

Describe in outline how you would measure and express the specific radioactivity of a polysaccharide.

A microsomal preparation was incubated under suitable conditions with ^{14}C-uridine diphosphate glucuronic acid (UDP-glucuronic acid) labelled in the glucuronic acid residue and ^{3}H-uridine diphosphate N-acetylglucosamine (UDP-N-acetylglucosamine) labelled in the acetyl residue, the two materials being added simultaneously at a final concentration of each of 0·5 mM. The incubation mixture had a volume of 0·05 ml and on measurement of radioactivity the whole mixture gave 10^6 counts/min ^{3}H and 4×10^5 counts/min ^{14}C. After incubation, a polysaccharide was isolated and purified after dilution with carrier material, transferred quantitatively to a planchet and counted on the same counter as above. The material had 18,200 counts/min ^{3}H and 6700 counts/min ^{14}C.

Calculate the specific activities of UDP-N-acetylglucosamine and UDP-glucuronic acid and the relative proportions of N-acetyl-glucosamine and glucuronic acid incorporated into the polysaccharide.

(University of Manchester, Biological Chemistry, Second B.Sc. Examination, 1968.)

11.36. A preparation of t-RNA is incubated with ATP, Mg^{++} ions, phosphoenol pyruvate, pyruvate kinase, a dialysed preparation of amino acyl transferases, 9·0 μg of unlabelled DL-alanine and 1·0 μCurie of [^{14}C] L-alanine of specific activity 100 m-Curie per mmole. Samples, each containing 10μg of t-RNA, are taken at the following times, treated with 5 per cent. trichloroacetic acid, filtered on a membrane filter and washed with 5 per cent. trichloroacetic acid. The radioactivity on each filter is measured :

Time of sampling (min)	0	5	10	20	30
Radioactivity (counts/min) after background correction	10	500	1300	1320	1280

Describe briefly what is happening during these procedures. The efficiency of counting is 25 per cent., the molecular weight of t-RNA is 25,000 and 1 Curie of radioactivity is $3·7 \times 10^{10}$ disintegrations per sec. What percentage of the t-RNA is L-alanyl t-RNA ?

(University of Newcastle upon Tyne, Biochemistry Honours Course, Part I, 1970.)

11.37. An oligosaccharide X was isolated from a partial enzyme digest of glycogen. Periodate oxidation of oligosaccharide X yielded 4 moles of formic acid, 1 mole of formaldehyde and consumed 9 moles of periodate per mole of oligosaccharide X. Exhaustive methylation and acid hydrolysis yielded 2,3,4,6-tetramethyl glucose, 2,3-dimethyl glucose, 2,3,6-trimethyl glucose, and 1,2,3,6-tetramethyl glucose. Suggest a formula for oligosaccharide X and explain your conclusions.

The glycogen molecule can be labelled with radioactivity by incubation of liver slices with [3-^{14}C] glucose. The specific activity of the maltose obtained from this radioactive glycogen and of the periodate oxidized maltose were $50,000 \pm 200$ counts/min/mg and $25,000 \pm 200$ counts/min/mg respectively. Has there been any randomization of the labelling from the 3 position to other positions of the glucose molecule during biosynthesis of the glycogen ?

(University of Newcastle upon Tyne, Biochemistry Honours Course, Part I, 1970.)

APPENDIX 1

SYMBOLS

A	area, absorbancy (extinction)
a_i	activity of ion i
C	concentration
c	concentration, velocity of light
Ci	curie
D	diffusion coefficient
E	activation energy, electrode potential, extinction (absorbancy)
F	Faraday
F	Helmholtz free energy, molar frictional coefficient
f	molecular frictional coefficient
f_i	molar activity coefficient of ion i
G	Gibbs free energy
g	gram
g	gravity, centrifugal
H	heat content (enthalpy)
h	hour
h	Planck's constant (6.63×10^{-34} Js)
I	ionic strength, light intensity
K	equilibrium constant, extinction coefficient
K_a	thermodynamic equilibrium constant (activities)
K_a	acidic dissociation constant
$K_{app.}$	apparent dissociation constant
K_b	basic dissociation constant
K_c	equilibrium constant (concentrations)
K_i	enzyme-inhibitor dissociation constant (inhibitor constant)
K_m	Michaelis constant
K_s	enzyme-substrate dissociation constant (substrate constant)
k	rate constant, Boltzmann's constant (1.38×10^{-23} JK^{-1})
l	length
M	moles per litre (molar)
M	molecular weight
M_n	number-average molecular weight
M_w	weight-average molecular weight
M_z	Z-average molecular weight

m	milli (10^{-3}), metre
N	Avogadro's number ($6 \cdot 02 \times 10^{23}$ mol^{-1})
N	normality (gram-equivalents per litre)
n	refractive index
P	pressure, probability or steric factor
pf_i	$-\log f_i$
pH	$-\log (H^+)$
pI	isoelectric point
pK_a	$-\log K_a$
Q	heat absorbed, Q_V at constant volume, Q_p at constant pressure
Q_x	metabolic quotient: μl of gas X/mg dry weight of biological material/h
Q_{10}	temperature coefficient
R	gas constant ($8 \cdot 314$ $JK^{-1}mol^{-1}$)
R.Q.	respiratory quotient
r	radius
rH	$-\log (H_2)$
S	Svedberg unit (10^{-13}s)
S	entropy
s	second
s	sedimentation coefficient
T	absolute temperature
t	time
$t_{\frac{1}{2}}$	half-life period
U	internal energy
V	volume, velocity of reaction
v	volume
$\bar{v}$	partial specific volume
W	work done by system
Z	molecular collision rate
z	valency
α	degree of dissociation, solubility coefficient, angle of rotation
β	buffer value
γ_i	molal activity coefficient of ion i
Δ	measurable increment
δ	infinitesimal increment
ϵ	molar extinction coefficient

$\epsilon_{spec.}$	specific extinction coefficient
η	viscosity coefficient
$[\eta]$	intrinsic viscosity
η_r	relative viscosity coefficient
$\eta_{sp.}$	specific viscosity
λ	wavelength
μ	micro (10^{-6})
ν	frequency
Π	osmotic pressure
ρ	density
Σ	sum
τ	turbidity
ϕ	volume fraction
ω	angular velocity
$^{\circ}$	standard state
$\ddagger$	transition state (activated complex)
( )	activity
[ ]	concentration

APPENDIX 2

PREFIXES FOR SI UNITS

Fraction	Prefix	Symbol	Multiple	Prefix	Symbol
10^{-1}	deci	d	10	deka	da
10^{-2}	centi	c	10^2	hecto	h
10^{-3}	milli	m	10^3	kilo	k
10^{-6}	micro	μ	10^6	mega	M
10^{-9}	nano	n	10^9	giga	G
10^{-12}	pico	p	10^{12}	tera	T
10^{-15}	femto	f			
10^{-18}	atto	a			

APPENDIX 3

THE GAS CONSTANT

THE universal gas equation is expressed by the relationship

$$PV = RT$$

and the value of R obviously depends on the units used to express P and V.

Where P is expressed in atmospheres and V in litres, at N.T.P. $R = \dfrac{1 \times 22 \cdot 4}{273}$

$= 0 \cdot 08204$ litre atmospheres per mole per degree.

However, if P is expressed in dynes per cm^2 and V in ml.

$$P = 76 \times 13 \cdot 59 \times 981 \text{ dynes/cm}^2$$

and $R = \dfrac{76 \times 13 \cdot 59 \times 981 \times 22400}{273} = 8 \cdot 314 \times 10^7 \text{ ergs/mole/degree}$

$$= 8 \cdot 314 \text{ joules/mole/degree (J mol}^{-1}\text{K}^{-1}).$$

Furthermore, since $4 \cdot 185 \times 10^7$ ergs $= 1$ calorie

$R = \dfrac{8 \cdot 314}{4 \cdot 184} = 1 \cdot 987$ calories per mole per degree.

Quite often in osmotic pressure measurements the pressure is determined in cm of water instead of mercury, since this means a larger value to read experimentally. In these cases

$$R = \dfrac{76 \times 13 \cdot 59 \times 22400}{273} = 8 \cdot 471 \times 10^4 \text{ ml cm H}_2\text{O/mole/degree.}$$

APPENDIX 4

THE GRAPHICAL SOLUTION OF PROBLEMS

MANY of the problems in this book demand graphical solution, and a few notes on this topic may be of value.

Choice of scale is very important, especially where the slope of the line is required. Generally the scale should be chosen so that the angle of the slope is approximately 45°, since this usually offers maximum accuracy. The scale selected should be large enough for the error in setting out the points to be small in comparison with the random measuring error, i.e. the spread of the points on either side of the correct line.

From a set of experimental observations of equal weight the slope of the straight line may be obtained by the *method of least squares*.

Suppose that a series of observations of two related quantities (x, y) are made, say

$$(x_1 y_1) \ldots \ldots (x_r y_r) \ldots \ldots (x_n y_n)$$

Suppose that the expected theoretical relationship between y and x is linear, thus

$$y = mx + c.$$

The problem is to find (using only the observed data) the most probable values of (m, c). Graphically the situation presented is to find the 'best' straight line passing through a scatter of points. The most common method of obtaining such values for (m, c) is the method of least squares. In this the sum of the squares of the y residuals is minimized. [For any observed pair $(x_r y_r)$, the y residual is the difference between y_r (the observed y) and $mx_r + c$ (the value of y on the line at $x = x_r$).]

Therefore we minimize

$$Y = [y_1 - (mx_1 + c)]^2 + \ldots \ldots + [y_n - (mx_n + c)]^2$$

$$= \sum_{r=1}^{n} [y_r - (mx_r + c)]^2$$

As different lines, i.e. different values of (m, c), are taken so Y varies. An application of the calculus yields the result that the minimum value of Y occurs when

$$\frac{dY}{dm} = 0 = \frac{dY}{dc}$$

Thus, differentiating,

$$\sum_{r=1}^{n} x_r(y_r - mx_r - c) = 0 = \sum_{r=1}^{n} (y_r - mx_r - c)$$

It follows that two simultaneous algebraic equations for which the solution for (m, c) fits the best line are

$$\left(\sum_{r=1}^{n} x_r^2\right) m + \left(\sum_{r=1}^{n} x_r\right) c = \sum_{r=1}^{n} x_r y_r \ldots \ldots (1)$$

$$\left(\sum_{r=1}^{n} x_r\right) m + (n)c = \sum_{r=1}^{n} y_r \ldots \ldots (2)$$

Each of these coefficients is easily calculated from the observational data. There is no need to plot the points in order to draw the line but they may be plotted for other purposes, e.g. to check for scatter about the line.

Note.—If equations (1) and (2) are written in the form

$$Am + Bc = D \ldots \ldots (1a)$$
$$Bm + nc = C \ldots \ldots (2a)$$

where

$$A = \sum_{r=1}^{n} x_r^2, \quad B = \sum_{r=1}^{n} x_r, \quad C = \sum_{r=1}^{n} y_r$$

and

$$D = \sum_{r=1}^{n} x_r y_r$$

the solution of these equations will be as follows :

Slope of best line $= m = \dfrac{nD - BC}{nA - B^2}$

y intercept of best line $= c = \dfrac{AC - BD}{nA - B^2}$

APPENDIX 5

ANGULAR VELOCITY

THE calculation of molecular weights from sedimentation data demands a knowledge of the angular velocity of the centrifuge in radians per second. If the angle described in time t is θ, the angular velocity ω is given by

$$\omega = \frac{\theta}{t}$$

and where θ is in radians and t in seconds ω is obtained in radians per second.

Since a radian is the angle subtended by an arc equal to the radius of a circle, the circumference of which is $2\pi r$, it follows that one circumference or revolution is equal to 2π radians. Hence multiplication of the speed of the centrifuge in revolutions per second by 2π gives the angular velocity in radians per second.

APPENDIX 6

SOLUTION OF QUADRATIC EQUATIONS

THE two roots of the general quadratic equation

$$ax^2 + bx + c = 0$$

are given by the formula

$$x = \frac{-b \pm \sqrt{(b^2 - 4ac)}}{2a}$$

APPENDIX 7

SOLUTION OF SIMULTANEOUS EQUATIONS BY DETERMINANT METHODS

See A. C. AITKEN (1959), *Determinants and Matrices*, 9th ed. Edinburgh: Oliver & Boyd.

To illustrate the determinant method for the solution of simultaneous equations consider the following set of equations:

$$a_{11}x_1 + a_{12}x_2 + a_{13}x_3 = h_1$$
$$a_{21}x_1 + a_{22}x_2 + a_{23}x_3 = h_2$$
$$a_{31}x_1 + a_{32}x_2 + a_{33}x_3 = h_3$$

The solutions for x_1, x_2, and x_3 are quotients having as their denominator (D) the following expression:

$$D = \begin{vmatrix} a_{11} & a_{12} & a_{13} \\ a_{21} & a_{22} & a_{23} \\ a_{31} & a_{32} & a_{33} \end{vmatrix} = a_{11} \begin{vmatrix} a_{22} & a_{23} \\ a_{32} & a_{33} \end{vmatrix} - a_{12} \begin{vmatrix} a_{21} & a_{23} \\ a_{31} & a_{33} \end{vmatrix} + a_{13} \begin{vmatrix} a_{21} & a_{22} \\ a_{31} & a_{32} \end{vmatrix}$$

where a determinant of the type:

$$\begin{vmatrix} a_{22} & a_{23} \\ a_{32} & a_{33} \end{vmatrix}$$

yields on expansion:

$$a_{22} a_{33} - a_{23} a_{32}.$$

The numerator (N_1) of the quotient for the solution for x_1 will be by Cramer's Rule

$$N_1 = \begin{vmatrix} h_1 & a_{12} & a_{13} \\ h_2 & a_{22} & a_{23} \\ h_3 & a_{32} & a_{33} \end{vmatrix}$$

Solutions for x_2 and x_3 may be obtained in an analogous manner.

In manipulating determinants it should be remembered that interchanging a row or column by one place alters the sign of the resultant expression, i.e.

$$\begin{vmatrix} a & b & c \\ d & e & f \\ g & h & i \end{vmatrix} = - \begin{vmatrix} a & c & b \\ d & f & e \\ g & i & h \end{vmatrix}$$

$$= \begin{vmatrix} c & a & b \\ f & d & e \\ i & g & h \end{vmatrix}$$

APPENDIX 8

THE INTERNATIONAL SYSTEM OF UNITS (SI)

The name Système Internationale d'Unités (abbreviation SI) has been adopted by the General Conference on Weights and Measures for a coherent system based on six units : metre, kilogram, second, ampere, kelvin and candela. (A system of units is said to be coherent when the units for all derived physical quantities are obtained from the basic units by multiplication or division without the introduction of any numerical factors [including powers of ten].) The mole has also been recommended for inclusion in the SI as the basic unit amount of substance.

The SI names and symbols for the basic units are as follows :

Basic physical quantity	Name of basic SI unit	Symbol
length	metre	m
mass	kilogram	kg
time	second	s
electric current	ampere	A
thermodynamic temperature	kelvin	K
amount of substance	mole	mol
luminous intensity	candela	cd

There are also two supplementary dimensionless units :

Physical quantity	Name of unit	Symbol
plane angle	radian	rad
solid angle	steradian	sr

The SI units for derived physical quantities are those constructed by multiplication and division of the basic units.

Thus :

The basic SI unit of energy : $kg \times m^2 \times s^{-2}$ is given the special SI name joule and the symbol J.

Definitions of the basic SI units

METRE.—The metre is the length equal to $1\ 650\ 763 \cdot 73$ wavelengths in vacuum of the radiation corresponding to the transition between the levels $2p_{10}$ and $5d_5$ of the krypton-86 atom.

KILOGRAM.—The kilogram is the unit of mass ; it is equal to the mass of the international prototype of the kilogram. (This prototype is in the custody of the Bureau International des Poids et Mesures at Sèvres, France.)

SECOND.—The second is the duration of $9\ 192\ 631\ 770$ periods of the radiation corresponding to the transition between the two hyperfine levels of the ground state of the caesium-133 atom.

AMPERE.—The ampere is that constant current which, if maintained in two straight parallel conductors of infinite length, of negligible circular cross-section, and placed 1 metre apart in a vacuum, would produce between these conductors a force equal to 2×10^{-7} newton per metre of length.

KELVIN.—The kelvin, unit of thermodynamic temperature, is the fraction 1/273·16 of the thermodynamic temperature of the triple point of water.

CANDELA.—The candela is the luminous intensity, in the perpendicular direction, of a surface of 1/600 000 square metre of a black body at the temperature of freezing platinum under a pressure of 101 325 newtons per square metre.

MOLE.—The mole is the amount of substance which contains as many elementary units as there are atoms in 0·012 kilogram of carbon-12. The elementary unit must be specified and may be an atom, a molecule, an ion, a radical, an electron, a photon, etc., or a specified group of such entities.

For further information concerning the SI the reader is referred to the following books :

The International System (SI) Units. (1964). British Standard 3762.
The Use of SI Units. (1967). British Standards Institution, PD 5686.
McGLASHAN, M. L. (1968). *Physico-Chemical Quantities and Units.* Royal Institute of Chemistry Monographs for Teachers, No. 15.

APPENDIX 9

INTERNATIONAL ATOMIC WEIGHTS

	Symbol	Atomic Number	Atomic Weight
Aluminium . .	Al	13	26·97
Antimony . .	Sb	51	121·76
Argon . . .	A	18	39·944
Arsenic . . .	As	33	74·91
Barium . . .	Ba	56	137·36
Beryllium . .	Be	4	9·02
Bismuth . .	Bi	83	209·00
Boron . . .	B	5	10·82
Bromine . .	Br	35	79·916
Cadmium . .	Cd	48	112·41
Caesium . .	Cs	55	132·91
Calcium . . .	Ca	20	40·08
Carbon . .	C	6	12·01
Cerium . . .	Ce	58	140·13
Chlorine . .	Cl	17	35·457
Chromium . .	Cr	24	52·01
Cobalt . . .	Co	27	58·94
Copper . . .	Cu	29	63·57
Dysprosium . .	Dy	66	162·46
Erbium . . .	Er	68	167·2
Europium . .	Eu	63	152·0
Fluorine . .	F	9	19·00
Gadolinium . .	Gd	64	156·9
Gallium . . .	Ga	31	69·72
Germanium . .	Ge	32	72·60
Gold . . .	Au	79	197·2
Hafnium . .	Hf	72	178·6
Helium . . .	He	2	4·003

	Symbol	Atomic Number	Atomic Weight
Holmium . . .	Ho	67	164·94
Hydrogen . .	H	1	1·0081
Indium . . .	In	49	114·76
Iodine . . .	I	53	126·92
Iridium . . .	Ir	77	193·1
Iron . . .	Fe	26	55·84
Krypton . .	Kr	36	83·7
Lanthanum . .	La	57	138·92
Lead . . .	Pb	82	207·21
Lithium . . .	Li	3	6·940
Lutecium . .	Lu	71	175·00
Magnesium . .	Mg	12	24·32
Manganese . .	Mn	25	54·93
Mercury . . .	Hg	80	200·61
Molybdenum . .	Mo	42	95·95
Neodymium . .	Nd	60	144·27
Neon . . .	Ne	10	20·183
Nickel . . .	Ni	28	58·69
Niobium . . .	Nb	41	92·91
Nitrogen . . .	N	7	14·008
Osmium . . .	Os	76	190·2
Oxygen . . .	O	8	16·000
Palladium . .	Pd	46	106·7
Phosphorus . .	P	15	30·98
Platinum . . .	Pt	78	195·23
Potassium . .	K	19	39·096
Praseodymium . .	Pr	59	140·92
Protactinium . .	Pa	91	231
Radium . . .	Ra	88	226·05
Radon . . .	Rn	86	222
Rhenium . . .	Re	75	186·31
Rhodium . . .	Rh	45	102·91
Rubidium . .	Rb	37	85·48
Ruthenium . .	Ru	44	101·7
Samarium . .	Sm	62	150·43
Scandium . .	Sc	21	45·10
Selenium . . .	Se	34	78·96
Silicon . . .	Si	14	28·06
Silver . . .	Ag	47	107·880
Sodium . . .	Na	11	22·997
Strontium . . .	Sr	38	87·63
Sulphur . . .	S	16	32·06
Tantalum . . .	Ta	73	180·88
Tellurium . . .	Te	52	127·61
Terbium . . .	Tb	65	159·2
Thallium . . .	Tl	81	204·39
Thorium . . .	Th	90	232·12
Thulium . . .	Tm	69	169·4
Tin . . .	Sn	50	118·70
Titanium . . .	Ti	22	47·90
Tungsten . . .	W	74	183·92
Uranium . . .	U	92	238·07
Vanadium . .	V	23	50·95
Xenon . . .	Xe	54	131·3
Ytterbium . .	Yb	70	173·04
Yttrium . . .	Y	39	88·92
Zinc . . .	Zn	30	65·38
Zirconium . .	Zr	40	91·22

ANSWERS

FOUR or five figure logarithmic tables have been used, where occasion demands, in the solution of the problems.

Where a problem requires graphical solution there is, of course, the possibility of deviation from the recorded answer due to personal assessment of the best straight line through the experimental points. The answer recorded in such cases is, wherever possible, the mean of two individual determinations. Furthermore, it is possible to solve some problems by alternative methods and the answers do not always correspond exactly.

CHAPTER I

1.1. 600; 2S, 3P.
1.2. Horse: Fe 16,669; S-S 16,878; S 16,446. 1 Fe, 1 S-S, 2S. Pig: Fe 13,960; S 13,362. 1 Fe, 2 S.
1.3. Tryptophan 36,447, 3 mol.; tyrosine 36,216, 9 mol.; β-hydroxyglutamic acid 36,244, 4 mol.
1.4. 19,866; 18,697; 16,729; 21,920; 36,700.
1.5. Fe 33,238; S 33,405; arginine 32,856. 2 Fe, 5 S, 8 arginine.
1.6. 1 Fe combines with $0.944 \approx 1$ mole oxygen.
1.7. 740; 780.
1.8. 1,041; 1,430.
1.9. 48; 47.
1.10. 414.
1.11. 678.
1.12. Acid 1,041, acidic dye 961·3; base 1,429, basic dye 1,429.
1.13. M_n 46,750; M_w 57,170.
1.14. 2·629 atm.
1.15. $59.98 \approx 60$.
1.16. 62,320.
1.18. A 59,990; B 56,990; greater solvation effect with B.
1.20. Mean value 74,600.
1.21. *ca.* 6·7.
1.22. 69,900.
1.23. 66,800.
1.24. 66,700.
1.25. 2·08 days, 13,600; 10·2 days, 6,100.
1.26. Mean value 5,930; individual values, 5,900, 5,960, 5,930, 5,930.
1.27. 1,475,000; 1,060,000; 1,291,000; 1,235,000.
1.28. 184,000 before, 98,000 after treatment.
1.29. 34,800.
1.30. 61,500.
1.31. 270,000.
1.32. 35,210; 60,970 cm.^{3}
1.33. By activity: 0·0214, 0·0217, 0·0220; by nitrogen: ——, 0·0212, 0·0218 cm²/day. Average of all determinations 0·0216 giving $r = 2·67 \times 10^{-7}$ cm and $V = 48,300$ cm.^{3}
1.34. Volume of water of hydration per g dry trypsin: 0·50, 0·49, 0·50, 0·55, 0·72 ml (in order of increasing trypsin concentration).
1.35. Reduced osmotic pressure 60,000; Kunitz average 46,870.
1.36. Saccharose: 0·00578, 0·0104; 0·0120, 0·0216; 0·0313, 0·0564; 0·0731, 0·132; 0·127, 0·228; 0·204, 0·367. Glucose: 0·0138, 0·0248; 0·0289, 0·0520; 0·0702, 0·126; 0·138, 0·248; 0·200, 0·360; 0·270, 0·486 (Kunitz value given first in each case; values in order of increasing sugar concentration).

1.37. Volume of water of hydration per g HbCO: 0·14, 0·18, 0·27, 0·43, 0·55 ml, in order of increasing HbCO concentration. The effect of concentration on the degree of hydration is much less pronounced if the Kunitz equation is used in its full form. The corresponding values are then 0·13, 0·12, 0·14, 0·20 and 0·22 ml/g HbCO.

1.38. 180.

1.39. 73,000.

1.40. $M = S^{\frac{3}{2}} \times k$ (constant); $M_{60S} = 465k$; $M_{40S} = 253k$; $M_{60S} + M_{40S} = 718k$. Compare with $M_{80S} = 716k$.

1.41. (1)

	I	II	III	IV	V	VI
λ_0	4·58	8·57	4·30	8·13	2·89	2·90
$10^{-2}M$	0·65	1·22	0·61	1·16	0·41	0·41
(2a)	0·41	1·14	0·71	1·09	0·28	0·21

(2c) $\beta = 0·14$.

1.42. (a) 67,300; (b) In 'non-diffusible' compartment K^+, 1·246 M; Cl^- 0·246 M. In other compartment K^+, 0·554 M; Cl^-, 0·554 M.

1.43. 65,580.

1.44. Zero time correction Δt, 2·4 min; D_{corr}, 4·19 × $10^{-7}cm^2sec^{-1}$.

1.45. 65,330.

1.46. 0·002 cm^3 per g dry protein.

1.47. 21·2s; 21·0s; 17·7s; 18·3s.

1.48. 8·35 cm. (A–T base pair, M 586; G–C base pair, M 576; assuming rat liver contains some 40% G–C).

1.49. Specific for cysteine SH. Ficin has 1 SH per molecule. Activity correlates with SH content. Incorporation of mercury suggests mercaptide bond formation. Value of reagent is its specificity, its colour and its possible utilization in studies of the structure of the active centre.

1.50. M_{NaCl}, 1·48×10^5; M_{H_2O}, 1·68×10^4.

1.51. 25,150.

1.54. 13,430.

1.55. 79,930.

1.56. $S = kM^\alpha$, where α has the values: native, 0·33; alkaline denatured, 0·37; neutral denatured, 0·57. For random coil, $\alpha = 0·50$.

1.57. $M = 23,500$.

1.58. In 2M denaturing agent $M = 67,100$; in 7M denaturing agent $M = 66,560$.

1.61. Enzyme alone, low concn., $M = 46,000$; alone, high concn., $M = 106,000$; low concn. plus soybean factor, $M = 166,000$; in 1% dodecyl sulphate, $M = 10,000$.

1.62. Enzyme consists of two chains of approx. $M = 12,250$ linked by disulphide bridges. Ala and ileu occupy N-termini. Only one of ser involved in active site.

1.64. (a) $M = 39,700$; (b) Min $M = 21,167$.

1.66. 0·313 mg/ml.

CHAPTER II

2.1. 1 × 10^{-6}; 3·16 × 10^{-10} g/l.

2.2. 0·869.

2.3. (a) 2·7; (b) 11; 3·65.

2.4. 3·8 × 10^{-2} M.

2.5. (a) 0·1; (b) 3; (c) 0·1; (d) 0·12; (e) 0·45; (f) 3; (g) 0·075. (1) I = M; (2) I = 3M; (3) I = 4M; (4) I = 6M; (5) I = 15M.

2.6. 4·24 × 10^{-4}.

2.7. (a) f_{H^+} 0·869; a_{H^+} 8·69 × 10^{-3}; f_{SO_4} = 0·569; a_{SO_4} = 2·85 × 10^{-3}. (b) $f_{Na^+} = f_{Cl^-} = 0·91$; $a_{Na^+} = a_{Cl^-} = 1·898 × 10^{-3}$.

2.8. 0·025; $f_{Mg^{++}}$ 0·483; $a_{Mg^{++}}$ 3·62 × 10^{-3}; f_{Cl^-} 0·834; a_{Cl^-} 8·34 × 10^{-3}; f_{SO_4} = 0·483; a_{SO_4} = 1·21 × 10^{-3}.

2.9. $[H_2CO_3]/[HCO_3^-] = 0·082/1·0$.

2.10. $1 \cdot 61 \times 10^{-7}$.
2.11.   Oxyhaemoglobin. [acid]/[salt]: haemoglobin $6 \cdot 03/1 \cdot 0$; oxyhaemoglobin $0 \cdot 17/1 \cdot 0$.
2.12.   (a) $6 \cdot 00$; (b) $6 \cdot 90$; (c) $5 \cdot 41$; (d) $5 \cdot 05$; (e) $6 \cdot 02$; (f) $6 \cdot 12$; (g) $5 \cdot 65$; (h) $5 \cdot 12$.
2.13.   (a) $5 \cdot 60$; (b) $5 \cdot 60$; (c) $3 \cdot 05$; (d) $7 \cdot 30$; (e) $5 \cdot 70$; (f) $5 \cdot 69$; (g) $3 \cdot 31$.
2.14. β $0 \cdot 079$. Maximum buffering range ca. pH $3 \cdot 0$-$5 \cdot 70$.
2.15. pK_{a_2} $6 \cdot 82$; |pH, β|$6 \cdot 82$, $0 \cdot 55$|$6 \cdot 0$, $0 \cdot 23$|$7 \cdot 5$, $0 \cdot 28$|.
2.16. $7 \cdot 41$.
2.17.   Arterial: p_{CO_2} $38 \cdot 3$ mm; $[CO_2]$ $1 \cdot 15$ mM; $[HCO_3^-]$ $25 \cdot 24$ mM.
        Venous: p_{CO_2} $47 \cdot 1$ mm; $[CO_2]$ $1 \cdot 42$ mM; $[HCO_3^-]$ $26 \cdot 31$ mM.
2.18.   Oxygenated: $12 \cdot 5$ mM; non-oxygenated: $26 \cdot 6$ mM.
2.19.   Oxygenated: $19 \cdot 85$ mM; non-oxygenated: $28 \cdot 96$ mM.
2.20.   Between pH $6 \cdot 1$ and $9 \cdot 0$ oxyhaemoglobin is the stronger acid; between pH $4 \cdot 5$ and $6 \cdot 1$ haemoglobin is stronger. Above pH 9 and below pH $4 \cdot 5$ there is no difference.
2.21. $21 : 1$; $11 : 1$; $2 \cdot 82 : 1$; $1 \cdot 83 : 1$; $1 \cdot 08 : 1$.
2.22. H_2CO_3: $1 \cdot 997$, $5 \cdot 07$, $4 \cdot 87$, $4 \cdot 61$, $4 \cdot 44$, $4 \cdot 28$ mM. pK_{a_1}': $6 \cdot 267$, $6 \cdot 204$, $6 \cdot 177$, $6 \cdot 155$, $6 \cdot 134$, $6 \cdot 100$. $pK_{a_1}' = -0 \cdot 53\sqrt{I} + 6 \cdot 327$; $\log f_{HCO_3^-} = -0 \cdot 53\sqrt{I}$.
2.23.   Compartment A: | Na^+, Cl^- | $0 \cdot 508$, $0 \cdot 498$ | $1 \cdot 333$, $0 \cdot 333$ | $1 \cdot 125$, $0 \cdot 125$ | $1 \cdot 000098$, $0 \cdot 000098$ |.   Compartment B: $Na^+ = Cl^-$, $0 \cdot 502$, $0 \cdot 667$, $0 \cdot 375$, $0 \cdot 0099$.   All concentrations mM.
2.24.   Average value $6 \cdot 60$. Result probably rather high since limiting solubility imposes very low concentration of sulphadiazine for the experiment.
2.25.   (a) $3 \cdot 997$ litres N-HCl; (b) 4; (c) $3 \cdot 162 \times 10^{-4}$.
2.26. $2 \cdot 97 \times 10^{-7}$ M.
2.27. $pKa_1 = 2 \cdot 35$; $pKa_0 = 9 \cdot 69$.
2.28.   (a) pH $5 \cdot 5$: A, $10+$; B, $18+$
        pH $8 \cdot 0$: A, $6+$; B, $8 \cdot 5+$.
2.29.   (a) 11 mM; (b) 27.

CHAPTER III

3.1. $-2,388 \cdot 2$ kcal.
3.2. $-2,693 \cdot 4$ kcal.
3.3. $-325 \cdot 13$ kcal; $-68 \cdot 31$ kcal.
3.4. $-691,040$ cal.
3.5. $-144,620$ cal.
3.6. $-3,350$ cal.
3.7. $1 \cdot 50$ kcal; $1 \cdot 165$ kcal; $28 \cdot 7\%$, $0 \cdot 05\%$. This is rather an extreme case but serves to emphasize the inaccuracy of the method.
3.8. $-57 \cdot 9$ kcal; $-10 \cdot 3$ kcal.
3.9. $-154 \cdot 90$ kcal.
3.10. $-2,854$ cal.
3.11.   (a) $48 \cdot 3$; (b) $60 \cdot 2$; (c) $75 \cdot 8$; (d) $66 \cdot 6$; (e) $56 \cdot 6$.
3.12.   15,000 cal energy bound in phosphoglycerylphosphate molecule.
3.13. K, $17 \cdot 52$, ΔG, -1770 cal.
3.14. $-3 \cdot 57$ kcal; $-7 \cdot 6$ kcal.

CHAPTER IV

4.1. $807 \cdot 2$.
4.2. $-78,594$ cal.
4.3. $-111,447$ cal.
4.4.   (a) 4,887; (b) $37 \cdot 15$; (c) $3 \cdot 13$.
4.5. $6 \cdot 76 \times 10^{12}$; $3 \cdot 85 \times 10^{-8}$ M; $7 \cdot 94 \times 10^{15}$; $1 \cdot 45 \times 10^{-8}$ M.
4.6. $17 \cdot 52$; -1770 cal.
4.7. $96 \cdot 2\%$ dihydroxyacetone phosphate; $3 \cdot 8\%$ glyceraldehyde phosphate.
4.8. $-3,720$ cal.
4.9. $-3,800$ cal/mole; $25°$: -826 cal/mole; $38°$: -703 cal/mole.

4.10. K_c (M × 10¹³): (a) 7·23; (b) 7·46; (c) 7·76. Average: 7·48; $\Delta G°$ +16,540 cal.

4.11. Reaction (3) K_{app}. (litres/mole): 0·227, 0·190, 0·195, 0·166. Average: 0·195. Reaction (1) K_{app} (litres/mole): 8·37 × 10³; $\Delta G°$, −5295 cal.

4.12. K_{app}. (litres/mole): 18·6, 17·7, 19·6, 21·5, 23·5. Average: 20·2. K_a(×10⁻⁸): 3·71, 3·53, 3·91, 4·29, 4·69. Average: 4·03. K_{app}. (pyruvate-oxaloacetate): 4·71 × 10⁻⁴.

4.13. pH 8·5: 0·0261, 0·0261, 0·0297, 0·0348. Average: 0·0292 M. $\Delta G°$, 2,185 cal; pH 5·8: 0·0122 M. $\Delta G°$, 2,725 cal.

4.14. 298·1°: K 1·334, $\Delta G°$ −171 cal.; 310·7°: K 0·921, $\Delta G°$ +51 cal. ΔH −5,405 cal.

4.15. Iso-citric/aconitic: 25°, 2·14; 38°, 1·54; citric/aconitic: 25°, 31·35; 38°, 20·72.

4.16. K values: 30 min, 6·81; 40 min, 6·78; 50 min, 6·82; equilibrium attained within 30 min.

4.17. [2-oxoglutarate] = [alanine] = 237 μl/ml solution, and [glutamate] [pyruvate] = 193 μl/ml solution.

4.18. pH 6·5: 1·29, 0·503; pH 6·0: 1·33, 0·622; pH 7·5: 1·38, 0·619.

4.19. (a) 6·26, 6·14, 6·27, 6·13 × 10⁻³, average value 6·20 × 10⁻³. (b) Not necessarily, but this form of mass law equation is probably correct in general, e.g. first power of all species. Could be Prot⁻ + Ca^{2+} ⇌ CaProt⁺, but not 2 Prot⁼ + Ca^{2+}. (c) Equilibrium not attained. Final concentrations: Ca^{2+}, 2·10; CaProt, 2·37; Prot⁼, 6·99 m-moles/kg H_2O. (d) Not instantaneously in equilibrium; CaProt will dissociate. (e) Ca^{2+}, 1·39 × 10⁻³; CaProt, 1·72 × 10⁻³ m-moles/kg H_2O.

4.20. 11·10 kcal; 5·21 kcal; −0·32 volt.

4.21. K values: 41·1, 43·6, 21·9, 41·7, 45·1, average 38·68 moles⁻¹ litre. $\Delta G°$, −2179 cal; (a) 3·9%; (b) 85·1%. First order cleavage kinetics, second order condensation. If the low (third) value of K is rejected the average becomes 42·87 moles⁻¹ litre and $\Delta G°$, −2241 cal; (a) 85·8%; (b) 3·2%.

4.22. ΔH, 3987 cal; ΔG, −2410 cal.

4.23. −6875 cal/mole.

4.24. −1768 cal.

4.25. (a) and (b) 86·3.

CHAPTER V

5.1. $t_\frac{1}{2}$, 47 days; k, 0·0147 days⁻¹.

5.2. First order; k, 0·0142 min⁻¹.

5.3. Zero order.

5.4. (a) 8,100 cal/mole; (b) 12,200 cal/mole.

5.5. 18,200 cal/mole.

5.6. +67,700 cal; +213 cal/mole/degree.

5.7. Cat, 9,869 cal; human, 10,140 cal.

5.8. 13,460 cal.

5.9. 12,000 cal.

5.10. −20,200 cal.

5.11. 9,139 cal.

5.12. First order; k, 0·127 min⁻¹.

5.13. Both first order. Oxaloacetate: k, 0·00140 min⁻¹; oxalosuccinate: k, 0·0117 min⁻¹.

5.14. 70 kcal/mole. (From rate constants: 35·2, 5·63 × 10⁻⁴; 30·2, 0·925 × 10⁻⁴; 25·0, 0·13 × 10⁻⁴ sec⁻¹.)

5.15. First order, k = 14·2 × 10⁻⁶ sec⁻¹. Bimolecular; carry out reaction in dioxane containing low concentrations of water to demonstrate dependence on [H_2O]. E = 21,700 cal/mole; $\Delta G^{\ddagger}$ = 25,900 cal/mole; $\Delta H^{\ddagger}$ = 21,000 cal/mole; $\Delta S^{\ddagger}$ = −15·3 cal/mole/degree.

5.16. E = 12,000 cal/mole. $k(26°)$ = 13·2 flashes/min.

5.17. First order. k values (min⁻¹): 65·6°, 0·0566; 69·6°, 0·24; 71·5°, 0·52; 73·6°, 1·75. E, 97,450 cal/mole.

Q

5.18. (b) Apparent first order in early stages (15 min); expect unimolecular since simple decomposition. If unimolecular, rate constant will be independent of initial $[HbO_2]$. Rate of change of $[HbO_2]$ should be independent of total Hb concentration. If dithionite reacts only with free oxygen then, above a certain minimum concentration, the velocity of decomposition of HbO_2 should be independent of [dithionite]. If it reacts with HbO_2, rate of dissociation should increase as [dithionite] is increased.

5.19. $\Delta H^{\ddagger}$, 83,570 cal; $\Delta S^{\ddagger}$, 183·4 cal/mole/degree.

5.20. 8,654 cal.

5.21. (a) $a_t = \dfrac{a_0}{k_1 + k_2} (k_2 + k_1 e^{-(k_1 + k_2)t})$

$b_t = \dfrac{a_0}{k_1 + k_2} (k_1 - k_1 e^{-(k_1 + k_2)t})$

(b) $a_t = a_0 e^{-k_1 t}$

$b_t = \dfrac{k_1 a_0}{k_2 - k_1} (e^{-k_1 t} - e^{-k_2 t})$

$c_t = a_0 \left(1 - \dfrac{k_2}{k_2 - k_1} e^{-k_1 t} + \dfrac{k_1}{k_2 - k_1} e^{-k_2 t}\right)$

5.22. $E(Fe^{2+})$, 10,020 cal.; E (catalase), 1,590 cal.

CHAPTER VI

6.1. $5·62 \times 10^{-3}$ M.
6.2. Serine: $2·5 \times 10^{-3}$ M; threonine: $2·2 \times 10^{-3}$ M.
6.3. pH 7·6.
6.4. $2·3 \times 10^{-6}$ M.
6.5. $1·3 \times 10^{-4}$ M.
6.6. Oxidation $\approx 10^{-5}$ M; for phosphorylation, concentrations are too far from K_s to give reliable values; best value is obtained with highest sarcosome concentration and is $\approx 5 \times 10^{-5}$ M.
6.7. $7·81 \times 10^{-2}$ M; $0·1250$ moles/hour.
6.8. 15°: $1·01 \times 10^{-6}$ M; 22°: $1·61 \times 10^{-6}$ M.
6.9. K_s $1·25 \times 10^{-2}$ M; K_i $9·92 \times 10^{-3}$ M.
6.10. Hydrocinnamate, phenylacetate and phenylbutyrate competitive; benzoate non-competitive.
6.11. 2,724,000.
6.12. Competitive; $6·08 \times 10^{-6}$ M.
6.13. $6·0 \times 10^{-5}$ M.
6.14. B non-competitive; C competitive.
6.15. $1·66 \times 10^{-5}$ M.
6.16. $K_s = 1·9 \times 10^{-3}$ M (rejecting 0·32 mM Mg^{2+} value); competitive; average $K_i = 5·5 \times 10^{-4}$ M (rejecting 0·32 mM Mg^{2+} experiment.)
6.17. $0·001$ M.
6.18. (a) $60·6 \mu l/10$ min; (b) $90·9 \mu l/10$ min.
6.19. (a) 4; (b) $1·41 \times 10^{-5}$ M.
6.21. K_s, $8·70 \times 10^{-4}$ M; K_i, $2·64 \times 10^{-3}$ M (competitive inhibitor).
6.22. $\Delta H^{\ddagger}$, 1410 cal; $\Delta G^{\ddagger}$, 10,000 cal; $\Delta S^{\ddagger}$, −29·3 cal/mole/degree.
6.23. K_m values fall within the range 2·05 to $2·28 \times 10^{-2}$ M.
V values (in order of increasing pH): 0·29, 0·80, 1·22, 1·82, 2·00 and 2·22.
6.24. Competitive inhibition (for 1·26 mM malonate, $K_i = 1·88$ mM; for 1·95 mM malonate, $K_i = 1·54$ mM.)
6.25. K_m, 2·25 mM; K_i, 26·26 mM.
6.26. ES complex has a pK of approximately 7·0.
6.28. Competitive; K_i, 5 μM.

6.30. Non-competitive; K_i, 0·134 mM.
6.31. Expt. 1: v_t theory, 137; observed, 105.
       Expt. 2: v_t theory, 146; observed, 120.
6.32. K_m for X, 2·63 mM; for Y, 2·5 mM.
6.33. 0·36 mM.
6.34. V_{max}, 5×10^{-4} mole/sec; K_m 5·8 mM; k_{+2}, 12·5 sec^{-1}.
6.36. Competitive. $K_m = 0·856$ mM; $K_1 = 2·7$ mM.
6.37. $K_m(A) = k_{+3}/k_{+1}$; $K_m(B) = (k_{+3}+k_{-2})/k_{+2}$.
6.38. $K_m = 4·34 \times 10^{-4}$ M; $V = 0·0256$ μmole min^{-1}.
6.40. 2-SH per mol.
6.42. C non-competitive; D competitive.
6.45. Pure enzyme: $K_m = 0·83$ mM; $V = 4·93$ μmol/min/mg protein.
       Crude enzyme: $K_m = 1·51$ mM; $V = 1·92$ μmol/min/mg protein.
6.46. $K_1 = 0·16$ M if competitive; 0·32 M if non-competitive.
6.47. $K_m = 4·24 \times 10^{-3}$ M; $K_1 = 0·09$ M. Competitive with both acetic acid and sodium acetate.
6.50. $K_1 = 7·5$ mM.

CHAPTER VII

7.1. 4·01, 2·46, 2·31.
7.2. (a) 0·022; (b) 0·056; (c) 0·149; (d) 0·301; (e) 0·757; (f) 2·000.
7.3. $1·432 \times 10^3$; 43·9%.
7.4. $\epsilon_{spec.} = 88·83$.
7.5. (a) 0·0; (b) −1·15; (c) −5·35; (d) −8·08; (e) −15·15.
7.6. 700, 3,256.
7.7. 36·7.
7.8. NADH $3·33 \times 10^{-5}$ M; NAD$^+$ $1·67 \times 10^{-5}$ M.
7.9. *ortho* 0·14 g/100 ml, 4·9%; *meta* 1·55 g/100 ml, 53·8%; *para* 1·19 g/100 ml, 41·3%.
7.10. 87, 48·5, 8·5.
7.11. (1) $6·25 \times 10^3$; (2) $6·27 \times 10^3$; (3) $6·11 \times 10^3$; (4) $5·93 \times 10^3$ cm^2/mole. The authors attribute the low value for NADP-isocitrate to the presence of *ca.* 5% impurity in the preparation.
7.13. (a) 0·30μmole; (b) A, 0·144 mM; B, 0·149 mM; C, 0·180 mM.
7.14. NAD, $1·229 \times 10^{-5}$ M; NADH, $3·215 \times 10^{-5}$ M.
7.15. Tryptophan, 0·101 mM; tyrosine, 0·109 mM.
7.16. 7·1.
7.17. 10·05.
7.18. Single stranded.
7.19. (i) 0·630 (ii) 0·415.
7.20. (1) 1·054; (2) 1·137; (3) 0·961.
7.21. $E = 0·495$; $T = 32\%$.
7.24. $K_{diss} = 0·032$ M. (iii) 0·544 p.p.m. (v) $\Delta H = 11·2$ kcal/mole.
7.25. $a_0 = -647·8$; $b_0 = -93·7$. 14·9%.

CHAPTER VIII

8.1. (a) 1·61; (b) 1·59; (c) 1·43; (d) 1·70; (e) 1·55.
8.2. (a) 1·99; (b) 1·80; (c) 2·11; (d) 1·88.
8.3. (a) 1·55; (b) 1·82; (c) 1·46; (d) 1·37.
8.4. (a) 0·86; (b) 1·20; (c) 1·60; (d) 1·25; (e) 1·33; (f) 1·11; (g) 0·86; (h) 0·71; (j) 0·69.
8.5. O_2 159·6 mm; N_2 593·2 mm Hg.
8.6. $5·78 \times 10^{-4}$; $1·83 \times 10^{-3}$; $5·78 \times 10^{-3}$; $1·83 \times 10^{-2}$ M.
8.7. 76·9 μl/5 min.
8.8. $7·5 \times 10^{-3}$ M; 0·099% (w/v).
8.9. $3·33 \times 10^{-3}$ M.
8.10. Theoretical O_2 uptake for complete oxidation: 541, 1,082, 2,164, 4,328 μl. Per cent. theoretical O_2 uptake, endogenous not subtracted: 31·4,

30·4, 30·4, 29·9; endogenous subtracted: 21·6, 23·5, 25·6, 26·2. Endogenous suppressed completely in all except lowest concentrations.

8.11. 33·8% assimilated; endogenous respiration not suppressed by substrate.
8.12. Inhibits dissimilation of pyruvic acid. If KOH added to centre well, in presence of arsenite there should be no gas evolution.
8.13. 1·03, 1·57, 1·27, 5·00.
8.14. O_2 uptake: 200 μl; CO_2 evolution: 201 μl R.Q. 1·0.
8.15. Endogenous Q_{O_2}: 14·4, 7·45, 3·40, 1·65, 0·60. Glucose Q_{O_2}: 3·55, 4·75 6·05, 7·73, 9·07.
8.16. 4, 5-Dihydroxyphthalate: $-5·9$ μmoles O_2, $+11·8$ μmoles CO_2. Protocatechuate: $-5·98$ μmoles O_2, $+6·1$ μmoles CO_2. These values suggest a reaction sequence:
4, 5-Dihydroxyphthalate→protocatechuate + CO_2
Protocatechuate + O_2→3-oxoadipate + CO_2.
8.18. Parent 55%; mutant 11·9%.
8·19. 7·29.

CHAPTER IX

9.1. 18, 38, 80 min.
9.2. M.g.t. 35 min; lag 50 min.
9.3. Unaerated 40 min; aerated 32 min. Aeration began at 200 min.
9.4. M.g.t. 45 min.; 569 min.
9.5. Lag periods: control 400; malic 450; α-oxoglutaric 380; succinic 340; aspartic 260; glutamic 240 min.
9.6. M.g.t. of phenol and control cultures both 45 min, therefore no effect.
9.7. (a) 25; (b) 35; (c) 50 min.
9.8. 10-11 min; 0·5 g galactose/litre.
9.9. 450 mg glucose and 550 mg mannose/litre.
9.10. 104 min.
9.11. 1·527 mM; 4·722 mM. Must be at least 0·90 mM but precise concentration cannot be ascertained from the data given.
9.12. Triauxic growth. First growth cycle: lag, about 200 min; m.g.t., 57 min; stationary population, 2·513 × 10⁶ cells/ml. Second growth cycle: lag, 130 min; m.g.t., 80 min; stationary population, 19·95 × 10⁶ cells/ml. Third growth cycle: lag, 140 min; m.g.t., 90 min; stationary population, 126·2 × 10⁶ cells/ml.
9.13. 29·5. Probably thioclastic reaction, conserving 1ATP in addition to 2ATP from glycolysis.
9.14. 29·5.

CHAPTER X

10.1. (a) 5·1%; (b) 73·1%.
10.2. 1 : 2.
10.3. $-0·141$ V.
10.4. $+0·136$ V. $+0·139$ V.
10.5. $-0·402$ V.
10.6. 99·9% oxidized.
10.7. $n = 1$ for pH range 0-4; $n = 2$ for pH range 4-8; $E° = 0·386$ V.
10.8. $-14·88$, 26·04, 12·18; $+20,300$ cal, $-17,760$ cal, $-8,307$ cal.
10.9. 94·5%; r′H=16·1.
10.10. $-1,691$ cal; 60% increase.
10.11. $-26,900$ cal, $-23,530$ cal.
10.12. $+0·136$ V; $n = 2$.
10.13. $+0·796$ V.
10.14. $pK = 6·65$; $K = 2·2 × 10^{-7}$.
10.15. $E = 0·82 + 0·0148 \log p_{O_2}$; $p_{O_2} = 10^{-24}$ atmosphere.
10.16. $-0·18$ V.
10.17. pH 4·19; K, 3·85 × 10^{-5}.

10.18. 70% decrease.
10.19. 94·3%, 1·04.
10.20. *L. mesenteroides*: 104·3%, 0·86; *L. pentoaceticus*: 91·85%, 1·41; *L. plantarum*: 98·6%.
10.21. (*a*) 94·5% recovery; (*b*) O-R ratio = 1·04.
10.22. (*a*) 38; (*b*) 87. Glucose, 1·69 kcal/g; undecylic acid, 3·74 kcal/g.

CHAPTER XI

11.1. 0·0485 days^{-1}; 61·54; 37·93, 23·36, 8·68%.
11.2. 1,000.
11.3. 87·1 days; 0·00795 days^{-1}.
11.4. Haemoglobin, when incorporated into a red blood cell, does not take part in dynamic equilibria characteristic of other body proteins; it offers, therefore, a method for determining the life span of the red blood cell.
11.5. 0·870 g, 29%. N.B. The original article contains a typographical error and gives the excess density of water formed from leucine in the hydrolysate as 225 instead of 161 parts per million (personal communication, Dr. H. H. Ussing).
11.7. 3·418 and 3·026 g.
11.8. Tyrosine 3·0%, 11; phenylalanine 6·8%, 27; arginine 3·7%, 14; lysine 8·5%, 39; glutamic acid 8·6%, 39; aspartic acid 10·4%, 52; glycine 5·5%, 50; leucine 15·1%, 77.
11.10. Glutamic acid 19·1%, 55; aspartic acid 11·3%, 36; lysine 11·4%, 33; leucine 15·6%, 50; glycine 1·5%, 8.
11.11. Glycerophosphate. Average value: 6·75 h.
11.12. β-C 940; α-C 10; carboxyl-C 10 counts/min/μmole phenylalanine. Glucose has 2,040 counts/min/μmole and the β-C has about half this activity. 3-Carbon fragment could therefore have come from triosephosphate.
11.13. Glutamic acid 16·9%, 80; aspartic acid 9·71%, 51; tyrosine 4·78%, 18.
11.14. (*a*) 4·7%, 4·6%; (*b*) 6·2%, 6·1%.
11.15. 0·748 mCi.
11.16. 4·32 × 10^{-4} μCi; 7·29 × 10^{-12}.
11.17. (*a*) 13·32 × 10^6 counts/min; (*b*) $t_{\frac{1}{2}}$, 8·7 days.
11.18. Phospholipid, 0·417; RNA-P, 0·288.
11.19. Aspartate, C-1, 220; C-4, 530. Glutamate, C-1, 752. Arginine, C-1, 739; C-6, 800. Proline, C-5, 750. ^{14}C is confined to carboxyl or guanidyl (which fixes CO_2) groups and therefore results are consistent with suggestion that H^{14}COOH is utilized only after conversion to $^{14}CO_2$.
11.20. P/O ratios: 5·07, 4·73, 4·40, 3·05, 2·78.
11.21. Percentage distribution of total radioactivity

Time (sec)	0	5	10	15	20	30	60	105
Malate	0	23·1	23·6	24·7	29·0	6·3	6·8	2·3
Glycerate	0	59·0	46·2	24·7	13·2	12·5	9·7	13·7
Alanine	0	17·9	29·2	50·6	51·4	72·8	71·6	76·1
Lactate	0	0·0	0·0	0·0	6·4	8·4	11·9	7·9

Glycerate is the compound formed first from tartrate.

11.23. 4.
11.24. 30 per cent. pentose phosphate cycle.
11.25. ATP, 33,765 counts/min/μmole; terminal phosphates, 11,238 counts/min/μmole; ester phosphate, 11,289 counts/min/μmole. Total activity, 0·869 μCi/100 ml.
11.26. (*a*) 1·18 min. (*b*) 7·36 m-equiv./h/kg body weight. (*c*) 1·70 min. (*d*) 7·73 m-equiv./animal. (*e*) 10·59 litres.
11.27. ^{14}C, 1074 counts/min; ^{32}P, 1852 counts/min.
11.30. 1·014×10^8.

11.32. $x = s_2 t_1 / s_1 t_2$.

11.33.   (a) $7 \cdot 96 \times 10^{-3}$ days^{-1};   (b) $15 \cdot 15$ mg methionine.

11.34.   Specific activity 90μCi/mg;   Relative s.a. $0 \cdot 075$.

11.35.   Specific activities: UDP-glucuronic acid, $1 \cdot 6 \times 10^4$ counts/min/nmole; NAG, 4×10^4 counts/min/nmole.   Relative proportions $1 \cdot 086 : 1 \cdot 0$, i.e. $1 : 1$.

11.36. 35%.

11.37.   A branched chain of 7 glucose residues involving five [1→4] and one [1→6] linkage.   No randomization.

LOGARITHMIC TABLES

LOGARITHMS

	0	1	2	3	4	5	6	7	8	9	1 2 3	4 5 6	7 8 9
10	0000	0043	0086	0128	0170						5 9 13	17 21 26	30 34 38
						0212	0253	0294	0334	0374	4 8 12	16 20 24	28 32 36
11	0414	0453	0492	0531	0569						4 8 12	16 20 23	27 31 35
						0607	0645	0682	0719	0755	4 7 11	15 18 22	26 29 33
12	0792	0828	0864	0899	0934						3 7 11	14 18 21	25 28 32
						0969	1004	1038	1072	1106	3 7 10	14 17 20	24 27 31
13	1139	1173	1206	1239	1271						3 6 10	13 16 19	23 26 29
						1303	1335	1367	1399	1430	3 7 10	13 16 19	22 25 29
14	1461	1492	1523	1553	1584						3 6 9	12 15 19	22 25 28
						1614	1644	1673	1703	1732	3 6 9	12 14 17	20 23 26
15	1761	1790	1818	1847	1875						3 6 9	11 14 17	20 23 26
						1903	1931	1959	1987	2014	3 6 8	11 14 17	19 22 25
16	2041	2068	2095	2122	2148						3 6 8	11 14 16	19 22 24
						2175	2201	2227	2253	2279	3 5 8	10 13 16	18 21 23
17	2304	2330	2355	2380	2405						3 5 8	10 13 15	18 20 23
						2430	2455	2480	2504	2529	3 5 8	10 12 15	17 20 22
18	2553	2577	2601	2625	2648						2 5 7	9 12 14	17 19 21
						2672	2695	2718	2742	2765	2 4 7	9 11 14	16 18 21
19	2788	2810	2833	2856	2878						2 4 7	9 11 13	16 18 20
						2900	2923	2945	2967	2989	2 4 6	8 11 13	15 17 19
20	3010	3032	3054	3075	3096	3118	3139	3160	3181	3201	2 4 6	8 11 13	15 17 19
21	3222	3243	3263	3284	3304	3324	3345	3365	3385	3404	2 4 6	8 10 12	14 16 18
22	3424	3444	3464	3483	3502	3522	3541	3560	3579	3598	2 4 6	8 10 12	14 15 17
23	3617	3636	3655	3674	3692	3711	3729	3747	3766	3784	2 4 6	7 9 11	13 15 17
24	3802	3820	3838	3856	3874	3892	3909	3927	3945	3962	2 4 5	7 9 11	12 14 16
25	3979	3997	4014	4031	4048	4065	4082	4099	4116	4133	2 3 5	7 9 10	12 14 15
26	4150	4166	4183	4200	4216	4232	4249	4265	4281	4298	2 3 5	7 8 10	11 13 15
27	4314	4330	4346	4362	4378	4393	4409	4425	4440	4456	2 3 5	6 8 9	11 13 14
28	4472	4487	4502	4518	4533	4548	4564	4579	4594	4609	2 3 5	6 8 9	11 12 14
29	4624	4639	4654	4669	4683	4698	4713	4728	4742	4757	1 3 4	6 7 9	10 12 13
30	4771	4786	4800	4814	4829	4843	4857	4871	4886	4900	1 3 4	6 7 9	10 11 13
31	4914	4928	4942	4955	4969	4983	4997	5011	5024	5038	1 3 4	6 7 8	10 11 12
32	5051	5065	5079	5092	5105	5119	5132	5145	5159	5172	1 3 4	5 7 8	9 11 12
33	5185	5198	5211	5224	5237	5250	5263	5276	5289	5302	1 3 4	5 6 8	9 10 12
34	5315	5328	5340	5353	5366	5378	5391	5403	5416	5428	1 3 4	5 6 8	9 10 11
35	5441	5453	5465	5478	5490	5502	5514	5527	5539	5551	1 2 4	5 6 7	9 10 11
36	5563	5575	5587	5599	5611	5623	5635	5647	5658	5670	1 2 4	5 6 7	8 10 11
37	5682	5694	5705	5717	5729	5740	5752	5763	5775	5786	1 2 3	5 6 7	8 9 10
38	5798	5809	5821	5832	5843	5855	5866	5877	5888	5899	1 2 3	5 6 7	8 9 10
39	5911	5922	5933	5944	5955	5966	5977	5988	5999	6010	1 2 3	4 5 7	8 9 10
40	6021	6031	6042	6053	6064	6075	6085	6096	6107	6117	1 2 3	4 5 6	8 9 10
41	6128	6138	6149	6160	6170	6180	6191	6201	6212	6222	1 2 3	4 5 6	7 8 9
42	6232	6243	6253	6263	6274	6284	6294	6304	6314	6325	1 2 3	4 5 6	7 8 9
43	6335	6345	6355	6365	6375	6385	6395	6405	6415	6425	1 2 3	4 5 6	7 8 9
44	6435	6444	6454	6464	6474	6484	6493	6503	6513	6522	1 2 3	4 5 6	7 8 9
45	6532	6542	6551	6561	6571	6580	6590	6599	6609	6618	1 2 3	4 5 6	7 8 9
46	6628	6637	6646	6656	6665	6675	6684	6693	6702	6712	1 2 3	4 5 6	7 7 8
47	6721	6730	6739	6749	6758	6767	6776	6785	6794	6803	1 2 3	4 5 5	6 7 8
48	6812	6821	6830	6839	6848	6857	6866	6875	6884	6893	1 2 3	4 4 5	6 7 8
49	6902	6911	6920	6928	6937	6946	6955	6964	6972	6981	1 2 3	4 4 5	6 7 8

By permission of Messrs. Macmillan & Co., Ltd.

LOGARITHMS

	0	1	2	3	4	5	6	7	8	9	1 2 3	4 5 6	7 8 9
50	6990	6998	7007	7016	7024	7033	7042	7050	7059	7067	1 2 3	3 4 5	6 7 8
51	7076	7084	7093	7101	7110	7118	7126	7135	7143	7152	1 2 3	3 4 5	6 7 8
52	7160	7168	7177	7185	7193	7202	7210	7218	7226	7235	1 2 2	3 4 5	6 7 7
53	7243	7251	7259	7267	7275	7284	7292	7300	7308	7316	1 2 2	3 4 5	6 6 7
54	7324	7332	7340	7348	7356	7364	7372	7380	7388	7396	1 2 2	3 4 5	6 6 7
55	7404	7412	7419	7427	7435	7443	7451	7459	7466	7474	1 2 2	3 4 5	5 6 7
56	7482	7490	7497	7505	7513	7520	7528	7536	7543	7551	1 2 2	3 4 5	5 6 7
57	7559	7566	7574	7582	7589	7597	7604	7612	7619	7627	1 2 2	3 4 5	5 6 7
58	7634	7642	7649	7657	7664	7672	7679	7686	7694	7701	1 1 2	3 4 4	5 6 7
59	7709	7716	7723	7731	7738	7745	7752	7760	7767	7774	1 1 2	3 4 4	5 6 7
60	7782	7789	7796	7803	7810	7818	7825	7832	7839	7846	1 1 2	3 4 4	5 6 6
61	7853	7860	7868	7875	7882	7889	7896	7903	7910	7917	1 1 2	3 4 4	5 6 6
62	7924	7931	7938	7945	7952	7959	7966	7973	7980	7987	1 1 2	3 3 4	5 6 6
63	7993	8000	8007	8014	8021	8028	8035	8041	8048	8055	1 1 2	3 3 4	5 5 6
64	8062	8069	8075	8082	8089	8096	8102	8109	8116	8122	1 1 2	3 3 4	5 5 6
65	8129	8136	8142	8149	8156	8162	8169	8176	8182	8189	1 1 2	3 3 4	5 5 6
66	8195	8202	8209	8215	8222	8228	8235	8241	8248	8254	1 1 2	3 3 4	5 5 6
67	8261	8267	8274	8280	8287	8293	8299	8306	8312	8319	1 1 2	3 3 4	5 5 6
68	8325	8331	8338	8344	8351	8357	8363	8370	8376	8382	1 1 2	3 3 4	4 5 6
69	8388	8395	8401	8407	8414	8420	8426	8432	8439	8445	1 1 2	2 3 4	4 5 6
70	8451	8457	8463	8470	8476	8482	8488	8494	8500	8506	1 1 2	2 3 4	4 5 6
71	8513	8519	8525	8531	8537	8543	8549	8555	8561	8567	1 1 2	2 3 4	4 5 5
72	8573	8579	8585	8591	8597	8603	8609	8615	8621	8627	1 1 2	2 3 4	4 5 5
73	8633	8639	8645	8651	8657	8663	8669	8675	8681	8686	1 1 2	2 3 4	4 5 5
74	8692	8698	8704	8710	8716	8722	8727	8733	8739	8745	1 1 2	2 3 4	4 5 5
75	8751	8756	8762	8768	8774	8779	8785	8791	8797	8802	1 1 2	2 3 3	4 5 5
76	8808	8814	8820	8825	8831	8837	8842	8848	8854	8859	1 1 2	2 3 3	4 5 5
77	8865	8871	8876	8882	8887	8893	8899	8904	8910	8915	1 1 2	2 3 3	4 4 5
78	8921	8927	8932	8938	8943	8949	8954	8960	8965	8971	1 1 2	2 3 3	4 4 5
79	8976	8982	8987	8993	8998	9004	9009	9015	9020	9025	1 1 2	2 3 3	4 4 5
80	9031	9036	9042	9047	9053	9058	9063	9069	9074	9079	1 1 2	2 3 3	4 4 5
81	9085	9090	9096	9101	9106	9112	9117	9122	9128	9133	1 1 2	2 3 3	4 4 5
82	9138	9143	9149	9154	9159	9165	9170	9175	9180	9186	1 1 2	2 3 3	4 4 5
83	9191	9196	9201	9206	9212	9217	9222	9227	9232	9238	1 1 2	2 3 3	4 4 5
84	9243	9248	9253	9258	9263	9269	9274	9279	9284	9289	1 1 2	2 3 3	4 4 5
85	9294	9299	9304	9309	9315	9320	9325	9330	9335	9340	1 1 2	2 3 3	4 4 5
86	9345	9350	9355	9360	9365	9370	9375	9380	9385	9390	1 1 2	2 3 3	4 4 5
87	9395	9400	9405	9410	9415	9420	9425	9430	9435	9440	0 1 1	2 2 3	3 4 4
88	9445	9450	9455	9460	9465	9469	9474	9479	9484	9489	0 1 1	2 2 3	3 4 4
89	9494	9499	9504	9509	9513	9518	9523	9528	9533	9538	0 1 1	2 2 3	3 4 4
90	9542	9547	9552	9557	9562	9566	9571	9576	9581	9586	0 1 1	2 2 3	3 4 4
91	9590	9595	9600	9605	9609	9614	9619	9624	9628	9633	0 1 1	2 2 3	3 4 4
92	9638	9643	9647	9652	9657	9661	9666	9671	9675	9680	0 1 1	2 2 3	3 4 4
93	9685	9689	9694	9699	9703	9708	9713	9717	9722	9727	0 1 1	2 2 3	3 4 4
94	9731	9736	9741	9745	9750	9754	9759	9763	9768	9773	0 1 1	2 2 3	3 4 4
95	9777	9782	9786	9791	9795	9800	9805	9809	9814	9818	0 1 1	2 2 3	3 4 4
96	9823	9827	9832	9836	9841	9845	9850	9854	9859	9863	0 1 1	2 2 3	3 4 4
97	9868	9872	9877	9881	9886	9890	9894	9899	9903	9908	0 1 1	2 2 3	3 4 4
98	9912	9917	9921	9926	9930	9934	9939	9943	9948	9952	0 1 1	2 2 3	3 4 4
99	9956	9961	9965	9969	9974	9978	9983	9987	9991	9996	0 1 1	2 2 3	3 3 4

AUTHOR INDEX

A

Adair, G. S., 54, 55
Aitken, A. C., 444
Alberty, R. A., 97, 232]
Alexander, A. E., 50, 51
Altman, C., 194, 197, 198, 232
Andrews, P., 38, 40, 51
Anson, M. L., 34, 58, 59, 177
Archibald, W. J., 23, 24, 25
Aronoff, S., 425
Arrhenius, S., 166, 171
Atkinson, D. E., 121, 122, 126
Atkinson, E., 149
Avison, A. W., 117
Avogadro, A., 2

B

Ball, E. G., 369, 370, 371
Bannerjee, R., 368
Barcroft, J., 292
Barker, H. A., 299, 308
Barker, S. B., 266
Bauchop, T., 123, 126, 319, 328
Beaufay, H., 28, 51
Beaven, G. H., 283
Beer, 260-268
Benzinger, T. H., 108, 109, 126
Berkeley, Earl of, 54
Berthet, J., 28, 51
Bigeleisen, J., 424, 425
Blackwood, A. C., 367
Block, R. J., 51
Bodansky, O., 177
Bonnichsen, R., 59
Borsook, H., 142
Bowen, T. J., 51, 61
Boyer, P. D., 232
Bray, H. G., 174, 232
Briggs, G. E., 185, 187, 189, 209
Brigham, E. H., 236
Brönsted, J. N., 71, 77, 97
Brown, H. D., 108, 126
Bull, H. B., 51, 97
Burk, D., 189, 205, 206, 207, 210, 215, 217, 218, 232
Burk, N. F., 54
Burris, R. H., 292, 306
Burton, K., 117, 134, 143, 147
Butler, J. A. V., 111, 126

C

Calvet, 108
Calvin, M., 401
Campbell, J. J. R., 308

Carpenter — E

Carpenter, P. G., 59
Chance, B., 250, 277
Chang, S. F., 63
Changeux, J-P., 225, 232
Chao, F. C., 59
Chapman, L. M., 52
Charlwood, P. A., 57
Chase, A. M., 236
Cheldelin, V. H., 425
Chervenka, C. H., 51
Clark, W. M., 85, 97, 102, 361, 367
Cleland, W. W., 195, 202, 203, 232, 233, 350
Clifton, C. E., 299, 309
Cohn, E. J., 52
Colowick, S. P., 146
Coolidge, T., 52
Cordes, E. H., 232
Creeth, J. M., 56
Crosbie, G. W., 425

D

Da Costa, E., 147
Dagley, S., 240, 309, 330, 331, 433
Daly, C., 100
Dann, W. J., 178
Dawes, E. A., 174, 235, 240, 299, 309, 330, 331
Debye, P., 45, 47, 72, 73
de Duve, C., 28, 30, 51
Dickens, F., 303
Dill, D. B., 100
Dixon, M., 142, 166, 174, 189, 212, 213, 214, 215, 216, 232, 292, 303, 306, 353, 355, 367
Djerassi, C., 283
Donnan, F. G., 13, 94-97
Doty, P., 282, 283
Doudoroff, M., 123, 405, 408, 410, 417, 421

E

Edsall, J. T., 2, 51, 97
Eggleston, L. V., 443
Eichelberger, L., 147
Einstein, A., 41
Elkins-Kaufman, E., 237
Elsden, S. R., 123, 126, 319, 328
Elsworth, R., 321, 326, 327, 328
Embden, G., 123
Entenman, C., 403, 425
Entner, N., 123, 405, 408, 410, 417, 421
Eriksson-Quensel, I-B., 56
Evans, E. A., 141

SUBJECT INDEX

A

Absolute reaction rates, theory of, 167-173
Absorbance, decadic, 261
  Napierian, 261
Absorption coefficient, 261,
  molar, 262
Absorption spectrum, 260
  of nicotinamide-adenine dinucleotide, 274
  of nicotinamide-adenine dinucleotide phosphate, 274
Acid-base equilibria, 71-94
Activation energy, 166-172
  determination of, 166
Activity, 72, 114, 135, 361
  optical, 273
Activity coefficient 72, 73, 136, 361
Allosterism, 224-230
  heterotropic, 225
  homotropic, 225
Amino acid :
  decarboxylases, 296
  diffusion coefficient of, 31
  dipolar ion form, 83-91
Ampholyte, definition, 71
Analysis :
  elementary and amino acid, 2-6
  end-group, 6
  isotope dilution, 391-395
  photometric, 256-283
Angular velocity, 18, 443
Answers to problems, 448
Approach to sedimentation equilibrium, 23-25
Archibald method, 23-25
Arrhenius equation, 166, 171
Atom per cent., excess, 374, 375
Atomic number, 373
Atomic weights, table, 446, 447
Avogadro number, 31, 440
Axial ratio, 32, 44

B

Background count, 383
Bacterial growth, 312-328
  continuous culture, 320-328
  diauxie phenomenon, 320, 321
  doubling time, 314
  exponential phase, 313-316
  growth cycle, 312
  lag phase, 316-318
  mean generation time, 154, 313
  molar growth yield, 123, 319
  specific growth rate, 315
  stationary phase, 312
  stationary population, 318
  total growth, 318
  yield coefficient (constant), 319
Bacteriophage, 275
Bactogen, 326
Band width, spectral, 268, 269
Barcroft-Haldane differential manometer, 292
Beer's Law, 260, 262
  deviations from, 263
Bicarbonate-CO_2 buffers, 303-306
Biological oxidation-reduction systems, 336, 337, 358
Bond energy, definitions, 119
Bond, high energy, 116-120
  hydrogen, 275
Briggs-Haldane equation, 187
Buffer solutions, 80-82
  bicarbonate-CO_2, 303-306
  retention of CO_2 by, 302
Buffer value, 81

C

Calibration curves, for colorimeter, 266, 267
Calorimetry, 108
Carbon dioxide production, measurement, 300-302
Carbon dioxide retention by buffers, 302
Carbonic acid, dissociation, 304, 305
Catalytic centre activity, 231
Cellulose, chain length, 6
Chemostat, 326
Coincidence correction, 383
Colligative properties of solutions, 1, 3, 11
Collision theory of reactions, 167
  comparison with transition state theory, 171, 172
Colorimeter, Duboscq, 264, 265
  photoelectric, 265, 266
  calibration curves, 266, 267
Colorimetry, 264-267
Consecutive reactions, 159
  reversible, 161
Conservation equation, 186
Continuous culture of micro-organisms, 320-328
Cotton effect, 280
Coupled reactions, 137
Crystal analysis, 259
Curie, definition, 378

Printed by T. & A. Constable Ltd., Edinburgh